Intermediary Xenobiotic Metabolism in Animals

Intermediary Xenobiotic Metabolism in Animals: Methodology, Mechanisms and Significance

Edited by

D.H. Hutson
Shell Research Ltd, Sittingbourne

J. Caldwell
St. Mary's Hospital Medical School, London

G.D. Paulson
United States Department of Agriculture
State University Station, Fargo
North Dakota 58105, USA

Taylor & Francis
London · New York · Philadelphia
1989

UK	Taylor & Francis Ltd, 4 John St, London WC1N 2ET
USA	Taylor & Francis Inc, 242 Cherry St, Philadelphia, PA 19106-1906

Copyright © Taylor & Francis 1989

All rights reserved. No part of this publication may be reproduced, stored in a retrieval system, or transmitted, in any form or by any means, electronic, electrostatic, magnetic tape, mechanical photocopying, recording or otherwise, without the prior permission of the copyright owner.

British Library Cataloguing in Publication Data

Intermediary xenobiotic metabolism in animals:
   methodology, mechanisms and significance.
   1. Animals. Xenobiotics. Metabolism I. Hutson, D.H.
   II. Caldwell, John, Paulson, G.D
   591.1′33

ISBN 0-85066-764-X

Printed in Great Britain by Taylor & Francis (Printers) Ltd Basingstoke, Hampshire

CONTENTS

Preface · vii

1. Sites of Metabolism

Xenobiotic metabolism: an introduction · 3
 J. Caldwell

Metabolism of xenobiotics in the gastro-intestinal tract · 13
 A.G. Renwick and C.F. George

Metabolism of xenobiotics in the respiratory tract · 41
 J.A. Bond and A.R. Dahl

The absorption and disposition of xenobiotics in skin · 65
 S.A. Hotchkiss and J. Caldwell

Biotransformation and bioactivation of xenobiotics by the kidney · 81
 M.W. Anders

Bioactivation and detoxication of organophosphorus insecticides in rat brain · 99
 J.E. Chambers, C.S. Forsyth and H.W. Chambers

2. Intermediary Xenobiotic Metabolites and their Significance

Metabolites resulting from oxidative and reductive processes · 119
 P.E. Levi and E. Hodgson

Bioactivation resulting from hydrolytic processes · 139
 W.C. Dauterman

N-Acetyltransferase polymorphism and arylamine toxicity: relationship to O-acylation reactions · 151
 T.J. Flammang, G. Talaska, D.Z.J. Chu, N.P. Lang and F.F. Kadlubar

Sulphate ester and glucuronic acid conjugates as intermediates in xenobiotic metabolism · 179
 D.H. Hutson

Metabolites derived from glutathione conjugation · 205
 J.E. Bakke

Reactive intermediates and their reaction with macromolecules · 225
 D.R. Hawkins

3. Methods for Studying Intermediary Xenobiotic Metabolism

Chemical modulation of xenobiotic metabolism *in vivo*   245
   D.J. Jollow, V.F. Price, S.A. Roberts and S.L. Longacre

Germfree rats in the study of the metabolism of xenobiotic compounds   263
   P. Goldman

NMR and the study of xenobiotic metabolism: an introduction   277
   T.A. Moore and P.G. Morris

The use of autoradiography as a tool to study xenobiotic metabolism   295
   I. Brandt and E.B. Brittebo

In situ perfusion and collection techniques for studying xenobiotic metabolism in animals   315
   K.L. Davison

Substrate delivery as a critical element in the study of intermediary metabolites of lipophilic xenobiotics *in vitro*   335
   T. Nakatsugawa, J. Timoszyk and J.M. Becker

Stable isotopes in metabolism studies: use of isotope effects   355
   N. Kurihara

In vivo detection of free radical metabolites as applied to carbon tetrachloride and related hydrocarbons   375
   K.T. Knecht, H.D. Connor, L.B. LaCagnin, R.G. Thurman and R.P. Mason

Index   383

PREFACE

The study of intermediary xenobiotic metabolism is essential to our understanding of the response of animals to chemical challenge. It is the objective of this Book to collect together some recent practical experiences of working in the area and to stimulate further such research.

The term 'Intermediary Xenobiotic Metabolism' refers to the formation of intermediate products which may not be obvious from the structures of the terminal metabolites of a xenobiotic. Intermediary metabolites are thus those metabolites formed after the administration of a xenobiotic but often not eliminated as such from the animal. These intermediates may be unstable transition states, which collapse to other products, or they may be chemically reactive species which interact spontaneously with biological nucleophiles and electrophiles. They may be products which act as substrates for secondary metabolism. Classical examples are the arene oxides (observed in the urine as phenols, dihydrodiols or mercapturic acids), carbinolamines (intermediates in $\underline{N}$-dealkylations) and glutathione conjugates (observed in the urine as mercapturic acids). Some lipophilic metabolites which are stored in the body for relatively long periods can also be classified as intermediary metabolites because they must be metabolized prior to elimination. Also included are those metabolites normally regarded as terminal products but which have a degree of reactivity which may have been partly realised in the animal. The reactivity of ester glucuronides in the acylation of proteins is such an example. The site of formation of a reactive metabolite may be a critical determinant of the site of toxic action. Thus research on these metabolites must take into account the capacities of the various organs and tissues for their formation and removal.

The material is arranged in three sections: (i) an introductory chapter followed by five chapters on sites (organs) of metabolism, (ii) six chapters on types of metabolites generated by specific biotransformation pathways and (iii) eight chapters on methods for studying intermediary xenobiotic metabolism.

The chapters are based on papers presented at a Symposium forming part of the Third Chemical Congress of North America (Toronto, June 1988). The editors would like to take this opportunity of acknowledging the support of the Congress organization and thanking the American Chemical Society (Agrochemicals Division) and the International Society for the Study of Xenobiotics for their financial co-sponsorship.

D.H. Hutson	Sittingbourne, UK	
J. Caldwell	London, UK	
G.D. Paulson	Fargo, ND, USA	October, 1988

1. Sites of Metabolism

XENOBIOTIC METABOLISM: AN INTRODUCTION

John Caldwell

Department of Pharmacology and Toxicology,
St. Mary's Hospital Medical School,
London W2 1PG, UK.

INTRODUCTION

Humans and other animals live in an environment which abounds in xenobiotics, that is compounds which are foreign to the normal energy-yielding metabolism of the body. Exposure to these xenobiotics may occur <u>deliberately</u>, as in the cases of drugs, food additives, cosmetics, etc., <u>accidentally</u>, as in the cases of food contaminants and pesticides and <u>coincidentally</u>, as in the cases of industrial chemicals and other environmental pollutants. These xenobiotics may enter the body in a variety of ways. The most important of these are the gastrointestinal tract, which is important for substances present in our foods, the skin, which is the principle barrier between the internal milieu and our external environment and the respiratory tract, which is important for volatile compounds and other materials present in aerosols or on dust particles. In addition, in cases of the deliberate use of drugs and accidental exposure to other chemicals, a variety of parenteral (or injection) routes of administration may also be involved. It is the purpose of this presentation to give an introductory overview of the life cycle of a xenobiotic in the organism, which encompasses its absorption, distribution, metabolism and excretion and to comment upon factors influencing these various processes and their biological significance.

ABSORPTION

The processes of absorption are those which lead to the entry of a xenobiotic into the systemic circulation of the body. In mechanistic terms this may be achieved in one of two ways. The xenobiotic may enter by passive diffusion across lipid membranes of cells or be a substrate for an active transport process (Benet and Sheiner, 1985). In quantitative terms, passive diffusion is by far the more important, occurring in the cases of the skin, the respiratory tract and for the vast majority of substances absorbed from the gastrointestinal tract. Active transport processes are found in the gastrointestinal tract and are important for the absorption of xenobiotics (particularly drugs) which may be analogues of nutrients (Myrhe et al., 1982).

DISTRIBUTION

Once a xenobiotic has entered the systemic circulation it will be carried throughout the body and its distribution into the different tissues of the body will reflect tissue haemodynamics and the participation of three different processes (Benet and Sheiner, 1985), namely (i) any specific tissue uptake mechanisms which may act upon the compound in question, (ii) passive diffusion across lipid membranes and (iii) protein binding. Specific uptake mechanisms are frequently determinants of the tissue distribution of compounds which are analogues of hormones or nutrients. Specific protein binding is particularly important in terms of binding of xenobiotics by plasma proteins but may also involve proteins within the tissues of the body. The importance of protein binding is seen in its influence upon passive diffusion across membranes. For the vast majority of compounds, their tissue concentrations will be determined by haemodynamic considerations which will influence their passive diffusion across cellular membranes. It is important to appreciate that the equilibria governing such passive diffusion are those reached between the free fractions of a compound, i.e. those fractions which are not subject to protein binding, rather than between the total quantity of compound present on each side of the lipid membrane under consideration (see Levy and Shand, 1984).

METABOLISM

We use the term metabolism to describe the chemical fate of a xenobiotic in the organism and we may distinguish four different chemical possibilities for this (Williams, 1959). These are that the compound may be:

1) eliminated unchanged from the body,
2) retained unchanged in the body,
3) undergo enzymic metabolism, and,
4) undergo spontaneous chemical transformations.

Of these various options it is enzymic metabolism which is by far the most important. This process is often also referred to as biotransformation.

Three classes of compounds may be distinguished by their resistance to enzymic metabolism in the animal body (Renwick, 1983). These are (i) highly polar compounds such as strong carboxylic acids, sulphonic and sulphamic acids and quaternary amines, (ii) volatile compounds and (iii) nonpolar lipophilic compounds. Highly polar compounds are in general absorbed into the body extremely poorly and their physicochemical characteristics lead to their facile excretion if they do enter the systemic circulation. Volatile compounds may, by virtue of their volatility, be lost readily from the lungs and hence may not be available for metabolism. Some nonpolar lipophilic compounds may be resistant to metabolism as, for example, when a high degree of halogen substitution prevents oxygenation. This resistance to metabolism coupled with their extreme lipophilicity will lead them to be retained in the animal body rather than be excreted.

The non-enzymic spontaneous chemical transformations which xenobiotics may undergo in the animal body are frequently overlooked (Testa, 1983). However, these reactions can be important both in quantitative and functional terms. They include reactions with endogenous nucleophiles such as glutathione and other thiols of tissue macromolecules, or with endogenous electrophiles such as aldehydes and ketones (leading to the formation of Schiff's bases). Xenobiotic quinones may participate in redox cycling with a variety of endogenous molecules, while a number of xenobiotics may undergo chemical breakdown and/or rearrangement by reactions such as hydrolysis and hydrolytic ring opening which may be promoted by metals and other ions present in body fluids.

The enzymic metabolism of xenobiotics is characteristically a biphasic process in which the xenobiotic undergoes first a Phase 1 or functionalization reaction of oxidation, reduction or hydrolysis giving rise to a metabolite which may subsequently undergo conjugation in Phase 2 (Williams, 1959). Conjugation refers to the linkage of a xenobiotic or its metabolites to an endogenous molecule, termed the conjugating agent (Caldwell, 1982). While the biphasic process describes the fate of the majority of xenobiotics, it is important to appreciate that some xenobiotics undergo only functionalization reactions while others have in their structures functional groupings appropriate for conjugation directly. The metabolites of a xenobiotic may thus include the unchanged compound, Phase 1 metabolites, conjugates of the xenobiotic itself and conjugates of the Phase 1 metabolites.

The functionalization reactions of oxidation may involve carbon, nitrogen, sulphur and phosphorus atoms in a molecule. The oxidation of these centres may be carried out by a variety of enzymes, of which the cytochrome P-450 system (Boobis et al., 1985) and the flavoprotein oxidases (Ziegler, 1980) are by far the most important. Reduction of carbonyl, nitro and olefin groups may involve the participation of both tissue (Hewick, 1982) and gut microflora enzymes (Goldman, 1982); the latter are generally more important. The facile hydrolysis of esters, amides, epoxides and glycosides by a range of tissue enzymes is well established (Heymann, 1982). Conjugation reactions may be divided into two distinct groups depending on the source of energy for the synthesis (Caldwell, 1980). In most cases the energy derives from the activated endogenous conjugation agent and this occurs with the glucuronic acid, sulphate, methylation and acetylation mechanisms. However, in the case of glutathione and amino acid conjugations, the energy is derived from prior metabolic activation of the xenobiotic.

Most of the important xenobiotic metabolizing enzymes consist of families of related enzymically active proteins under separate regulatory control each with overlapping, often very marked, substrate selectivities (Jakoby, 1980). This leads to the systems generally appearing to be non-selective and results in their ability to catalyze the transformations of a wide range of substrates. The subcellular distributions of the principle xenobiotic metabolizing enzymes are listed in Table 1.

CONSEQUENCES OF METABOLISM

The processes of metabolism are critical determinants of the disposition of a xenobiotic in the body. In a small number of cases, xenobiotics may be retained in the body as a consequence of their metabolism. This may be due to the formation of highly reactive intermediates in metabolism which may undergo covalent binding to tissue macromolecules such as proteins or lipids or as a result of their ability to be incorporated into macromolecules by normal biosynthetic processes; this is particularly the case for acidic xenobiotics which may be incorporated into various pathways of lipid metabolism including fatty acid synthesis, triglyceride synthesis and esterification with cholesterol (Caldwell and Marsh, 1983; Quistad and Hutson, 1986). Lastly, a small number of metabolites may be so lipophilic as to be retained in the tissues of the body. A good example of this is seen in the case of the methyl sulphones

TABLE 1 - Subcellular distribution of xenobiotic metabolizing enzymes

Mitochondria	Amino acyl transferases
Endoplasmic reticulum	Cytochrome P-450 oxygenases Flavoprotein oxidases Esterases Epoxide hydrolase Glucuronyl transferases
Cytoplasm	Glutathione transferases Sulphotransferases Epoxide hydrolases $\underline{N}$-Acetyl transferases Methyl transferases

(drawn from Gibson and Skett (1986))

produced by the transformation of glutathione conjugates of haloaromatic compounds (Bakke, 1986).

In the vast majority of cases the processes of metabolism lead to the elimination of xenobiotics from the body. The most important routes by which xenobiotics and their metabolites are lost from the body are by renal elimination, leading to their voiding in the urine (Pritchard and James, 1982), and excretion by the liver into the bile leading to loss in the faeces (Smith, 1973). In both cases, the physicochemical changes occurring as a consequence of metabolism, i.e. increasing water solubility and polarity and, in the case of the bile, molecular weight, serve to enhance the affinity of the xenobiotic for elimination by these pathways. In addition, with volatile compounds exhalation is an important route of loss, while there are several minor routes by which xenobiotics may be eliminated. These include loss across the skin in the sweat and other secretions and in the other secretions of the body such as saliva, breast milk, etc.

The great majority of the organs and tissues of the body have some xenobiotic metabolizing capacity but this is highly variable (Wolf, 1984). It is most convenient to consider the organ distribution of xenobiotic metabolism in three ways. Firstly, high metabolizing capacity is seen in the gut wall, the skin and the lung, all organs which constitute 'portals of entry' of xenobiotics into the body from the external environment. The metabolic capacity of these tissues serves to restrict entry of the xenobiotic into the systemic circulation. Secondly, the liver (Peterson and Holtzmann, 1980) and kidney (Jones et al., 1980) have high metabolic activity. Both of these organs are involved in the excretion of xenobiotics from the body and, in the case of the liver, it may also contribute to protection from xenobiotics absorbed from the gastrointestinal tract, since it lies between the gut and the systemic circulation. Lastly, a variety of so-called target organs for toxicity can be marked by their particular metabolic capacities towards small numbers of xenobiotics. In quantitative terms the contribution of these target organs to xenobiotic metabolism is small but it is often an important determinant of a highly localized adverse response to a xenobiotic (Cohen, 1986). There occurs a frequently complex interplay between the various organs of the body in terms of their contributions to xenobiotic metabolism. The best documented example of this is seen with the enterohepatic circulation (Smith, 1973) in which a compound is metabolized and eliminated in the bile. Its metabolites are then re-absorbed from the gastrointestinal tract leading to both further processing and prolonged exposure of tissues to the drug and its metabolites. The consequence of this may be that the pattern of metabolites seen in the bile is very often different from that seen in the urine (Caldwell et al., 1980). This process may be further complicated by the potential contribution of the kidney to the nature of urinary metabolites which is seen in the cases of the transformation of glutathione conjugates (Tate, 1980; Tateishi and Shimizu, 1980) and in the formation of amino acid conjugates of acidic xenobiotics (Caldwell et al., 1980). For compounds whose toxicity is mediated through metabolites, there may occur a two stage metabolic activation in which the first step is performed in the liver followed by a second activation within the target organ for toxicity (Caldwell et al., 1986).

The processes of xenobiotic metabolism are of great significance for the effects which a xenobiotic may exert in the animal body. These may be summarized very briefly in terms of the Phase 1 and 2 reactions as follows. The processes of functionalization result in relatively minor changes to the structure and physicochemical properties of the xenobiotic. This often results in a retention of activity in the metabolites and sometimes may result in a chemical activation (Anders, 1985). In contrast, the conjugation or Phase 2 reactions bring about major changes to the structure and physicochemical properties of the xenobiotic which generally result in a loss of activity and also serve to facilitate elimination from the body (Caldwell, 1982). Only relatively rarely is metabolic activation the consequence of a conjugation reaction, although well-documented examples do exist (Caldwell, 1982; Mulder et al., 1986) and several cases are cited later in this Volume. Metabolism thus determines both the nature of the molecules present in the body after the administration of a xenobiotic and the time courses of their concentrations (their kinetics), with the result that the overall profile of biological activity of the xenobiotic will be greatly influenced by its metabolic transformations. Accordingly, factors able to influence metabolic transformation are of considerable importance, and a selection of examples of these factors are listed in Table 2.

TABLE 2 - Some factors affecting xenobiotic metabolism

Physiological	Endocrine influences
	Ontogenesis and senescence
Exogenous	Induction
	Inhibition
Pathological	
Genetic	Interspecies
	Intraspecies

(drawn from Gibson and Skett (1986))

CONCLUSIONS

The foregoing brief survey has served to sketch the major features of the life cycle of a xenobiotic in the organism and has indicated the significance of the various processes involved for the biological activity of the xenobiotic. Absorption, distribution, metabolism and excretion operate together to determine the nature of compounds present in the body following exposure to a xenobiotic and the concentration-time relationships for each compound in the various tissues and organs of the body. Overall, therefore, these processes represent a major influence upon the profile of biological activity of a xenobiotic, showing the importance of considering "what the body does to the xenobiotic" when asking "what does the xenobiotic do to the body?".

The chapters which follow in this section exemplify in greater detail the roles in xenobiotic metabolism of the gastro-intestinal tract (and its contents), the respiratory tract, the skin, the kidney and the brain. Section II of the book emphasizes the role of metabolic transformation in the generation of intermediary, often chemically reactive, metabolites and the importance of such reactions. The last eight chapters (Section III) describe some techniques available for the study of this important aspect of xenobiotic metabolism.

REFERENCES

Anders, M.W., 1985, (editor) Bioactivation of Foreign Compounds (New York: Academic Press).

Bakke, J.E., 1986, Catabolism of glutathione conjugates. In Xenobiotic Conjugation Chemistry, edited by G.D. Paulson, J. Caldwell, D.H. Hutson and J.J. Menn, ACS Symposium Series 299 (Washington, DC: American Chemical Society), pp. 301-321.

Benet, L.Z. and Sheiner, L.B., 1985, Pharmacokinetics: the dynamics of drug absorption, distribution and elimination. In The Pharmacological Basis of Therapeutics, edited by A.G. Gilman, I.S. Goodman, T.W. Rall and F. Murad (New York: Macmillan), pp. 3-34.

Boobis,A.R., Caldwell,J., De Matteis,F. and Elcombe,C.R., 1985, (editors) Microsomes and Drug Oxidations (London:Taylor and Francis).

Caldwell, J., 1980, In Concepts in Drug Metabolism, Part A, edited by P. Jenner and B. Testa (New York: Marcel Dekker), pp. 211-237.

Caldwell, J., 1982, Conjugation reactions in foreign compound metabolism: definition, consequences and species variations. Drug Metabolism Reviews, 13, 745-777.

Caldwell, J. and Marsh, M.V., 1983, Inter-relationships between xenobiotic metabolism and lipid biosynthesis. Biochemical Pharmacology, 32, 1667-1672.

Caldwell, J., Idle, J.R. and Smith, R.L., 1980, The amino acid conjugations. In Extrahepatic Metabolism of Drugs and Other Foreign Compounds, edited by T.E. Gram (New York: SP Medical and Scientific), pp. 453-477.

Caldwell, J., Sangster, S.A. and Sutton, J.D., 1986, The role of metabolic activation in target organ toxicity. In Target Organ Toxicity, edited by G.M. Cohen (Boca Raton, FL: CRC Press), pp. 37-54.

Cohen, G.M., 1986, (editor) Target Organ Toxicity, Vol. 1, Boca Raton, FL: CRC Press).
George,C.F. and Shand,D.G., 1982, Presystemic drug metabolism in the liver. In Presystemic Drug Elimination, edited by C.F.George, D.G.Shand and A.G.Renwick (London:Butterworth) pp.69-77.

Gibson, G.G. and Skett, P., 1986, <u>Introduction to Drug Metabolism</u>, (London: Chapman and Hall).

Goldman, P., 1982, Role of the intestinal microflora. In <u>Metabolic Basis of Detoxication</u>, edited by W.B. Jakoby, J.R. Bend and J. Caldwell (New York: Academic Press), pp. 323-337.

Hewick, D.S., 1982, Reductive metabolism of nitrogen-containing functional groups. In <u>Metabolic Basis of Detoxication</u>, edited by W.B. Jakoby, J.R. Bend and J. Caldwell (New York: Academic Press), pp. 151-170.

Heymann, E., 1982, Hydrolysis of carboxylic esters and amides. In <u>Metabolic Basis of Detoxication</u>, edited by W.B. Jakoby, J.R. Bend and J. Caldwell (New York: Academic Press), pp. 229-245.

Jakoby, W.B. (editor), 1980, <u>Enzymatic Basis of Detoxication</u>, Vols. I and II (New York: Academic Press).

Jones, D.P., Orrenius, S. and Jakobson, S.W., 1980, Cytochrome P-450-linked monooxygenase systems in the kidney. In <u>Extrahepatic Metabolism of Drugs and Other Foreign Compounds</u>, edited by T.E. Gram (New York: SP Medical and Scientific), pp. 123-158.

Levy, R. and Shand, D.G. (editors), 1984, Symposium: Clinical Implication of Drug-Protein Binding. Clinical Pharmacokinetics, $\underline{9}$, suppl. 1, 1-104.

Mulder, G.J., Meerman, J.H. and van den Goorbergh, A., 1986, Bioactivation of xenobiotics by conjugation. In <u>Xenobiotic Conjugation Chemistry</u>, edited by G.D. Paulson, J. Caldwell, D.H. Hutson and J.J. Menn, ACS Symposium Series 299 (Washington, DC: American Chemical Society), pp. 282-300.

Myrhe, K., Rugstad, H.E. and Hansen, T., 1982, Clinical pharmacokinetics of methyldopa. Clinical Pharmacokinetics, $\underline{7}$, 221-233.

Peterson, F.J. and Holtzmann, J.L., 1980, Drug metabolism in the liver - a perspective. In <u>Extrahepatic Metabolism of Drugs and Other Foreign Compounds</u>, edited by T.E. Gram (New York: SP Medical and Scientific), pp. 1-121.

Pritchard, J.B. and James, M.O., 1982, Metabolism and urinary excretion. In <u>Metabolic Basis of Detoxication</u>, edited by W.B. Jakoby, J.R. Bend and J. Caldwell (New York: Academic Press), pp. 339-357.

Quistad, G.B. and Hutson, D.H., 1986, Lipophilic xenobiotic conjugates. In <u>Xenobiotic Conjugation Chemistry</u>, edited by G.D. Paulson, J. Caldwell, D.H. Hutson and J.J. Menn, ACS Symposium Series 299 (Washington, DC: American Chemical Society), pp. 204-213.

Renwick, A.G., 1983, Unmetabolized compounds. In *Biological Basis of Detoxication*, edited by J. Caldwell and W.B. Jakoby (New York: Academic Press), pp. 151-179.
Smith, R.L., 1973, *The Excretory Function of Bile*, (London: Chapman and Hall).
Tate, S.S., 1980, Enzymes of mercapturic acid formation. In *Enzymatic Basis of Detoxication*, Vol. II, edited by W.B. Jakoby (New York: Academic Press), pp. 95-120.
Tateishi, M. and Shimizu, H., 1980, Cysteine conjugate β-lyase. In *Enzymatic Basis of Detoxication*, Vol. II, edited by W.B. Jakoby (New York: Academic Press), pp. 121-130.
Testa, B., 1983, Nonenzymatic biotransformation. In *Biological Basis of Detoxication*, edited by J. Caldwell and W.B. Jakoby (New York: Academic Press), pp. 137-150.
Williams, R.T., 1959, *Detoxication and Mechanisms*, 2nd edition (London: Chapman and Hall).
Wolf, C.R., 1984, Distribution and regulation of drug-metabolizing enzymes in mammalian tissues. In *Foreign Compound Metabolism*, edited by J. Caldwell and G.D. Paulson (London: Taylor and Francis), pp. 37-49.
Zeigler, D.M., 1980, Microsomal flavin-containing monooxygenase: oxygenation of nucleophilic nitrogen and sulfur compounds. In *Enzymatic Basis of Detoxication*, Vol. I, edited by W.B. Jakoby (New York: Academic Press), pp. 201-227.

METABOLISM OF XENOBIOTICS IN THE GASTRO-INTESTINAL TRACT

Andrew G. Renwick and Charles F. George

Clinical Pharmacology,
Medical and Biological Sciences Building,
Bassett Crescent East,
Southampton, S09 3TU, UK

INTRODUCTION
The gut is a major route of entry into the body for many foreign compounds and represents the first potential site for metabolism. The gastro-intestinal tract is frequently regarded as a simple organ involved in the digestion and absorption of nutrients. However, when considering the metabolism of foreign compounds it provides a diversity and complexity not found in organs such as the liver. An indication of this complexity is given in Table 1; clearly it is impossible to cover such a large subject adequately and in depth within a single chapter. Therefore the following account concentrates on those morphological or constitutional elements that are characteristic of and unique to the gastro-intestinal tract. These concepts are illustrated with data for a limited number of substrates, principally levodopa, sulphinpyrazone and sulindac.

METABOLISM BY THE INTESTINAL WALL
The wall of the upper intestine is the first major metabolic barrier for compounds that are resistant to acid catalyzed decomposition and digestive enzymes present in the lumen of the upper intestine. Thus, most studies on drug metabolism by the intestinal wall have concentrated on the activity in the duodenum and ileum.

In vitro metabolism
A number of in vitro techniques have been employed (Table 2) which involve widely differing degrees of structural integrity.
Everted intestinal sacs are one of the earliest and simplest techniques to have been applied to studies on the absorption and metabolism of foreign compounds (Barr and

TABLE 1 - Important characteristics of the gastrointestinal tract with respect to the metabolism of xenobiotics

Gut lumen	- extremes of physiological pH - host digestive enzymes - intestinal microflora
Gut wall	- large surface area for absorption - transport systems for absorption - major site of first-pass inactivation
Functional heterogeneity	- different regions of the gut have different functions (e.g. stomach, colon etc.) - different functions result in different metabolic capacities with respect to xenobiotics

TABLE 2 - In vitro methods for investigating xenobiotic metabolism by the intestinal wall

1. Everted intestinal sacs - allow presentation of the foreign compound from mucosal or serosal surface.
2. Tissue culture - preparations of varying degrees of complexity/integrity, e.g. explants, slices or isolated cells.
3. Cell fractionation - preparations of microsomes, cytosol etc.
4. Brush border vesicles - used primarily for studies on transport processes

Riegelman, 1970). Major advantages of this technique are the structural integrity of the wall architecture and the ability to present the drug from the mucosal surface. In contrast, the other methods shown in Table 2 can provide more detailed and specific information relating to the enzymes involved and their distribution along the intestine. Published information derived by in vitro techniques was reviewed by George (1981) and Caldwell and Varwell Marsh (1982). Biotransformation reactions detected include a wide range of both phase 1 (oxidation, reduction and hydrolysis) and phase 2 (conjugation) processes (Caldwell and Varwell Marsh, 1982). The distribution of drug metabolizing activity is not uniform either along the intestinal wall or within a specific region of the intestine. The duodenum and jejunum have the highest activity for both cytochrome P-450 mediated oxidations (Wattenberg, 1972) and for conjugations such as glucuronidation (Hartiala, 1973).

The decarboxylation of levodopa to dopamine is an important example of the generation of a biologically active intermediary metabolite. Dopamine is subsequently oxidised by mono-amine oxidase (MAO) to dihydroxyphenylacetic acid (Figure 1). Because dopamine is the pharmacologically active species, considerable attention has been focused on the enzymes L-aromatic amino acid decarboxylase (LAAD) and MAO. LAAD activity is present in large amounts in the intestinal mucosa with the highest activity in the jejunum of dog (Sasahara et al., 1981a) and rat (Table 3).

TABLE 3 - Levodopa metabolism by the wall of the gastrointestinal tract of the rat

Region of gastro-intestinal tract	% levodopa remaining unmetabolised	
	Normal	+ Benserazide (0.8mM)
Stomach	79 ± 17	90 ± 5
Duodenum	58 ± 17	90 ± 5
Jejunum	10 ± 4	87 ± 4
Ileum	26 ± 24	91 ± 3
Caecum	54 ± 22	84 ± 14
Colon	22 ± 11	89 ± 5

The results are the mean ± SD of the % of the added [^{14}C]levodopa (0.1μCi; 0.01 μmol) remaining after aerobic incubation at 37°C with an homogenate containing 60 mg of tissue. The nature of the radioactivity was defined by tlc. (From Wood et al., unpublished).

The distribution of xenobiotic metabolizing enzyme activity is not uniform within the epithelial cells of the small intestine. Dubey and Singh (1988a) have described a method of differential isolation of cells along the villus to crypt surface of the epithelium. Cells from the upper and mid-villus regions had more cytochrome P-450 mediated oxidation activity than crypt region cells using benzo[a]-pyrene and 7-ethoxycoumarin as substrates. The activity in all regions was increased by pretreatment of the animals with 3-methylcholanthrene (orally) or phenobarbitone

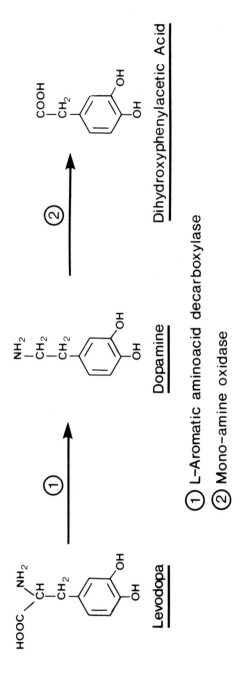

Figure 1. The metabolism of levodopa to dopamine. Other important pathways of metabolism, for example O-methylation, are not discussed in this chapter

(by intraperitoneal injection) (Dubey and Singh, 1988a). A similar distribution was found for the phase 2 enzymes, UDP-glucuronyltransferase and glutathione-S-transferase, with the highest activities in cells isolated from the upper villus region (Dubey and Singh, 1988b). The concentration of the cofactor UDPGA was also highest in the villus tip cells, but these cells contained the lowest amounts of glutathione. Clearly, the greatest xenobiotic metabolizing activity is provided at the very tips of the villi of the duodenum providing the possibility of an effective immediate metabolic barrier to the entry of potentially toxic xenobiotics.

An important feature of the epithelium of the upper intestine is the presence of micro-villi which form the brush border. A number of techniques for the isolation of the brush border as vesicles have been described (Forstner et al., 1968; Schmitz et al., 1973; Hopfer et al., 1973; Kessler et al., 1978; Ganapathy et al., 1981) although in most studies these preparations have been used to investigate transport and uptake processes. We have recently used such a preparation (Figure 2) to study the transport of levodopa by rat intestinal brush border vesicles (Wood et al., unpublished). The "uptake" of [^{14}C]levodopa into the vesicles was much slower than reported for other amino acids. The uptake was not affected by the presence of other amino acid substrates for the transport process or by benserazide, an inhibitor of LAAD, but was inhibited by the addition of ascorbate. The radioactivity was found to be covalently bound to the membrane vesicles; this observation and the inhibition by ascorbate suggests the generation of a reactive intermediary metabolite of levodopa. The enzymatic basis for the formation of this reactive intermediate of levodopa by intestinal brush border vesicles is not known and is currently under investigation. Active covalently binding metabolites of α-methyldopa (Dybing et al., 1976) and other catechols (Ito et al., 1988) have been reported to be formed by cytochrome P-450 mediated oxidation and by tyrosine oxidase, and probably involve an o-quinone intermediate (Figure 3). The data on levodopa indicate that there may be metabolic activity in the microvilli of the epithelial cells. This raises the intriguing possibility that xenobiotic metabolizing enzymes may be the first intracellular components encountered by absorbed foreign compounds.

In vivo metabolism

The detection of a specific metabolic reaction in vitro should not be equated automatically with a significant

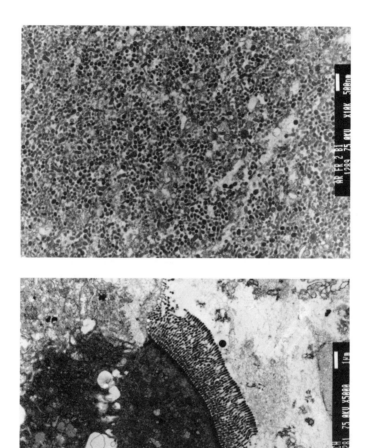

Figure 2. The preparation of brush border vesicles from rat small intestine. a) initial crude homogenate showing a section of brush border membrane (x 5,000 approx.). b) resuspended vesicle pellet (x 10,000 approx.). Electron micrographs were prepared by Dr. J.T.R. Fitzsimons, Neurophysiology, University of Southampton.

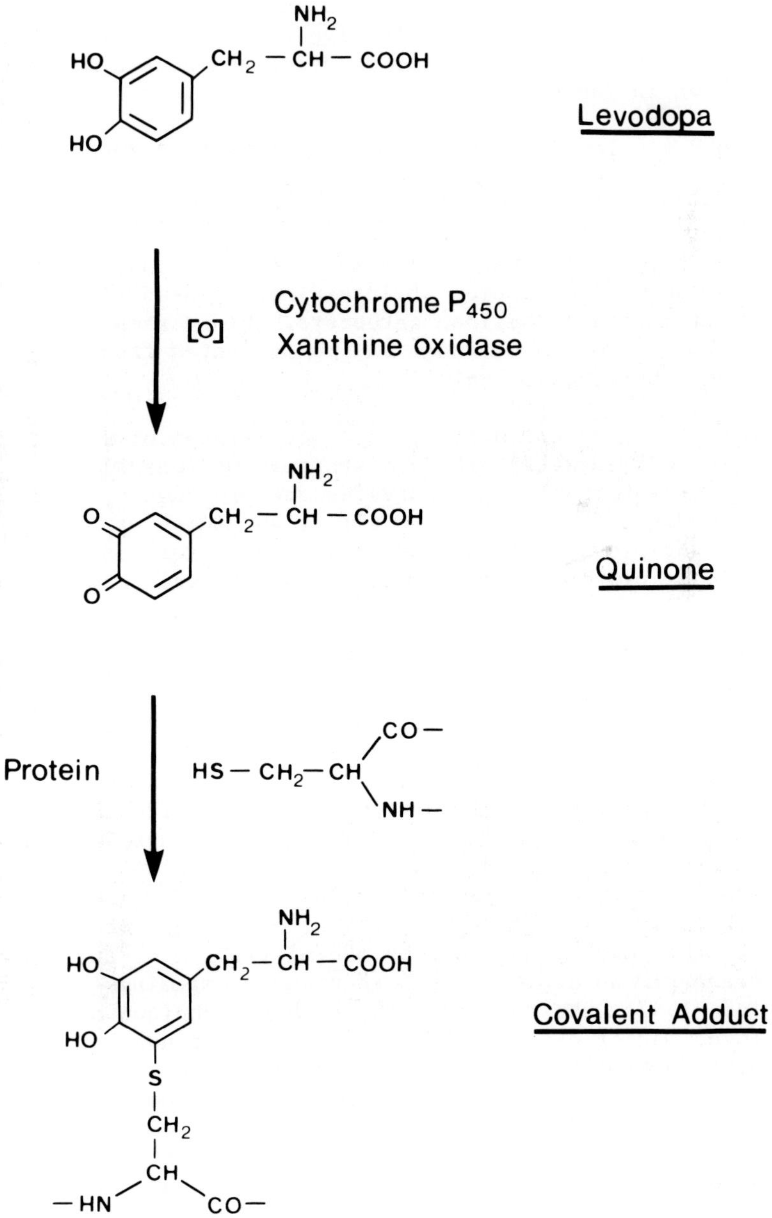

Figure 3. The metabolic oxidation of levodopa to a covalently binding species. Based on Dybing et al., 1976 and Ito et al., 1988

contribution to the overall fate of a compound. The true
importance of the gut wall to the fate of orally
administered foreign compounds can only be derived by a
combination of in vitro and in vivo techniques. The
principal in vivo methods for studying xenobiotic metabolisr
are shown in Table 4.

TABLE 4 - In vivo methods for investigating xenobiotic metabolism by the intestinal wall

---

1. In situ intestinal loop - xenobiotic is added to the lumen of a section of intestine isolated either by ligation or balloon catheters. Blood samples, or total venous drainage may be collected from the hepatic portal vein.

2. Route-dependent differences - comparison of metabolite profiles after oral and systemic peripheral administration demonstrates the presence of first-pass metabolism due to either gut lumen, gut wall, gut flora or liver (see Figure 4). Administration by infusion into the hepatic portal vein can define the contribution of the liver

---

The most important technique which provides direct
evidence of gut wall metabolism is the in situ intestinal
loop preparation (Barr and Riegelman, 1970). In addition to
the technical demands of surgery, a problem with this
technique is that sympathetic drive may decrease intestinal
blood flow. This could reduce the venous effluent flow and
decrease the rate of transfer through the epithelium,
thereby overestimating the extent of metabolism at this
site. This may be overcome by administration of an
α-adrenergic antagonist, such as phenoxybenzamine (Ilett and
Davies, 1982). Most studies using this technique have
employed larger experimental animals such as dogs or
rabbits. Few data have been published in man, because
cannulation of the hepatic portal vein is seldom performed.
Adventitious studies in patients with portal cannulae have
demonstrated that the gut wall may be an important site both
of oxidation (e.g. N-dealkylation of flurazepam; Mahon et
al., 1977) and of conjugation (e.g. sulphate conjugation of
ethinyloestradiol; Back et al., 1982) but not all substrates
are metabolized (e.g. imipramine; Dencker et al., 1976).

The use of oral and intravenous doses to define the extent of first pass metabolism cannot differentiate between gut lumen, gut wall and liver (Figure 4). A difference in the pattern of plasma or urinary metabolites after oral and

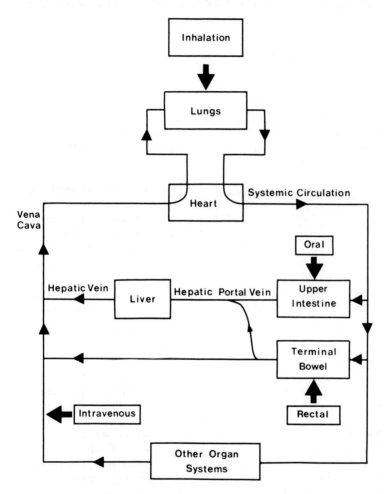

Figure 4. The position of the intestine within the general circulation. Unlike other routes of administration, foreign compounds taken orally have to pass the metabolic barriers of the intestine and liver prior to reaching the general circulation.

intravenous administration indicates the possibility of metabolism in the gut wall but additional data are necessary. Comparison of the areas under the plasma concentration-time curves for levodopa and its metabolites in dogs following intra-duodenal and intravenous administration showed a route dependent difference. Comparison with data obtained after intra hepatic-portal administration provided further evidence that the gut wall was a major site of first pass decarboxylation in this species (Sasahara et al., 1981a). Another example is the β-adrenoceptor agonist isoprenaline since about 80% is excreted as the sulphate conjugate of the parent drug in urine after oral administration, but only 1% after intravenous dosing (Connolly et al., 1972). Since the hepatic portal blood flow represents about 20% of cardiac output it is unlikely that such a dramatic difference could be due to hepatic metabolism (see Figure 4). Definitive proof, however, came from *in situ* loop preparations in the dog which showed that the sulphate conjugate was the major constituent in the portal venous blood following intraluminal instillation (George et al., 1974). Isoprenaline also illustrates an important aspect of the use of route of administration to define the role of the gut wall in first pass metabolism. A route of administration difference is dependent on limited amounts of the compound reaching sites of metabolism in the intestine following systemic administration. This was demonstrated directly for isoprenaline since no sulphate conjugate was detected in the venous effluent of the *in situ* dog intestine preparation when the drug was delivered by the arterial supply (Ilett et al., 1980).

Information on the metabolic capacity of the intestine can be obtained by comparison of the systemic profile of parent drug and its metabolites after the compound is instilled into isolated loops from different regions of the intestine. This *in vivo* technique has been used in the dog to show that the greatest absorption of intact levodopa occurred following intra-duodenal instillation, but that the greatest formation of circulating metabolites of dopamine occurred following intra-jejunal administration (Sasahara et al., 1981b). The results from this *in vivo* study were in agreement with the distribution of LAAD activity measured *in vitro*. It also showed that metabolism in the gut was a major determinant of the fate of levodopa *in vivo* because, if the liver were the main site of metabolism, the profile of parent drug to metabolites would have been independent of the region of the intestine from which the drug was absorbed. This is not true for the buccal cavity and rectum which do not drain into the hepatic portal vein. A similar study in man found no difference in the area under the

plasma concentration-time curve for levodopa in humans
following instillation into the duodenum or ileum (Gundert-
Remy et al., 1983). However, this study was not directly
comparable to that in the dog because the patients had
received a peripheral decarboxylase inhibitor to abolish
intestinal decarboxylation. In addition only the proximal
site of infusion was limited, and thus drug infused into the
duodenum would also reach the ileum. A better technique for
such studies in man has been described by Vidon et al.
(1985) in which the proximal end of the intestinal segment
is occluded by a balloon. The drug solution is perfused
into the intestine immediately distal to the balloon and the
perfusate collected by another tube positioned 30 cm below
the balloon. Thus, only 30 cm of intestine is exposed to
the perfused drug. No difference in the apparent
conjugation of oxprenolol was found following its perfusion
into the jejunum and ileum using this technique (Godbillon
et al., 1987); however, few other metabolic data have been
reported using this method. Information on the role of the
stomach and upper intestine in the metabolism of foreign
compounds in man can be provided by studies in patients who
have undergone gastrectomy. Such individuals may be
provided with a jejunostomy postoperatively to facilitate
feeding. Comparison of the fate of compounds given to such
patients via the jejunostomy with that in normal individuals
given an oral dose may produce a useful model (Nelson et
al., 1986).

METABOLISM BY THE INTESTINAL FLORA

The intestinal flora represent a complex, dynamic
metabolizing system which is susceptible to considerable
external manipulation. The gut flora can be summarised as a
vast number of organisms, up to 10^{10} bacteria per gram of
gut contents, most of which are anaerobic organisms present
in the lower bowel. While the generalisation is a
reasonable description of the human flora there are large
interspecies differences both in the numbers of organisms
present and their distribution along the gut (Drasar, 1988).
Particularly pertinent to studies in laboratory animals is
the presence of high numbers of bacteria in the stomach and
upper intestine of mice and rats. The distribution in
rabbits more closely resembles that in man, possibly due to
the lower gastric pH. Despite the more uniform distribution
of bacteria along the gut of the rat, nearly all the
microbial xenobiotic metabolizing activity is found in the
caecum and colon (e.g. azo- and nitro-reductase, β-glucur-
onidase, cyclamate hydrolysis and sulphoxide reduction;
Renwick, 1982; Renwick et al., 1982). Because of this

distribution within the gut, xenobiotic substrates reach the gut flora either due to incomplete absorption of an oral dose or following biliary excretion or secretion into the gut lumen from a systemic dose. This has important implications when considering methods to define the role of the gut flora in vivo.

Methods for studying the role of the gut flora in the disposition of xenobiotics have been reviewed recently (Coates et al., 1988). In this chapter these methods will be discussed in relation to the information they have provided concerning the reduction of the sulphoxide groups of the drugs sulphinpyrazone and sulindac. These reactions are important since in each case the parent sulphoxide may be regarded as a prodrug because the sulphide (or thioether) metabolite is considerably more potent. In the case of sulphinpyrazone, the sulphide is an inhibitor of platelet aggregation, whilst the sulphide of sulindac is an anti-inflammatory agent. In both cases the sulphide is an intermediate metabolite which is considerably more lipid soluble than the corresponding sulphoxide and therefore is not excreted as such in the urine. These sulphide metabolites may undergo oxidation (either at the S atom to reform the sulphoxide, or at carbon atoms) or they may undergo conjugation.

In vitro metabolism

Preparations of intestinal bacteria for in vitro studies may be obtained readily by removal from the gut lumen directly or from freshly voided faeces (Table 5). Such preparations should be made under anaerobic conditions and oxygen removed from media to avoid loss of oxygen-sensitive strict anaerobes. It should be appreciated that the luminal or faecal flora may not be representative of the flora associated with the wall of the lower intestine.

The most direct method is a simple, short-term incubation of a mixed culture with the xenobiotic of interest. Comparison of the activity in such a preparation with that present in animal tissues such as liver can give an indication of the potential of the bacterial flora to effect a reaction. In the case of the reduction of sulphinpyrazone to its active sulphide (Figure 5) the caecal contents were the most active system detected in either rat or rabbit (Table 6), and long term incubation was unnecessary. A potential problem with long term incubation or continuous culture systems is the selective growth of some organisms; because of this the composition of the flora at the end of the incubation may be different from that first isolated (Coates et al., 1988).

TABLE 5 - <u>In vitro</u> methods for studying xenobiotic biotransformation by the gut flora

---

1. <u>Mixed cultures</u> - using samples of caecal contents or faeces, etc.

    a) Short term incubation - simple medium; little selective growth; limited sensitivity.

    b) Long term incubations - nutrient medium; selective growth likely; higher sensitivity.

2. <u>Pure cultures</u> - long term incubations in nutrient medium will give unambiguous results; no bacterial interactions possible.

3. <u>Bacterial enzymes</u> - sterile enzyme preparations (fecalase) have been proposed as <u>in vitro</u> activation systems for short term tests

---

TABLE 6 - <u>In vitro</u> reduction of sulphinpyrazone under anaerobic conditions

Preparation	Enzyme activity (μmol/g tissue/h)	
	Rat[a]	Rabbit[b]
Liver	0.002 ± 0.001	0.055 ± 0.033
Kidney; spleen;) duodenum wall;) caecum wall)	ND	ND
Duodenum contents	0.003 ± 0.002	ND
Caecum contents	0.269 ± 0.038	0.311 ± 0.156

ND - not detected; <0.001 μmol/g/h

a) from Renwick et al., 1982
b) from Strong et al., 1984a

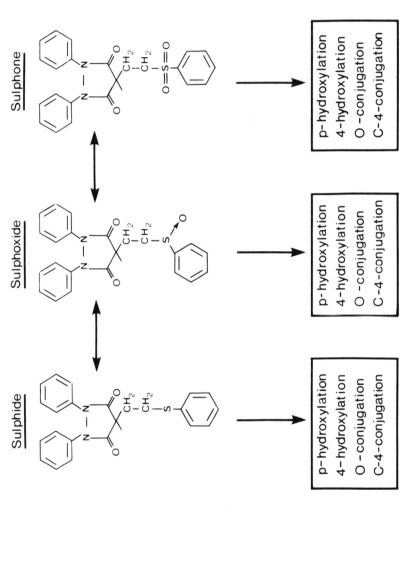

Figure 5. The interconversion of sulphinpyrazone (sulphoxide) with its oxidised and reduced forms. All three redox states undergo oxidation and conjugation reactions. The sulphide is a major circulating metabolite.

The use of pure cultures of isolated strains of bacteria can provide specific information concerning bacterial metabolism, but can be misleading. For example, ten-fold differences in enzyme activity between different organisms can pale into insignificance when considered in the context of up to 10^8-fold differences in the numbers of the different strains present. Closely related organisms may exhibit wide differences in metabolic activity so that misleading extrapolations may be made. An additional problem with this "pure" system is that interactions between organisms, which may be essential for certain reactions such as cyclamate hydrolysis (Renwick, 1988) are not possible. The finding that most strains of E. coli would reduce both sulphinpyrazone and sulindac (Table 7) whilst little activity was found in anaerobes is at variance with the in vivo data discussed below.

TABLE 7 - The reduction of sulphoxide-containing drugs by strains of bacteria isolated from human faeces

	Number of strains reducing >10% of substrate Number of strains studied	
	Sulphinpyrazone	Sulindac
Aerobes		
Escherichia coli	46/53	47/47
Enterobacter	5/13	13/13
Klebsiella spp.	3/14	14/14
Enterococci	1/18	2/14
Anaerobes		
Bacteroides spp.	0/43	31/46
Clostridia	0/22	14/14
Bifidobacteria	0/10	0/10
Gram positive rods	0/31	2/31

Data from Strong et al., 1987

In vivo metabolism

There are 3 principal methods available for studying the role of the gut flora in xenobiotic metabolism in vivo (Table 8).

TABLE 8 - In vivo methods for studying xenobiotic biotransformations by the gut flora

---

1. Route of administration - oral/systemic differences in metabolism; applicable to incompletely absorbed compounds only.

2. Oral antibiotic treatment - short term suppression of most bacteria; mixtures of antibiotics may be necessary; possible interference with absorption or systemic disposition of compound.

3. Germfree animals - unambiguous answers possible; special facilities are required for long term experiments; other anatomical and physiological differences may obscure findings.

Differences in metabolism after intravenous, oral or rectal administration (Figure 4) can provide strong evidence for the involvement of the gut/liver axis, but do not differentiate between these sites. For the gut flora to produce a route of administration dependent difference in metabolism it is essential that the orally administered dose undergoes incomplete absorption so that much of the substrate available for microbial metabolism is from the oral dose. If a compound is well absorbed and most of the substrate is derived from compound re-entering the gut lumen, e.g. in bile, then no clear route difference will be seen. This is shown clearly in the data obtained after sulphinpyrazone was given to rabbit, rat and man (Figure 5). For the rabbit, the area under the plasma concentration-time curve (AUC) of sulphinpyrazone was significantly lower after its oral administration (Figure 6; columns 2 and 3) compared with intravenous administration (Figure 6; column 1) indicating incomplete absorption of the parent drug. The AUC of the sulphide metabolite showed the opposite of this (Figure 6) indicating that the unabsorbed parent compound was converted to the additional sulphide detected. Data obtained after oral (Figure 7; columns 3 and 4) and systemic (Figure 7; columns 1 and 2) administration to rats showed

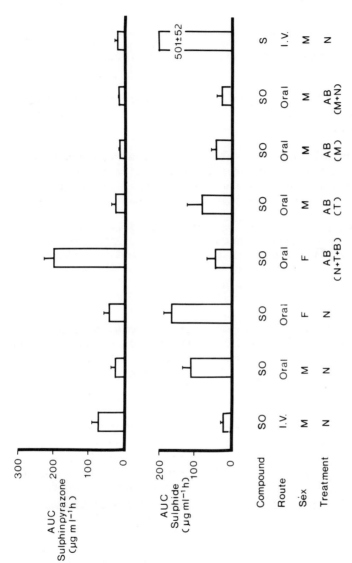

Figure 6. The metabolic interconversion in vivo of sulphinpyrazone and its sulphide in rabbits given a dose of 10 mg/kg. (Adapted from Strong et al., 1984a and 1987). The data are the means with SD given by a vertical bar line. AUC, area under the plasma concentration-time curve; SO, sulphinpyrazone; S, sulphide metabolite of sulphinpyrazone; IV, intravenous; M, male; F, female; N, normal; AB, antibiotic treated; (N), neomycin; (T), tetracycline; (B) bacitracin; (M) metronidazole

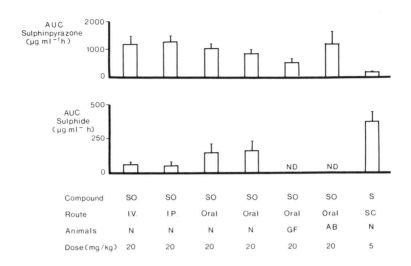

Figure 7. The metabolic interconversion in vivo of sulphinpyrazone and its sulphide in rats. (Adapted from Renwick et al., 1982). The data are the means with SD given by a vertical bar line. ND, not detectable; SC, subcutaneous; IP, intraperitoneal; GF, germfree; For other abbreviations see Figure 6.

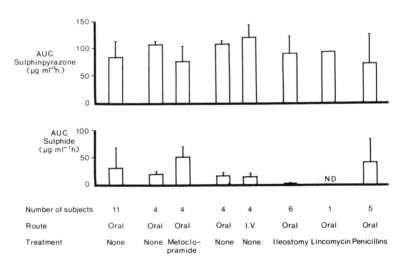

Figure 8. The metabolism of sulphinpyrazone to its active sulphide analogue in man (Adapted from Strong et al., 1984b). The results are the means with SD given by a vertical bar line. For abbreviations see Figures 6 and 7.

similar but less pronounced differences to those found in rabbits. In contrast there was no significant difference in the AUC of sulphinpyrazone in man between oral and intravenous (Figure 8; columns 4 and 5) dosing. This indicates that the drug was essentially completely absorbed in man and that the substrate for reduction was probably derived from the systemic drug. Thus no route-dependent difference was found in the AUC of the sulphide. However, when subjects were given metoclopramide, which increases upper intestinal motility, there was a decrease in AUC of the parent compound compared to control data (Figure 8; columns 2 and 3) and a significant increase in the AUC of the sulphide metabolite.

The use of oral antibiotics to suppress the intestinal flora provides a powerful technique for studying the role of the gut flora *in vivo*. A number of different regimens are available employing single or multiple antibiotics to give effective suppression of the gut flora (Coates et al., 1988). Unambiguous results are possible in animals using a mixture of antibiotics; for example, formation of sulphinpyrazone sulphide in rats was completely abolished by prior treatment with neomycin, bacitracin and tetracycline, with little apparent effect on the AUC of the parent drug (Figure 7; column 6). However, when a comparable protocol was used in rabbits (Figure 6; column 4) the AUC of the sulphide was reduced compared to normal (Figure 6; column 3) but not abolished. Of greater importance was the greatly increased AUC of sulphinpyrazone itself showing that this antibiotic cocktail decreased the systemic clearance of the drug in rabbits. Since many of the pathways of metabolism of the sulphide are similar to those of the sulphoxide (Figure 5) it is likely that the AUC of the sulphide may also have been exaggerated. Studies with single antibiotics or simple mixtures (Figure 6; columns 5-7) allowed suppression of gut flora metabolism without a marked systemic effect. However, with metronidazole, the AUC of the parent compound was reduced compared to controls suggesting interference with absorption, a phenomenon which has been reported previously with the use of antibiotics in metabolism studies (Gardner and Renwick, 1978; Renwick, 1982). Thus a small decrease in the formation of a postulated microbial metabolite following oral antibiotics could arise simply from reduced absorption and therefore a decreased amount of substrate available for tissue metabolism. Although broad spectrum antibiotics have been given to man specifically to study the influence of the gut flora on xenobiotic metabolism (Renwick, 1988) such clinically non-essential use is questionable. Studies on the fate of sulphinpyrazone in patients receiving antibiotics therapeutically showed no consistent effect of penicillins (Figure 8; column 8) although 1 patient produced no sulphide at all. Of greater interest was a single patient who had received lincomycin for a number of years and showed no sulphide formation, despite a normal AUC for the parent drug (Figure 8; column 7).

Studies in conventional, germfree and ex germfree animals provide the most powerful method for proving the role of the gut flora in the fate of xenobiotics. This approach requires special facilities for long term studies which has limited its widespread use. For example cyclamate metabolism has not been studied in germfree animals because of the long time required for induction of metabolizing activity and its variability in conventional rats (Renwick, 1988). However, germfree animals have been used successfully in short term metabolism studies to provide clear evidence for the involvement of the gut flora in the overall fate of xenobiotics (e.g. Gardner and Renwick, 1978; Renwick et al., 1982; Mikov et al., 1988). The data on the formation of the sulphide metabolite of sulphinpyrazone by germfree rats (Figure 7; column 5) are a good example of two facets of such studies. Firstly, the complete absence of detectable sulphide provides very clear evidence of the essential role played by the gut flora in the generation of this intermediate metabolite. However, the AUC for the parent compound was also reduced significantly compared to normal rats. This could arise from the one or more of other important differences between conventional and germfree animals. These include a thinner intestinal wall, massive caecal enlargement, the persistence of digestive enzymes into the lower gut, differences in intestinal transit time, altered absorption and changes in the cardiovascular and circulatory system in the germfree rat (Coates et al., 1988; Heneghan, 1988). If a small difference in metabolism is found between germfree and conventional rats, this could arise from factors other than direct bacterial metabolism. Therefore, subtle differences should be viewed as suggestive evidence rather than definitive proof.

Comparable studies in "germfree" humans are not ethically possible, although very rare neonatal immune deficiencies require that some babies are maintained in "germfree" conditions. Because most of the anaerobic gut flora of man are associated with the colon, patients who have undergone colectomy with the establishment of an ileostomy represent an interesting model since drugs given orally pass through the upper intestine only. The ileostomy effluent of such patients is not sterile, but contains large numbers of aerobic organisms (Draser and Hill, 1974). Ileostomy effluent did not reduce sulphinpyrazone to its intermediary sulphide metabolite *in vitro* (Table 9), and the AUC for the sulphide metabolite *in vivo* in ileostomy patients (Figure 8; column 6) was thirty-fold lower than that in normal subjects (Strong et al., 1984b). Since the AUC of the parent drug was similar to that in normal subjects, the potential for tissue metabolism was not affected and it was possible to conclude that the anaerobic gut bacteria are essential for the reduction of sulphinpyrazone in man.

TABLE 9 - Sulphoxide reduction *in vitro* under anaerobic conditions

Preparation	Activity (μmol/g/h)	
	Sulphinpyrazone	Sulindac
Rabbit liver		
10,000 g supernatant	0.08 $\pm$ 0.05	0.20 $\pm$ 0.11
microsomes	ND	ND
cytosol	ND	0.31 $\pm$ 0.24
Human - faeces (normal)	0.48 $\pm$ 0.19	1.30 $\pm$ 0.64
Human - faeces (patients receiving metronidazole)	0.16 $\pm$ 0.17	0.61 $\pm$ 0.31
Human - ileostomy effluent	ND	ND

The data are the mean $\pm$ SD
ND - not detectable
Data from Strong et al., 1984a, 1986 and 1987

The *in vivo* and *ex vivo* data on sulphinpyrazone reduction in man (suppression by lincomycin and metronidazole; almost complete absence in ileostomy patients) indicate that the anaerobic bacteria are probably responsible. However, the opposite conclusion was reached by studies with pure strains of intestinal bacteria (Table 7). This apparent discrepancy may be due to an interaction between different types of organisms, in that the aerobes possess the enzyme responsible for reduction and that the anaerobes provide a suitably anoxic environment *in vivo* for this activity to be expressed.

The potential value of studies in ileostomy patients was reinforced by data obtained with the drug sulindac. This drug also undergoes intermediary metabolism to an active sulphide or thioether, which is not an important urinary metabolite (Figure 9). Sulindac is readily reduced by human intestinal bacteria (Tables 7 and 9), but is also a substrate for reduction in the liver, kidneys and possibly other tissues (Anders et al., 1981; Ratnayake et al., 1981; Yoshihara and Tatsumi, 1985; Tables 7 and 9). Therefore the in vitro data suggested the possibility of gut flora involvement but there was no evidence for this in vivo. For example the time to peak plasma concentrations of the sulphide metabolite of sulindac almost coincided with that of the parent compound, whilst the peak of the sulphide of sulphinpyrazone, formed by the gut flora, was delayed by 8-24h compared with the parent drug. When sulindac was given to patients with an ileostomy (Strong et al., 1985) the plasma concentration-time curve of the parent drug was almost identical to that found in normal volunteers (Figure 10). The initial peak of the sulphide was also similar, confirming that absorption and tissue metabolism were unaffected. In patients with ileostomy the initial plasma concentrations of the sulphide decreased rapidly so that levels were undetectable by 24h. In contrast, in normal subjects there was a prolonged slow elimination phase to 55h after the dose (Figure 10). This late phase (12 hours onwards) accounted for 55% of the total AUC of the sulphide in normal subjects but only 7% in ileostomy patients. Since sulindac is excreted in the bile (Dobrinska et al., 1983) it is likely that the late phase in normal subjects resulted from the bacterial reduction of sulindac in the terminal gut (Strong et al., 1985).

CONCLUSIONS

The intestine is a potent but complex site of foreign compound metabolism. The principal components are the gut wall itself and the gut lumen with its bacterial flora. However, both components show a non-uniform distribution of enzyme activity along the intestinal tract. Thus a multiplicity of in vitro and in vivo methods may be necessary to define the importance of the gut in the presystemic or systemic metabolism of foreign compounds following their oral or parenteral administration.

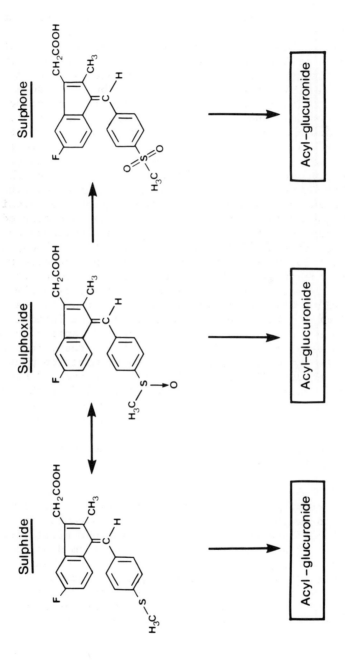

Figure 9. The interconversion of sulindac (sulphoxide) with its oxidised and reduced forms. All three forms undergo acyl-conjugation. The sulphide is the pharmacologically active species.

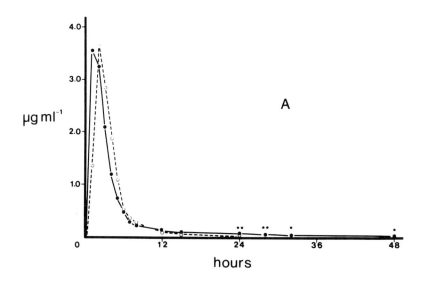

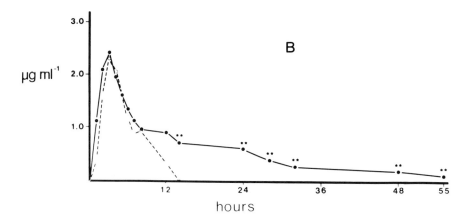

Figure 10. Plasma concentration-time curves for sulindac (A) and its sulphide metabolite (B) in normal subjects (● - ●) and patients with ileostomy (o--o) given a single oral dose of sulindac (200 mg). (Adapted from Strong et al., 1985). The asterisks indicate statistically significant differences between normal subjects compared to patients with an ileostomy (* $P<0.05$, ** $P<0.01$).

REFERENCES

Anders, M.W., Ratnayake, J.H., Hanna, P.E. and Fuchs, J.A., 1981, Thioredoxin dependent sulfoxide reduction by rat renal cytosol. Drug Metabolism and Disposition, 9, 307-310.

Back, D.J., Breckenridge, A.M., MacIver, M., Orme, M., Purba, H.S., Rowe, P.H. and Taylor, I., 1982, The gut wall metabolism of ethinyloestradiol and its contribution to the pre-systemic metabolism of ethinyloestradiol in humans. British Journal of Clinical Pharmacology, 13, 325-330.

Barr, W.H. and Riegelman, S., 1970, Intestinal drug absorption and metabolism I: Comparison of methods and models to study physiological factors of in vitro and in vivo intestinal absorption. Journal of Pharmaceutical Science, 59, 154-163.

Caldwell, J. and Varwell Marsh, M., 1982, Metabolism of drugs by the gastrointestinal tract. In *Presystemic Drug Elimination*, edited by C.F. George, D.G. Shand and A.G. Renwick (London: Butterworths), pp. 29-42.

Coates, M.E., Drasar, B.S., Mallett, A.K. and Rowland, I.R., 1988, Methodological considerations for the study of bacterial metabolism. In *Role of the gut flora in toxicity and cancer*, edited by I.R. Rowland (London: Academic Press), pp. 1-21.

Connolly, M.E., Davies, D.S., Dollery, C.T., Morgan, C.D., Paterson, J.W. and Sandler, M., 1972, Metabolism of isoprenaline in dog and man. British Journal of Pharmacology, 46, 458-472.

Dencker, H., Dencker, S.J., Green, A. and Nagy, A., 1976, Intestinal absorption, demethylation and enterohepatic circulation of imipramine. Clinical Pharmacology and Therapeutics, 20, 584-586.

Dobrinska, M.R., Furst, D.E., Spiegal, T., Vincek, W.C., Tompkins, R., Duggan, D.E., Davies, R.O. and Paulus, H.E., 1983, Biliary secretion of sulindac and metabolites in man. Biopharmaceutics and Drug Disposition, 4, 347-358.

Drasar, B.S., 1988, The bacterial flora of the intestine. In *Role of the gut flora in toxicity and cancer*, edited by I.R. Rowland (London: Academic Press), pp. 23-38.

Draser, B.S. and Hill, M.J., 1974, *Human Intestinal Flora* (London: Academic Press).

Dubey, R.K. and Singh, J., 1988a, Localisation and characterisation of drug-metabolising enzymes along the villus-crypt surface of the rat small intestine I. Monooxygenases. Biochemical Pharmacology, 37, 169-176.

Dubey, R.K. and Singh, J., 1988b, Localisation and characterisation of drug-metabolising enzymes along the villus-crypt surface of the rat small intestine II. Conjugases. Biochemical Pharmacology, 37, 177-184.

Dybing, E., Nelson, S.D., Mitchell, J.R., Sasame, H.A. and Gillette, J.R., 1976, Oxidation of α-methyldopa and other catechols by cytochrome P-450 generated superoxide anion: possible mechanism of methyldopa hepatitis. Molecular Pharmacology, 12, 911-920.

Forstner, G.G., Sabesin, S.M. and Isselbacher, K.J., 1968, Rat intestinal microvillus membranes. Purification and biochemical characterisation. Biochemical Journal, 106, 381-390.

Ganapathy, V., Mendicino, J.R. and Leibach, F.H., 1981, Transport of glycyl-L-proline into intestinal and renal brush border vesicles from rabbit. Journal of Biological Chemistry, 256, 118-124.

Gardner, D.M. and Renwick, A.G., 1978, The reduction of nitrobenzoic acids in the rat. Xenobiotica, 8, 679-690.

George, C.F., 1981, Drug metabolism by the gastrointestinal mucosa. Clinical Pharmacokinetics, 6, 259-274.

George, C.F., Blackwell, E.W. and Davies, D.S., 1974, Metabolism of isoprenaline in the intestine. Journal of Pharmacy and Pharmacology, 26, 265-267.

Godbillon, J., Vidon, N., Palma, R., Pfeiffer, A., Franchisseur, C., Bovet, M., Gossett, G., Bernier, J.J. and Hirtz, J., 1987, Jejunal and ileal absorption of oxprenolol in man: influence of nutrients and digestive secretions on jejunal absorption and systemic availability. British Journal of Clinical Pharmacology, 24, 335-341.

Gundert-Remy, U., Hildebrandt, R., Stiehl, A., Weber, E., Zurcher, G. and Da Prada, M., 1983, Intestinal absorption of levodopa in man. European Journal of Clinical Pharmacology, 25, 69-72.

Hartiala, K., 1973, Metabolism of hormones, drugs and other substrates by the gut. Physiological Reviews, 53, 496-534.

Heneghan, J.B., 1988, Alimentary tract physiology: interactions between the host and its microbial flora. In Role of the gut flora in toxicity and cancer, edited by I.R. Rowland (London: Academic Press), pp. 39-77.

Hopfer, U., Nelson, K., Perrotto, J. and Isselbacher, K.J., 1973, Glucose transport in isolated brush border membrane from rat small intestine. Journal of Biological Chemistry, 248, 25-32.

Ilett, K.F. and Davies, D.S., 1982, In vivo studies of gut wall metabolism. In Presystemic Drug Elimination, edited by C.F. George, D.G. Shand and A.G. Renwick (London: Butterworths), pp. 43-65.

Ilett, K.F., Dollery, C.T. and Davies, D.S., 1980, Isoprenaline conjugation: a "true first pass effect" in the dog intestine. Journal of Pharmacy and Pharmacology, 32, 362.

Ito, S., Kato, T. and Fujita, K., 1988, Covalent binding of catechols to proteins through the sulphydryl group. Biochemical Pharmacology, 37, 1707-1710.

Kessler, M., Acuto, O., Storelli, C., Murer, H., Muller, M. and Semenza, G., 1978, A modified procedure for the rapid preparation of efficiently transporting vesicles from small intestinal brush border membranes. Their use in investigating some properties of D-glucose and choline transport systems. Biochimica et Biophysica Acta., 506, 136-154.

Mahon, W.A., Inaba, T. and Stone, R.M., 1977, Metabolism of flurazepam by the small intestine. Clinical Pharmacology and Therapeutics, 22, 228-233.

Mikov, M., Caldwell, J., Dolphin, C.T. and Smith, R.L., 1988, The role of intestinal microflora in the formation of the methylthio adduct metabolites of paracetamol. Studies in neomycin-pretreated and germ-free mice. Biochemical Pharmacology, 37, 1445-1449.

Nelson, E.B., Abernethy, D.R., Greenblatt, D.J. and Ameer, B., 1986, Paracetamol absorption from a feeding jejunostomy. British Journal of Clinical Pharmacology, 22, 111-113.

Ratnayake, J.H., Hanna, P.E., Anders, M.W. and Duggan, D.E., 1981, Sulfoxide reduction. In vitro reduction of sulindac by rat hepatic cytosolic enzymes. Drug Metabolism and Disposition, 9, 85-87.

Renwick, A.G., 1982, First-pass metabolism within the lumen of the gastrointestinal tract. In Presystemic Drug Elimination, edited by C.F. George, D.G. Shand and A.G. Renwick (London: Butterworths), pp. 3-28.

Renwick, A.G., 1988, Intense sweeteners and the gut microflora. In Role of the gut flora in toxicity and cancer, edited by I.R. Rowland (London: Academic Press), pp. 175-206.

Renwick, A.G., Evans, S.P., Sweatman, T.W., Cumberland, J. and George, C.F., 1982, The role of the gut flora in the reduction of sulphinpyrazone in the rat. Biochemical Pharmacology, 31, 2649-2656.

Sasahara, K., Nitanai, T., Habara, T., Kojima, T., Kawahara, Y., Morioka, T. and Nakajima, E., 1981a, Dosage form design for improvement of bioavailability of levodopa IV: possible causes of low bioavailability of oral levodopa in dogs. Journal of Pharmaceutical Science, 70, 730-733.

Sasahara, K., Nitanai, T., Habara, T., Morioka, T. and Nakajima, E., 1981b, Dosage form design for improvement of bioavailability of levodopa V: Absorption and metabolism of levodopa in intestinal segments of dogs. Journal of Pharmaceutical Science, 70, 1157-1161.

Schmitz, J., Preiser, H., Maestracci, D., Ghosh, B.K., Cerda, J.J. and Crane, R.K., 1973, Purification of human intestinal brush border membranes. Biochimica et Biophysica Acta., 323, 98-112.

Strong, H.A., Renwick, A.G. and George, C.F., 1984a, The site of reduction of sulphinpyrazone in the rabbit. Xenobiotica, 14, 815-826.

Strong, H.A., Oates, J., Sembi, J., Renwick, A.G. and George, C.F., 1984b, Role of the gut flora in the reduction of sulfinpyrazone in humans. Journal of Pharmacology and Experimental Therapeutics, 230, 726-732.

Strong, H.A., Warner, N.J., Renwick, A.G. and George, C.F., 1985, Sulindac metabolism: the importance of an intact colon. Clinical Pharmacology and Therapeutics, 38, 387-393.

Strong, H.A., Angus, R., Oates, J., Sembi, J., Howarth, P., Renwick A.G. and George, C.F., 1986, Effects of ischaemic heart disease, Crohn's disease and antimicrobial therapy on the pharmacokinetics of sulphinpyrazone. Clinical Pharmacokinetics, 11, 402-410.

Strong, H.A., Renwick, A.G., George, C.F., Liu, Y.F. and Hill, M.J., 1987, The reduction of sulphinpyrazone and sulindac by intestinal bacteria. Xenobiotica, 17, 685-696.

Vidon, N., Evard, D., Godbillon, J., Rongier, M., Duval, M., Schoeller, J.P., Bernier, J.J. and Hirtz, J., 1985, Investigation of drug absorption from the gastrointestinal tract of man II. Metoprolol in the jejunum and ileum. British Journal of Clinical Pharmacology, 19, 107S-112S.

Wattenberg, L.W., 1972, Dietary modification of intestinal and pulmonary aryl hydrocarbon hydroxylase activity. Toxicology and Applied Pharmacology, 23, 741-748.

Wood, N., Wride, T.J. and Renwick A.G. Unpublished observations. In Part in Wride T.J., M.Sc. Thesis, University of Southampton.

Yoshihara, S. and Tatsumi, K., 1985, Guinea pig liver aldehyde oxidase as a sulfoxide reductase: its purification and characterisation. Archives of Biochemistry and Biophysics, 242, 213-224.

METABOLISM OF XENOBIOTICS IN THE RESPIRATORY TRACT

James A. Bond and Alan R. Dahl

Inhalation Toxicology Research Institute,
Lovelace Biomedical and Environmental Research Institute,
P.O. Box 5890,
Albuquerque, New Mexico 87185, USA

INTRODUCTION

Organic compounds including polycyclic aromatic hydrocarbons (PAH), nitro-PAH, aromatic amines, aza-arenes, aliphatic compounds, aldehydes, and ketones have been detected and, in many instances, quantitatively measured in cigarette smoke and in the emissions, products, and by-products associated with energy production and use. Organic compounds, including those specifically mentioned, are also frequently encountered in industrial settings as intermediates for the production of hundreds of chemicals. Usually organic chemicals exist as components of chemical mixtures, either as mixtures of vapours or adsorbed onto respirable particles that can be either organic or inorganic in nature. Insufficient information is available on the potential human health risks associated with exposure to these chemicals. In particular, the fate and target organs for these chemicals after inhalation in man are often unknown.

Since inhalation is a likely route of exposure of people to many organic chemicals, it is important to determine the metabolic fate of these chemicals after inhalation. Many factors, both chemical and biological, can influence the fate of inhaled chemicals. The metabolic fate of chemicals is dependent upon dynamic processes which involve varying rates of uptake, distribution, metabolism (activation and/or inactivation), and excretion. Alterations in one or more of these processes can affect the fate of an inhaled chemical. Inhaled chemicals can deposit in various regions of the respiratory tract. Factors that determine deposition amount and site include chemical reactivity and water solubility for gases and vapours, and the shape, physical size, and density for particles (Task Group on Lung Dynamics, 1966;

Davies, 1985). The retention of organic compounds in the respiratory tract following deposition depends upon their reactivity and binding to respiratory tract tissue and their solubility in body fluids. These factors alone and in concert, influence the metabolic fate of inhaled chemicals.

This report will review an important factor which influences the metabolic fate of inhaled chemicals; that is, the capacity of various regions of the respiratory tract to metabolize xenobiotics. Since the respiratory tract is directly exposed to chemicals, the activation of chemicals by respiratory tract tissues may have an important role in the pathogenesis of carcinogen-induced lesions in these tissues. This review will briefly discuss the importance of deposition of inhaled chemicals in the respiratory tract and present current information on the metabolic capacity of various regions of the respiratory tract to metabolize chemicals.

DEPOSITION OF INHALANTS IN THE RESPIRATORY TRACT

The atmospheric concentration of a chemical by itself does not define the total dose, nor the specific sites of local doses, of a chemical delivered to the respiratory tract. For particles, it is generally the particle size that influences the specific sites of deposition for the chemical in the respiratory tract. The importance of particle size in influencing the extent and loci of deposited particles has been a topic of numerous reviews and will not be emphasized here. Figure 1 depicts the predicted human deposition of particles ranging in aerodynamic size from submicron to micron (Task Group on Lung Dynamics, 1966). Particles deposit at various sites within the respiratory tract by several mechanisms (Figure 1), many of which have been reviewed (Schlesinger, 1988; Yeh et al., 1976). Regional deposition influences the subsequent pathways for removal of a deposited chemical and, as such is a major factor in determining the ultimate toxicological response to an inhaled chemical.

It is important to note that the particle size influences not only the sites of deposition in a given region of the respiratory tract but also influences the metabolic fate of the chemical. For example, inhalation of large particles (>5 μm) will result in the largest fraction of the inhaled chemical depositing in the nasal-pharyngeal region (Figure 1) (Task Group on Lung Dynamics, 1966). Cheng et al. (1988) recently reported that inspiratory deposition efficiencies in a human nasal cast were 16% and 40% for 0.01 μm and 0.005 μm particles, respectively. This would subsequently result in metabolism of the chemical by the nasal tissue enzymes.

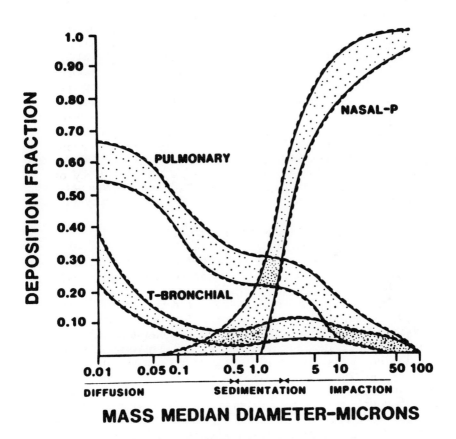

Figure 1. Predicted deposition of particles ranging in size from submicron to micron sizes at various sites within the human respiratory tract. Also shown are the mechanisms of deposition for the different particles sizes. (Modified from Task Group on Lung Dynamics, 1966)

On the other hand, inhalation of submicron particles (<1.0 μm) will result in a large fractional deposition not only in the nasal-pharyngeal region but also in the deep lung (i.e. alveoli) (Task Group on Lung Dynamics, 1966). Therefore, metabolism of the inhaled chemical would occur at both these sites of the respiratory tract and the metabolic fate of the chemical would be a function of its metabolism at both sites.

Inhaled vapours, even those with relatively high volatility and low water solubility, may be largely absorbed in the nasal cavity rather than in the lungs, although lung uptake may also be substantial (Stott and McKenna, 1984) (Figure 2). The deposition of vapours in the respiratory tract may differ among animal species (Morris et al., 1986). Dahl and Bechtold (1985) have shown that the initial deposition of vapours is greater in the anterior part of the nose than in the posterior area where the olfactory mucosa occur. Unless they are highly reactive with tissue, most vapours reentrain in the airstream after initial deposition for subsequent movement toward the gas exchange regions of the lung. For inhalation of many irritant vapours, however, it is the nose, particularly the anterior part, that is most prone to toxic effects (Buckley et al., 1984). Toxic effects appear therefore to be more related to the site of vapour deposition than to the intrinsic sensitivity of the

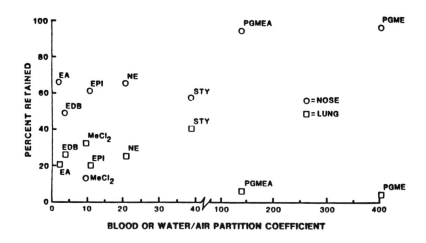

Figure 2. Retention of inhaled vapours by the nose and the lung as a function of blood or water/air partition coefficients. Adapted from data by Stott and McKenna, 1984. $MeCl_2$ = methylene chloride; EPI = epichlorhydrin; NE = nitroethane; STY = styrene; PGMEA = propyleneglycol monomethyl ether acetate; PGME = propyleneglycol monomethyl ether (from Dahl, 1988, with permission)

target tissue, since the olfactory tissue at the back of the nasal cavity is more sensitive than tissue in the anterior part of the nose (Miller et al., 1981; Appelman et al., 1982).

In summary, a major influence of particle size on the biological fate of an inhaled chemical is to affect the site of deposition in the respiratory tract. Since different portions of the respiratory tract have different abilities to remove and metabolize inhaled chemicals (see below), the eventual clearance and metabolism of a chemical to a reactive intermediate(s) will be closely related to particle size. For inhaled vapours, solubility, volatility, and reactivity will determine the site of interaction of the vapour with respiratory tract tissues.

DISTRIBUTION OF XENOBIOTIC METABOLIZING ENZYMES IN THE RESPIRATORY TRACT

The metabolism of chemicals has been studied in different preparations of respiratory tract tissues (including nasal tissue), cultured trachea and bronchus, the perfused lung, isolated lung cells, and pulmonary alveolar macrophages. In addition, there have been a few studies in which respiratory tract metabolism has been measured in the intact animal. Two animal species, the rat and dog, have been thoroughly investigated in terms of the ability of the respiratory tract to metabolize organic compounds (Bond et al., 1988a; Sabourin et al., 1988). These studies are particularly important in that they provide information on the regional metabolic capacity of the respiratory tract. Bond et al. (1988b) extended these studies to determine the net effect of respiratory tract metabolism on the regional "effective DNA dose" of an inhaled respiratory tract carcinogen. These studies are described below. Following this discussion, studies describing the metabolic capacity of the: 1) nasal tissue, 2) tracheal/bronchial tissue, 3) the pulmonary tissue, and 4) the pulmonary macrophage will be presented.

Bond et al (1988a) characterized the distribution of xenobiotic metabolizing enzymes in the respiratory tract of the dog. Substrates for different isozymes of cytochrome P-450, including benzo[a]pyrene (BaP), nitropyrene (NP), ethoxycoumarin, and ethoxyresorufin and selected Phase II enzymes, including UDP-glucuronyl transferase and glutathione transferase were measured. Specific regions of the dog respiratory tract that were sampled included the maxilloturbinate, nasoturbinate, ethmoid turbinate, larynx, proximal trachea, distal trachea, left principal bronchus (generation 1), and the pulmonary airways (generation 3-9 and generation 10-18). The data for BaP and NP were qualitatively similar in that there was higher metabolic activity in certain regions of the nasal tissue (e.g. ethmoid turbinates) and in the intrapulmonary airway generations 3-18 than in the major conducting airways (e.g. larynx, trachea, and bronchi) (Figure 3). The greatest ethoxycoumarin O-deethylase activity was in the nasal region with much less activity observed in the major airways or the pulmonary airways (Figure 3). The specific activity of ethoxycoumarin O-deethylase in the ethmoid turbinates was, in general, 5-10 times that observed for the other portions of the nasal cavity sampled. Only the ethmoid turbinates showed evidence of ethoxyresorufin metabolism. Both epoxide hydrolase and glutathione transferase activities were higher in the nasal cavity and pulmonary airways than in the major conducting airways. UDP-glucuronyl transferase was relatively evenly distributed throughout the respiratory tract. The data from these studies indicate that there is an uneven distribution of xenobiotic metabolizing enzymes, as demonstrated by quantitative differences in rates of metabolism of different substrates.

Sabourin et al (1988) measured the distribution of microsomal cytochrome P-450 isozymes 2, 4, 5, and 6 and the pulmonary FAD-containing monooxygenase in ten different anatomical regions of the rabbit respiratory tract using immunoblot analysis and enzymatic assays. The regions of the respiratory tract that were used for this work included the maxilloturbinates, ethmoturbinates, nasal mucosa, nasopharyngeal mucosa, laryngeal mucosa, trachea, carina, major bronchi, pleural and subpleural parenchyma, and parenchyma-containing small airways. Cytochrome P-450 isozymes 2 and 5 and the FAD-containing monooxygenase were detected by immunoblotting in all of the pulmonary and nasal samples, although levels in nasal tissues were generally much lower than those levels found in the lung (Table 1). Cytochrome P-450 isozyme 4, which is generally not present in extra-hepatic tissue, was detected in nasal ethmoturbinates and nasal mucosa. The presence of isozyme 3a in nasal samples was indicated by the presence of high rates of aniline hydroxylation. The data from these studies showed differences in profiles of monooxygenases throughout the respiratory tract. The enzymatic properties of the major bronchi were found to be similar to those of the pleura and parenchyma, and the nasopharyngeal mucosa, larynx, and trachea were found to contain low concentrations of P-450 and flavoprotein monooxygenase (FMO). Nasal tissues, however, are interesting in that they contain unique P-450 profiles and have enzymatic properties distinct

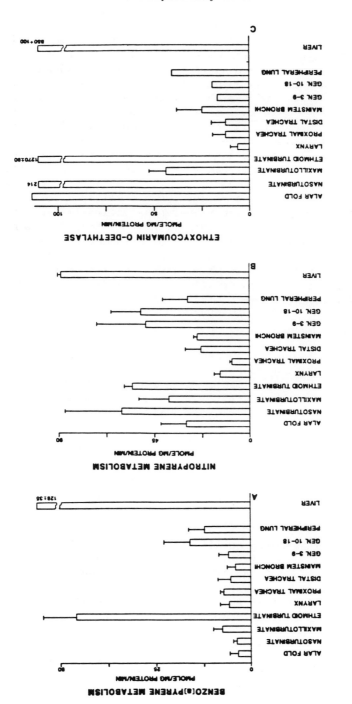

Figure 3. Distribution of cytochrome P-450-dependent metabolism of benzo[a]pyrene (A), nitropyrene (B) and 7-ethoxycoumarin (C) (from Bond et al., 1988a, with permission).

TABLE 1 - Quantitation of cytochrome P-450 isozymes and FAD-containing monooxygenase (FMO) in rabbit respiratory tract by immunoblotting[a,b]

Tissue	pmoles/mg microsomal protein				
	Form 2	Form 5	Form 4	Form 6	FMO
Maxilloturbinates	8	27	-	-	38
Ethmoturbinates	7	10	+	-	67
Nasal mucosa	10	13	+	-	43
Nasopharyngeal mucosa	10	10	-	-	62
Laryngeal mucosa	1	4	ND	ND	31
Trachea	13	6	ND	ND	32
Carina	9	9	+	+	45
Major bronchi	52	29	-	+	301
Pleural and sub-pleural parenchyma	124	73	-	+	360
Parenchyma containing small airways	180	85	-	+	576

[a]Data from Sabourin et al., 1988.

[b]For Form 4 and 6, qualitative observations were made with + = present, - = absent and ND = not determined. Values were obtained from measurements on one preparation of microsomes pooled from 15 rabbits.

from those of the rest of the respiratory tract.

Using electron microscopic immunochemistry, Serabjit-Singh et al. (1988) reported Clara cell and type II cell cytoplasmic labelling of anti-cytochrome P-450 isozyme 2 (anti-2), anti-cytochrome P-450 isozyme 5 (anti-5), and anti-NADPH cytochrome P-450 reductase (anti-R). This labelling was associated with the agranular endoplasmic reticulum of Clara cells and type II epithelial cells. The plasma membrane of the Clara cell, the tips of microvillae of the ciliated cell, secretory granules of the Goblet cell, and the plasma membrane and pinocytotic vesicles of the endothelial cell were labelled with anti-2 and anti-5, but not anti-R. The results from these studies point to the potential importance of extracellular metabolism of xenobiotics in the airways.

Bond et al (1988b) recently investigated the regional dosimetry of DNA adducts in the respiratory tract following exposure of rats to a carcinogenic concentration of inhaled diesel exhaust. These studies were designed to investigate the net effect of activating and inactivating enzymes on the "effective DNA dose" of an inhaled carcinogen in the respiratory tract. The maxilloturbinates, ethmoturbinates, trachea, left mainstem bronchus (airway generation 1), axial airway (airway generations 2-12), and peripheral lung tissue were dissected from the respiratory tract following exposure of rats to diesel exhaust for 12 weeks. DNA adducts were measured in these various regions of the respiratory tract using the ^{32}P-postlabelling assay (Randerath et al., 1985). DNA was isolated from the dissected samples and analyzed for the presence of adducts. Chromatographic maps of DNA adducts demonstrated unique patterns of DNA adducts for each of the regions. The highest level of total DNA adducts occurred in peripheral lung tissue (~ 20 adduct per 10^9 bases) (Figure 4). The levels of DNA adducts detected in the nasal tissues were about 1/4 to 1/5 of those in peripheral lung. There were less than 3 adducts per 10^9 bases in each of the regions of the major conducting airways (i.e. trachea, bronchi, axial airway). The data from this study indicate that higher levels of DNA adducts were present in tissues where exhaust-induced tumours had been shown to be located. These studies point to the importance of the balance of activating and inactivating enzymes in the production of reactive metabolites that can bind to DNA.

Nasal tissue metabolism

Until recently, nasal tissue has been generally neglected as a site of xenobiotic biotransformation (Dahl, 1986). Nasal tissue metabolism is particularly important because

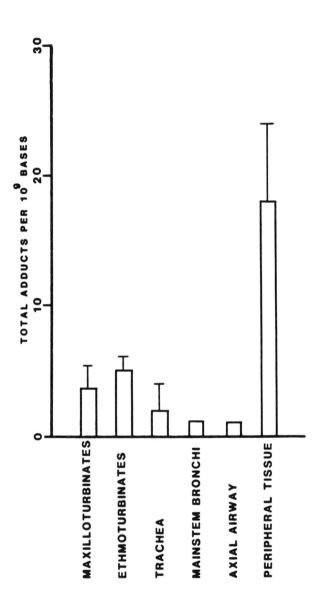

Figure 4. Distribution of DNA adducts in the respiratory tract of rats exposed to diesel exhaust for 12 weeks. (Bond et al., 1988b, with permission).

many environmental pollutants contain known carcinogens
adsorbed onto particles of sizes that deposit in the nasopharyngeal region of the respiratory tract (Natusch and
Wallace, 1974; Task Group on Lung Dynamics, 1966). Furthermore, chemicals inhaled in the vapour form can be absorbed
directly in the nasal region (see above). Known xenobiotic
metabolizing enzymes in the nasal cavity include cytochrome
P-450 and FMO, aldehyde dehydrogenases, epoxide hydro-
lyases, glutathione transferases, UDP-glucuronyl transferases, and carboxylesterases (Dahl, 1986; Baron et al.,
1986). In general, enzymes in the nasal tissue have metabolic activity (per g tissue) comparable to that in liver.

The nasal tissue metabolism of two PAHs, benzo[a]pyrene
(BaP) and nitropyrene (NP), has been thoroughly investigated
(Bond, 1983a, 1983b). Those in vitro studies using rat
nasal tissue homogenates indicated that nasal tissue
metabolized these compounds to phenols, quinones,
dihydrodiols, and tetrols. In vivo studies in hamsters have
also shown that BaP is metabolized by nasal tissue (Dahl et
al., 1985). The profile of BaP metabolites produced in
hamster noses (Figure 5) is nearly identical to that
measured using nasal tissue homogenates, suggesting that in
vitro models for nasal tissue metabolism of PAHs may predict
the metabolic profile in the intact animal. Petridou-
Fischer et al. (1988) recently reported that BaP was
metabolized in nasal tissue of dogs and monkeys. These
authors demonstrated that both dogs and monkeys were able to
metabolize BaP when it was instilled into either the ethmoid
or maxillary region. Similar studies were performed by
these authors using a representative methylenedioxyphenyl
compound, dihydrosafrole (Petridou-Fischer et al., 1987).
The data from those studies indicated that interspecies and
interregional differences occur in the metabolism of nasally
deposited dihydrosafrole in monkeys and dogs.

There have been only a few studies of the distribution of
metabolizing enzymes within the different regions of the
nasal tissue (see above discussion also). Dahl and
colleagues have demonstrated that for many substrates and
different species, the ethmoturbinates typically contain the
highest metabolic activity which was attributed to the very
high levels of cytochrome P-450 (Dahl, 1986). Casanova-
Schmitz et al. (1984) have shown that aldehyde
dehydrogenases are found primarily in the respiratory mucosa
of the anterior portion of the nose rather than in the
ethmoturbinates. Brittebo et al. (1983) have shown that
metabolites of some nitrosamines preferentially bind to the
olfactory and respiratory epithelia of rats. Immunohisto-

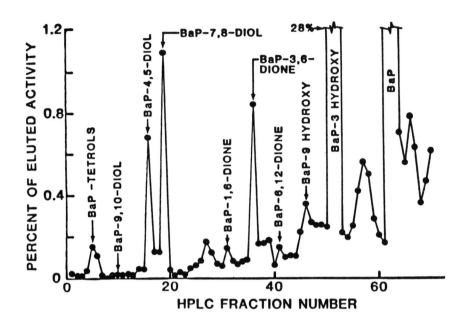

Figure 5. HPLC profile of BaP metabolites from hamster nasal metabolism (from Dahl et al., 1985, with permission)

chemical techniques were used to demonstrate that specific cytochrome P-450 isozymes and NADPH-cytochrome P-450 reductase were localized in specific cells of the respiratory and olfactory epithelia (Foster et al., 1986; Voigt et al., 1985; Baron et al., 1986). Antigens related to several rat hepatic enzymes (e.g. major isozymes of cytochrome P-450 induced in rat liver by β-naphthoflavone, 3-methylcholanthrene (3-MC), phenobarbital, and pregnenolone 16α-carbonitrile, NADPH-cytochrome P-450 reductase, epoxide hydrolase, and glutathione S-transferases B, C, and E) were observed within olfactory and respiratory epithelial cells as well as within Bowman's and seromucous glands in the lamina propria of these regions. Enzyme histochemistry was employed by Bogdanffy et al. (1986, 1987) to identify specific cell types within the respiratory and olfactory mucosa of rats and mice which contain aldehyde dehydrogenase activity (Figure 6) and carboxylesterase activity. These

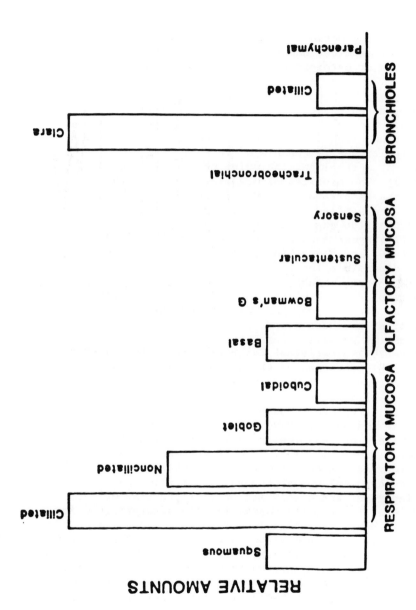

Figure 6. Distribution of aldehyde dehydrogenase in rat respiratory tract cells (Bogdanffy et al., 1986).

authors found that both the ciliated and nonciliated cells of the respiratory mucosa contained activities higher than those in other regions of the nasal tissue. Reed et al. (1986) confirmed the presence of a cytochrome P-450-dependent drug metabolizing system in the olfactory epithelium of mice, hamsters, and rats using several substrates of cytochrome P-450.

Tracheal and bronchial tissue metabolism

A considerable amount of research has been done on animal and human bronchial metabolism of chemicals, particularly PAH (Autrup, 1982). In general, the data indicate that bronchial tissue is capable of metabolizing PAHs such as BaP to several metabolites including phenols, dihydrodiols, and quinones. The profiles of organic-soluble metabolites of BaP in the respiratory tissues from humans and experimental animals are similar.

Evidence also suggests that bronchial tissue contains Phase II enzymes such as UDP-glucuronyl transferase, aryl sulphatase, and glutathione transferase. These enzymes are often responsible for the detoxification and elimination of reactive, toxic metabolites. Bronchial tissue also metabolizes PAHs to compounds that are capable of binding to DNA. The BaP-DNA adducts detected in human bronchial tissue are similar to those found in cultured tissue from experimental animals. Human bronchial epithelial cells also activate 7,12-dimethylbenz[a]anthracene (DMBA), 3-MC, and dibenz[a,h]anthracene to metabolites that covalently bind to DNA. In general, there is a positive correlation between the level of covalent binding of these PAHs to bronchial DNA and their carcinogenic potency in experimental animals (Huberman and Sachs, 1977). Human bronchial tissue can metabolize N-nitrosamines to metabolites that bind to DNA (reviewed in Harris, 1987). In some cases, DNA adduct formation was greater in bronchial tissue than in other tissues (e.g. colon, esophagus) (Table 2).

An important and highly metabolically active cell in bronchiolar tissue, the Clara cell, is a target cell for many toxicants. For example, it is well known that 4-ipomeanol is preferentially toxic to nonciliated bronchiolar epithelial (Clara) cells which contain high levels of cytochrome P-450 isozymes 2 and 5 (Table 3) (Domin et al., 1986). Using electron microscopic immunochemistry, Serabjit-Singh et al. (1988) reported Clara cell cytoplasmic labelling of anti-cytochrome P-450 isozyme 2 (anti-2), anti-cytochrome P-450 isozyme 5 (anti-5), and anti-NADPH cytochrome P-450 reductase (anti-R). This labelling was associated with the agranular endoplasmic reticulum of Clara

TABLE 2 - Metabolism of N-nitrosamines to form DNA adducts of human tissues[a,b]

N-Nitrosamine	Bronchus (%)	Colon (%)	Esophagus (%)
N-Nitrosodimethylamine	100	6	67
N-Nitrosodiethylamine	78	8	100
N-Nitrosopyrrolidine	100	52	ND[c]

[a]Data from Harris, 1987.

[b]The tissue type with the highest number of DNA adducts is given the value of 100 percent. Normal tissue explants from immediate autopsy donors were exposed to 100 μM N-nitrosamine for 24 hr.

[c]Not detected.

cells. The plasma membrane of the Clara cell, the tips of microvillae of the ciliated cell, secretory granules of the Goblet cell, and the plasma membrane and pinocytotic vesicles of the endothelial cell were labelled with anti-2 and anti-5, but not anti-R. The results from these studies point to the potential importance of extracellular metabolism of xenobiotic in the airways. Jones et al. (1983) reported that isolated rat Clara cells (50% pure) had high levels of UDP-glucuronyl transferase, glutathione transferase, and epoxide hydrolase activities. Specific activities of these enzymes were 189 $\pm$ 54 (SD), 630 $\pm$ 47, and 154 $\pm$ 19 pmole/10^6 cells/minute, respectively. In contrast, low enzyme activities were found for 7-ethoxycoumarin O-deethylase (13.3 $\pm$ 4.6 pmole/10^6 cells/minute) and aryl hydrocarbon hydroxylase (0.14 $\pm$ 0.03 pmole/10^6 cells/minute) activities.

Cohen and Moore (1976) and Cohen et al. (1979) observed that ethyl acetate-soluble metabolites from the culture media of rodent tracheal and bronchial cultures are quantitatively similar. Other investigators (Moore and Cohen, 1978; Cohen et al., 1979) have shown that BaP is metabolized to oxidative and conjugated metabolites in

TABLE 3 - Concentrations of microsomal cytochrome P-450 isozymes and NADPH-cytochrome P-450 reductase activities in samples from Clara cells, type II cells, alveolar macrophages, and whole lungs of rabbits[a]

Microsomal sample	Cytochrome P-450 isozymes[b]			Reductase[c]
	2	5	6	
Clara cell	318 ± 40	156 ± 17	1.9 ± 0.8	441 ± 73
Type II cell	244 ± 44	125 ± 35	1.7 ± 0.8	248 ± 73
Macrophage	14 ± 3	Trace	<0.1	80 ± 20
Whole lung	157 ± 45	78 ± 7	5.7 ± 2.7	131 ± 41

[a] Data from Domin et al., 1987.

[b] Results from Western blotting are given as the mean pmole/mg protein ± standard deviation (n = 3).

[c] Reductase activities are given as the mean units/mg ± standard deviation (n = 3).

cultured rodent trachea. Kaufman and coworkers (1973) demonstrated that these metabolites were covalently bound to tracheal DNA. Data obtained from these studies of rodent trachea showed BaP metabolite profiles similar to those found in short-term organ cultures of human bronchus. Coulombe et al. (1986) have shown that the nonciliated epithelial cells in rabbit trachea have a high content of the pulmonary cytochrome P-450 system. Their data suggest that the nonciliated cells can metabolize aflatoxin B_1. Aldehyde dehydrogenases occur in the Clara cells of the distal bronchioles but relatively little occurs in the tracheal epithelial cells (Bogdanffy et al., 1986). Serabjit-Singh et al. (1988) demonstrated labelling with anti-cytochrome P-450 isozyme 2 and anti-NADPH cytochrome P-450 reductase in the tracheal extracellular matrix suggesting the possibility that metabolism of inhaled xenobiotics may occur at the luminal surface of the trachea.

Pulmonary tissue metabolism

In terms of studies regarding respiratory tract metabolism, the lung has been the most extensively investigated. Several different preparations of lung tissue have been used to investigate pulmonary metabolism of xenobiotics. These include lung cell homogenates (Bond, 1983b), cultured type II alveolar cells (Sivarajah et al., 1983; Bond et al., 1983), Clara cells (Minchin and Boyd, 1983), lung slices (Stoner et al., 1978), the isolated perfused lung (Bond and Mauderly, 1984; Bond et al., 1985; Warshawsky et al., 1984), and the whole animal (for review see Bond et al., 1986). Metabolism of chemicals by these different lung preparations typically occurs at slower rates (i.e. amount per unit of incubation time) than observed in the liver.

Jones et al. (1983) measured several enzyme activities in isolated rat alveolar type II cells (80% pure with no detectable Clara cells). Specific activities for 7-ethoxycoumarin $\underline{O}$-deethylase, aryl hydrocarbon hydroxylase, UDP-glucuronyl transferase, glutathione transferase, and epoxide hydrolase were 0.7 ± 0.2 (SD), ≤ 0.02, 32 ± 10, 36 ± 28, and < 2 pmole/10^6 cells/minute, respectively. These specific activities are considerably lower than those reported by the same investigators for Clara cells. However, for both cell types, the conjugation enzymes have considerably higher specific activities than the cytochrome P-450 dependent reactions.

Burke et al. (1988) recently reported on the localization and distribution of UDP-glucuronyl transferases (UDPGT) in rat lungs. Antibodies to three rat hepatic microsomal UDPGTs were used and immunohistochemical staining methods were employed to localize the enzymes. All three UDPGT isozymes were detected in bronchial epithelial cells, Clara cells, ciliated bronchiolar epithelial cells, and type II cells. The intensity of staining for the different UDPGT isozymes differed among the cell types. For example, anti-3α-hydroxysteroid UDPGT staining was most intense in the alveolar wall, whereas anti-17β-hydroxysteroid UDPGT staining was most intense in bronchial epithelium and Clara cells.

Pulmonary macrophage metabolism

The role of pulmonary alveolar macrophages in metabolizing inhaled organic compounds has not been studied as extensively as that of other cells, but its importance should not be overlooked. Some particles deposited in lungs are phagocytized by macrophages. Some macrophages with engulfed particles remain in the lung for extended periods of time, and slow release of organic compounds and their metabolites from these macrophages may subject surrounding tissues to protracted exposures to potentially toxic or carcinogenic reactive metabolites.

Domin et al. (1986) demonstrated that pulmonary macrophages have very low concentrations of cytochrome P-450 isozymes 2, 5, and 6 compared to other cells in the respiratory tract (Table 3). Most of the research regarding pulmonary alveolar macrophages in humans and laboratory animals has used PAH as substrates (Autrup et al., 1978; Palmer and Creasia, 1984). BaP has been used as a model compound in these studies. Although the amount of PAH metabolized per unit of incubation time (i.e. metabolic rate) is lower in pulmonary alveolar macrophages than in other portions of the respiratory tract, macrophages nevertheless do activate BaP to reactive intermediates that bind to DNA (Autrup et al., 1978) (Table 4). These metabolites are released into the surrounding medium and it has been demonstrated that the metabolites formed by macrophages are capable of being taken up by surrounding respiratory tract tissue (Palmer et al., 1978). Limited data from a few studies of the capacity of pulmonary alveolar macrophages to metabolize particle-associated BaP or DMBA (Autrup et al., 1979; Bond et al., 1984; Palmer and Creasia, 1984) suggest that when BaP or DMBA is coated on a particle (for example, diesel exhaust or urban air particles), pulmonary alveolar macrophages can engulf the particle and metabolize it to several compounds including the proximate carcinogen.

TABLE 4 - Interaction between human alveolar macrophages (AM) and human bronchus during metabolism of benzo[a]-pyrene coated on Fe_2O_3 particles[a]

Group	Incubation time (hr)	pmole bound/10 mg DNA	Protein
Human bronchus + AM	6	250	1500
Human bronchus + AM	24	4600	10000
Human bronchus alone	24	120	890

[a]Data from Autrup et al., 1979

CONCLUSIONS

All segments of the respiratory tract from the nasal cavity to the periphery of the pulmonary compartment contain enzymes that metabolize xenobiotic compounds. These enzymes are capable of metabolizing some inhaled xenobiotic compounds to products that are less toxic than the inhalant. In other cases, however, the metabolites may be more toxic than the parent molecule. There are large differences in rates of metabolism at different sites in the respiratory tract within the same species and across species. In general, the data suggest that the nasal tissue and pulmonary tissue have a higher metabolic activity toward inhaled organic chemicals than do other regions of the respiratory tract (e.g. trachea, bronchi). While these different sites contain many of the enzymes necessary for overall metabolism of xenobiotics, the balance of "activating" and "inactivating" enzymes in the different regions of the respiratory tract is not clear from available data. Most of the available data concerns the Phase I enzymes (e.g. cytochrome P-450 monooxygenase) and there is little information on the Phase II enzymes (e.g. conjugation reactions). It is this balance of enzymes (Phase I and Phase II) that presumably determines whether a chemical will be toxic and/or carcinogenic in a particular tissue.

Acknowledgements. Portions of the research described in this article were sponsored by the United States Department of Energy Office of Health and Environmental Research under Contract DE-AC04-76EV01013 and conducted in facilities fully accredited by the American Association Accreditation of Laboratory Animal Care. The authors thank several of their colleagues for their review of this manuscript.

REFERENCES

Appleman, L.M., Woutersen, R.A. and Feron, V.J., 1982, Inhalation toxicity of acetaldehyde in rats: I, acute and subacute studies. Toxicology, 23, 293-307.

Autrup, H., 1982, Carcinogen metabolism in human tissues and cells. Drug Metabolism Reviews, 13, 603-646.

Autrup, H., Harris, C.C., Stoner, G.D., Selkirk, J.K., Schafer. P.W. and Trump, B.F., 1978, Metabolism of (^{3}H) benzo[a]pyrene by cultured human bronchus and cultured human pulmonary alveolar macrophages. Laboratory Investigation, 38, 217-224.

Autrup, H., Harris, C.C., Schafer, P.W., Trump, B.F., Stoner, G.D. and Hsu, I.C., 1979, Uptake of benzo[a]pyrene-ferric oxide particulates by human pulmonary macrophages and release of benzo[a]pyrene and its metabolites. Proceedings of the Society for Experimental Biology and Medicine, 161, 2800-2804.

Baron, J., Voigt, J.M., Whitter, T.B., Kawabata, T.T., Knapp, S.A., Guengerich, F.P. and Jakoby, W.B., 1986, Identification of intratissue sites for xenobiotic activation and detoxication. In *Biological Reactive Intermediates. III. Mechanisms of Action in Animal Models and Human Disease*, edited by J.J. Kocsis, D.J. Jollow, C.M. Witmer, J.O. Nelson and R. Snyder (New York: Plenum Press), Vol. 197, pp. 119-144.

Bogdanffy, M.S., Randall, H.W. and Morgan, K.T., 1986, Histochemical localization of aldehyde dehydrogenase in the respiratory tract of the Fischer-344 rats. Toxicology and Applied Pharmacology, 82, 560-567.

Bogdanffy, M.S., Randall, H.W. and Morgan, K.T., 1987, Biochemical quantitation and histochemical localization of carboxylesterase in the nasal passages of the Fischer-344 rat and B6C3F$_1$ mouse. Toxicology and Applied Pharmacology, 88, 183-194.

Bond, J.A., 1983a, Some biotransformation enzymes responsible for PAH metabolism in rat nasal turbinates; effects on enzyme activities of *in vitro* modifiers and intraperitoneal and inhalation exposure of rats to inducing agents. Cancer Research, 43, 4804-4811.

Bond, J.A., 1983b, Bioactivation and biotransformation of 1-nitropyrene in liver, lung, and nasal tissue of rats. Mutation Research, 124, 315-324.

Bond, J.A. and Mauderly, J.L., 1984, Metabolism and macromolecular covalent binding of ^{14}C-1-nitropyrene in isolated perfused and ventilated rat lungs. Cancer Research, 44, 3924-3929.

Bond, J.A., Mitchell, C.E. and Li, A.P., 1983, Metabolism and macromolecular covalent binding of benzo[a]pyrene in cultured Fischer-344 rat lung type II epithelial cells. Biochemical Pharmacology, 32, 3771-3776.

Bond, J.A., Butler, M.M., Medinsky, M.A., Muggenburg, B.A. and McClellan, R.O., 1984, Dog pulmonary macrophage metabolism of free and particle-associated (^{14}C)benzo[a]pyrene. Journal of Toxicology and Environmental Health, 14, 181-189.

Bond, J.A., Mauderly, J.L., Henderson, R.F. and McClellan, R.O., 1985, Metabolism of 1-(^{14}C)nitropyrene in respiratory tract tissue of rats exposed to diesel exhaust. Toxicology and Applied Pharmacology, 79, 461-470.

Bond, J.A., Sun, J.D., Dutcher, J.S., Mitchell, C.E., Wolff, R.K. and McClellan, R.O., 1986, Biological fate of inhaled organic compounds associated with particulate matter. In Aerosols, edited by S.D. Lee and P.J. Verkerk (Chelsea, MI: Lewis Publishers), pp. 487-502.

Bond, J.A., Harkema, J.R. and Russell, V.I., 1988a, Regional distribution of xenobiotic metabolizing enzymes in respiratory airways of dogs. Drug Metabolism and Disposition, 16, 116-124.

Bond, J.A., Wolff, R.K., Harkema, J.R., Mauderly, J.L., Henderson, R.F., Griffith, W.C. and McClellan, R.O., 1988b, Distribution of DNA adducts in the respiratory tract of rats exposed to diesel exhaust. Toxicology and Applied Pharmacology, in press.

Brittebo, E.B., Castonguay, A., Furuya, K. and Hecht, S.S., 1983, Metabolism of tobacco-specific nitrosamines by cultured rat nasal mucosa. Cancer Research, 43, 4343-4348.

Buckley, L.A., Jiang, X.Z., James, R.A., Morgan, K.T. and Barrow, C.S., 1984, Respiratory tract lesions induced by sensory irritants of the RD50 concentration. Toxicology and Applied Pharmacology, 74, 417-429.

Burke, J.P., Knapp, S.A., Voigt, J.M., Green, M.D., Tephly, T.R. and Baron, J., 1988, Localization and distribution of UDP-glucuronyltransferases (UDPGTs) in lungs of untreated rats. The Faseb Journal, 2, A1059.

Casanova-Schmitz, M., David, R.M. and Heck, H.D., 1984, Oxidation of formaldehyde and acetaldehyde by NADH-dependent dehydrogenases in rat nasal mucosal homogenates. Biochemical Pharmacology, 33, 1137-1142.

Cheng, Y.-S., Yamada, Y., Yeh, H.-C. and Swift, D.L., 1988, Diffusional Deposition of ultrafine aerosols in a Human Nasal Cast. Journal of Aerosol Science, in press.

Cohen, G.M. and Moore, B.P., 1976, Metabolism of (^{3}H)benzo[a]pyrene by different portions of the respiratory tract. Biochemical Pharmacology, 25, 1623-1629.

Cohen, G.M., Marchok, A.C., Nettesheim, P., Steele, V.E., Nelson, F., Huang, S. and Selkirk, J.K., 1979, Comparative metabolism of benzo[a]pyrene in organ and cell cultures derived from rat tracheas. Cancer Research, 39, 1980-1984.

Coulombe, R.A., Jr., Wilson, D.W., Hsieh, D.P.H., Plopper, C.G. and Serabjit-Singh, C.J., 1986, Metabolism of aflatoxin B_1 in the upper airways of the rabbit. Role of the nonciliated tracheal epithelial cell. Cancer Research, 46, 4091-4096.

Dahl, A.R., 1986, The role of nasal xenobiotic metabolism in toxicology. In Current Topics in Pulmonary and Toxicology, edited by Mannfred A. Hollinger (New York: Elsevier), Vol. 1, pp. 143-164.

Dahl, A.R., 1988, Metabolic characteristics of the respiratory tract. In Concepts in Inhalation Toxicology, in press.

Dahl, A.R. and Bechtold, W.E., 1985, Deposition and clearance of a water-reactive vapour, methylphosphonic difluoride (difluoro) inhaled by rats. Toxicology and Applied Pharmacology, 81, 58-66.

Dahl, A.R., Coslett, D.S., Bond, J.A. and Hesseltine, G.R., 1985, Exposure of the hamster alimentary tract to benzo[a]pyrene metabolites produced in the nose. Journal of the National Cancer Institute, 75, 135-139.

Davies, C.N., 1985, Absorption of gases in the respiratory tract. Annals of Occupational Hygiene, 29, 13-25.

Domin, B.A., Devereux, T.R. and Philpot, R.M. 1986, The cytochrome P-450 monooxygenase system of rabbit lung: Enzyme components, activities, and induction in the nonciliated bronchiolar epithelial (Clara) cell, alveolar type II cell, and alveolar macrophage. Molecular Pharmacology, 30, 296-303.

Foster, J.R., Elcombe, C.R., Boobis, A.R., Davies, D.S., Sesardic, A., McQuade, J., Robson, R.T., Haward, C. and Lock, E.A., 1986, Immunocytochemical localization of cytochrome P-450 in hepatic and extra-hepatic tissues of the rat with a monoclonal antibody against cytochrome P-450. Biochemical Pharmacology, 35, 4543-4554.

Harris, C.C., 1987, Biochemical and molecular effects of N-nitroso compounds in human cultured cells: An overview. IARC Scientific Publications 84, 20-25.

Huberman, E. and Sachs, L., 1977, DNA binding and its relationship to carcinogenesis by different polycyclic hydrocarbons. International Journal of Cancer, 19, 122-127.

Jones, K.G., Holland, J.F., Foureman, G.L., Bend, J.R. and Fouts, J.R., 1983, Induction of xenobiotic metabolism in Clara cells and alveolar type II cells isolated from rat lungs. Journal of Pharmacology and Experimental Therapeutics, 225, 316-319.

Kaufman, D.G., Genta, V.M., Harris, C.C., Smith, J.M. Sporn, M.B. and Saffiotti, U., 1973, Binding of ^{3}H-labelled benzo[a]pyrene to DNA in hamster tracheal epithelial cells. Cancer Research, 33, 2837-2841.

Miller, R.R., Ayres, J.A., Jersey, G.C. and McKenna, M.J. 1981, Inhalation toxicity of acrylic acid. Fundamental and Applied Toxicology, 1, 271-277.

Minchin, R.F. and Boyd, M.R., 1983, Localization of metabolic activation and deactivation systems in the lung: Significance to the pulmonary toxicity of xenobiotics. Annual Review of Pharmacology and Toxicology, 23, 217-238.

Moore, B.P. and Cohen, G.M., 1978, Metabolism of benzo-[a]pyrene and its major metabolites to ethyl acetate-soluble and water-soluble metabolites by cultured rodent trachea. Cancer Research, 38, 3066-3075.

Morris, J.B., Clay, R.J. and Cavanagh, D.G., 1986, Species differences in upper respiratory tract deposition of acetone and ethanol vapours. Fundamental and Applied Toxicology, 7, 671-680.

Natusch, D.S. and Wallace, J.R., 1974, Urban aerosol toxicity: The influence of particle size. Science (Washington, D.C.), 186, 695-699.

Palmer, W. and Creasia, D., 1984, Metabolism of 7,12-dimethylbenz[a]anthracene by alveolar macrophages containing ingested ferric oxide, aluminium oxide or carbon particles. Journal of Environmental Pathology and Toxicology, 5, 261-270.

Palmer, W.G., Allen, J.J. and Tomaszewski, J.E., 1978, Metabolism of 7,12-dimethylbenz[a]anthracene by macrophages and uptake of macrophage-derived metabolites by respiratory tissues in vitro. Cancer Research, 38, 1079-1084.

Petridou-Fischer, J., Whaley, S.L. and Dahl, A.R., 1987, In vivo metabolism of nasally instilled dihydrosafrole [1-(3,4-methylenedioxyphenyl)-propane] in dogs and monkeys. Chemico-Biological Interactions, 64, 1-12.

Petridou-Fischer, J., Whaley, S.L. and Dahl, A.R., 1988, In vivo metabolism of nasally instilled benzo[a]-pyrene in dogs and monkeys. Toxicology, 48, 31-40.

Randerath, K., Randerath, E., Agrawal, H.P., Gupta, R.C., Schurdak, M.E. and Reddy, M.V., 1985, Postlabelling methods for carcinogen-DNA adduct analysis. Environmental Health Perspectives, 62, 57-65.

Reed, C.J., Lock, E.A. and De Matteis, F., 1986, NADPH: Cytochrome P-450 reductase in olfactory epithelium. Relevance to cytochrome P-450-dependent reactions. Biochemical Journal, 240, 585-592.

Sabourin, P.J., Tynes, R.E., Philpot, R.M., Winquist, S. and Dahl, A.R., 1988, Distribution of microsomal monooxygenases in the rabbit respiratory tract. Drug Metabolism and Disposition, 16, 557-562.

Schlesinger, R.B., 1988, Deposition and clearance of inhaled particles. In Concepts in Inhalation Toxicology, in press.

Serabjit-Singh, C.J., Nishio, S.J., Philpot, R.M. and Plopper, C.G., 1988, The distribution of cytochrome P-450 monooxygenase in cells of the rabbit lung: An ultrastructural immunocytochemical characterization. Molecular Pharmacology, 33, 279-289.

Sivarajah, K., Jones, K.G., Fouts, J.R., Devereux, T., Shirley, J.E. and Eling, T.E., 1983, Prostaglandin synthetase and cytochrome P-450-dependent metabolism of ($\pm$)benzo(a)pyrene 7,8-dihydrodiol by enriched populations of rat Clara cells and alveolar type II cells. Cancer Research, 43, 2632-2636.

Stoner, G.D., Harris, C.C., Autrup, H., Trump, B.F., Kingsburg, E.W. and Myers, G.A., 1978, Explant culture of human peripheral lung. I. Metabolism of benzo(a)pyrene. Laboratory Investigation, 38, 685-692.

Stott, W.T. and McKenna, M.J., 1984, The comparative absorbtion and excretion of chemical vapours by the upper, lower and intact respiratory tract of rats. Fundamental and Applied Toxicology, 4, 594-602.

Task Group on Lung Dynamics, 1966, Deposition and retention models for internal dosimetry of the human respiratory tract. Health Physics, 12, 173-207.

Voigt, J.M., Guengerich, F.P. and Baron, J., 1985, Localization of a cytochrome P-450 isozyme (cytochrome P450 PB-B) and NADPH-cytochrome P-450 reductase in rat nasal mucosa. Cancer Letters, 27, 241-247.

Warshawsky, D., Bingham, E. and Niemeier, R.W., 1984, The effects of a carcinogen, ferric oxide, on the metabolism of benzo(a)pyrene in the isolated perfused lung. Journal of Toxicology and Environmental Health, 14, 191-209.

Yeh, H.C., Phalen, R.F. and Raabe, O.G., 1976, Factors influencing the deposition of inhaled particles. Environmental Health Perspectives, 15, 147-156.

THE ABSORPTION AND DISPOSITION OF XENOBIOTICS IN SKIN

Sharon A. Hotchkiss and John Caldwell

Department of Pharmacology and Toxicology,
St. Mary's Hospital Medical School,
Norfolk Place, London W2 1PG, UK

INTRODUCTION
  The skin, as a major interface between the body and its external environment, is constantly exposed to a wide range of chemicals, deliberately, coincidentally and accidentally. Deliberate exposure occurs to cosmetics, perfumes and topically applied drugs, coincidental exposure to household products and industrial chemicals, while accidental exposure may involve environmental pollutants. Historically the skin has been considered as an inert barrier preventing the loss of body constituents and the entrance of external agents. However, it is becoming increasingly apparent that the skin may be a highly permeable membrane which can exhibit considerable metabolic activity towards xenobiotic chemicals. Problems in dermal toxicology are of two types (1) the skin as a route of systemic exposure and (2) the skin as a target organ for local toxicity such as irritation, hypersensitivity, phototoxicity and skin carcinoma.
  The skin is commonly exposed to a wide range of chemicals which are known to cause systemic toxicity in animals. These include compounds such as the disinfectant hexachlorophene and certain fragrance agents such as musk ambrette. Local toxicity has been caused by a number of agents, including photoallergic contact dermatitis (musk ambrette), allergic contact dermatitis (ethylenediamine, para-aminobenzoic acid and nickel), contact urticaria (cinnamic acid and benzoic acid), skin cancer (benzo(a)pyrene, dimethylbenzanthracene, 3-methylcholanthrene), skin atrophy (steroids) and chloracne (TCDD).

THE SKIN AS A BARRIER

The skin comprises roughly 10% of the total body weight and receives 7.3% of the cardiac output (473 ml/min for a 70 kg man) (Shah, 1987). It is composed of two main layers, the epidermis and the dermis. The outermost layer, the stratum corneum, consisting of dead flattened keratinocytes, is the main barrier to skin penetration (Scheuplein and Blank, 1971; Scheuplein and Bronaugh, 1983) and exhibits "reservoir" characteristics (Dupuis et al., 1984; Rougier et al., 1985; Rougier et al., 1987). Below the stratum corneum is the viable tissue of the epidermis, and deeper still is the much thicker dermis which contains fibrous and elastic connective tissue, blood vessels and hair follicles. The skin is thus a highly structured barrier, organized to maintain the internal milieu and prevent the penetration of external xenobiotics. For a topically applied compound to be absorbed into the skin and thence into the systemic circulation, there are a number of diffusion and partitioning processes to be considered: partitioning into the lipophilic stratum corneum, diffusion across the stratum corneum, partitioning from the stratum corneum into the viable epidermis, diffusion across the epidermis and dermis, and passage into dermal blood vessels (Guy et al., 1987).

METHODOLOGY FOR STUDYING DERMAL ABSORPTION
In vivo studies

A wide range of animal species have been employed for percutaneous absorption studies but the general methodological approaches are the same. The skin is generally shaved (unless hairless mutant strains are used) and the compound of interest is applied to the shaved area. The site of application may be occluded to prevent removal or evaporation of the compound, or left open to the atmosphere. The penetration of the chemical may be assessed either by measuring blood levels or urinary excretion (Feldman and Maibach, 1970) and residual material is determined by excretion balance (Chidgey et al., 1987). Disadvantages of in vivo methodology include the complexity, expense and time consuming nature of these experiments, the removal of the compound from the site of application by the animal and the problems posed by irritant or toxic chemicals.

There has been some interest in surgical procedures whereby more representative in vivo models may be achieved. Such systems include human skin grafts onto athymic nude mice (Das et al., 1986), isolated perfused porcine skin flaps (Riviere et al., 1986) and rat - human skin flaps (Krueger et al., 1985). However, these methods are not easy to employ on a routine basis and are therefore not currently in widespread use.

<u>In vitro diffusion cells</u>

The difficulties inherent with <u>in vivo</u> work have lead to the development of several <u>in vitro</u> systems, in particular the use of diffusion cells (Franz, 1975; Bronaugh et al., 1982; Bronaugh and Stewart, 1984). There are a variety of models available but generally these are of two main types: the static diffusion cell and the flow-through diffusion cell. Both have two chambers (donor and receptor) made of an inert material such as teflon or glass. Pieces of excised skin are placed between the two chambers and the compound of interest is applied to the epidermis. The receptor fluid either bathes the dermis or flows continuously across the underside of the skin. The donor chamber may be open or closed and the receptor fluid is assayed for the absorbed chemical and/or its metabolites. The advantages of the flow-through system over its static counterpart include the minimization of sample collection problems, a reduction in problems due to the limited solubility of certain chemicals in the receptor fluid, and a more suitable model of <u>in vivo</u> blood flow. Recent work has shown that excised human abdominal skin in the flow-through diffusion system may provide a representative model of the <u>in vivo</u> absorption of certain compounds through skin (Bronaugh et al., 1985). We have found, using rodent skin, that the flow-through system gives an accurate reflection of the <u>in vivo</u> penetration of the simple fragrance compound, benzyl acetate (Hotchkiss et al., 1988a).

When using such systems, the integrity of the skin membrane should be established using tritiated water or cortisol (agents which do not penetrate intact skin to a significant degree) to ensure that no damage has occurred during handling. Full thickness skin or epidermis alone may be used, as it is argued that <u>in vivo</u>, dermal penetrants do not pass through the main thickness of the dermis, but are absorbed into the papillary microcirculation (Scheuplein and Blank, 1971). The epidermis can be separated from the dermis by immersion of the skin in water at $65^\circ C$ for 45 seconds or mechanically using a dermatome.

Various receptor fluids have been used with the main proviso that the fluid must not alter the permeability properties of the skin, and the penetrating molecule must be soluble in the receptor fluid. Saline is commonly employed and is suitable for water soluble molecules; for more lipophilic molecules, non ionic surfactant or alcohol mixtures, or serum have been employed (Bronaugh and Stewart, 1984; Riley and Kemppainen, 1985; Grissom et al., 1987).

FACTORS INFLUENCING SKIN PENETRATION

Percutaneous absorption studies have been carried out in a number of species, and general patterns of species variation in permeability have been observed. It would appear as a generalization that the skin of man is much less permeable than that of rodents, and that the squirrel and rhesus monkey best resemble man in their capacity to absorb topically applied compounds. Permeability is greatest in the rabbit and decreases in the order rabbit, rat, guinea pig, pig, monkey, man (Bartek et al., 1972). Table 1 gives comparative data for the topical absorption of a range of compounds in various laboratory species. There is considerable variation from one laboratory to the next, in the extent of percutaneous absorption found for chemicals within a particular animal species. Unfortunately, the systematic examination of these data is hampered by other variations in the application procedure such as the duration and vehicle of application.

TABLE 1 - Species differences in percutaneous absorption

Chemical	Absorption (% dose)					
	Rat	Rabbit	Pig	Squirrel Monkey	Rhesus monkey	Man
Caffeine	53.1	69.2	32.4		47.6	
DDT		46.3	43.4	1.5	10.4	
Lindane		51.2	37.6	16.0	9.3	
Cortisol					2.9	1.9
Testosterone	47.4	69.6	29.4		18.4	13.2
Benzoic acid					59.2	42.6

From Bartek et al., 1972; Wester and Noonan, 1980.

Other factors which have been found to affect absorption through skin include the physicochemical nature of the chemical itself (Jimbo, 1983), the vehicle of application (Mackee et al., 1945; Idson, 1975; Idson 1983; Jimbo et al., 1983; Bronaugh and Franz, 1986; Hotchkiss et al., 1988b), the extent of hydration and temperature of the skin (Behl et al., 1980), anatomical site (Feldmann and Maibach, 1967; Maibach et al., 1971; Wester et al., 1984), skin appendages (Kao et al., 1988), the concentration of the chemical applied (Wester and Maibach, 1976; Chidgey et al., 1987; Hotchkiss et al., 1988a), occlusion of the site of application (Bronaugh et al., 1985), skin damage and disease (Solomon and Lowe, 1979; Bronaugh et al., 1986; Smith, 1988) dermal blood flow, and bacterial degradation on the skin surface (Denyer et al., 1985).

METABOLISM IN THE SKIN

Metabolism in the skin includes bacterial degradation on the skin surface and enzymic metabolism in the viable epidermis and dermis. The skin contains all the drug metabolizing enzyme systems classically found in the liver (Table 2), including cytochrome P450 mixed function

TABLE 2 - Metabolic pathways present in skin

Metabolic pathway	Substrate
OXIDATION	
hydroxylation	benzo(a)pyrene
dealkylation	7-ethoxycoumarin
	ethoxyresorufin
epoxide formation	aldrin
	retinoic acid
REDUCTION	
ketone	testosterone
HYDROLYSIS	
esters	nitrate and aspirin esters
	betamethasone valerate
epoxides	styrene oxide
	benzo(a)pyrene 7,8-epoxide
GLUCURONIDATION	benzo(a)pyrene
	7-hydroxycoumarin
SULPHATION	dehydroepiandrosterone
	7-hydroxycoumarin
GSH CONJUGATION	styrene epoxide
ACETYLATION	aminofluorene
	p-aminobenzoic acid

oxygenases, UDP glucuronyltransferases (UDPGT), sulphotransferases and glutathione S-transferases. (Täuber, 1982; Noonan and Wester, 1983; Denyer et al., 1985). Classically, skin metabolism studies have been conducted using techniques previously applied to metabolism studies in other tissues i.e. skin homogenates (Golden et al., 1987), skin microsomes (Bickers et al., 1982; Moloney et al., 1982a; Rettie et al., 1986) and skin cell/organ cultures (Smith and Holland 1981; Don et al., 1987; Kao et al., 1984) together with other methods specifically applicable to the skin e.g. diffusion cells and culture of isolated hair follicles (Coomes et al., 1984; Hukkelhoven et al., 1983; Merk et al., 1987; Pohl et al., 1984). Additionally, in vivo studies may give evidence of a first pass effect in the skin (Wester et al., 1983).

Location of xenobiotic enzymes within the skin

Attempts have been made to localize the various enzymes within the structures of skin. Using corticosteroid 21-esters, esterase specific activity is 20 times greater in the epidermis than in the dermis, although total activities are similar (Täuber and Rost, 1987). The hydrolysis rate is greatest in the mouse, decreases in the rat and is lowest in man. Cytochrome P450 specific activity towards ethoxycoumarin is similar in the dermis and epidermis, with higher total activity in the dermis due to its greater mass (Finnen, 1987). The specific activities of cytochrome P450 and UDPGT are greatest in sebaceous cells followed by differentiated keratinocytes and then basal keratinocytes (Finnen, 1987). Aryl hydrocarbon hydroxylase specific activity is greater in the epidermis than the dermis with high activities in sebaceous cells and hair follicles (Akin and Norred, 1976; Thompson and Slaga, 1976; Mukhtar and Bickers, 1981; Bickers et al., 1982).

CONTRIBUTION OF SKIN TO THE WHOLE BODY METABOLISM OF XENOBIOTICS

Activities of the various skin enzymes have been compared to those of the liver (Rettie et al., 1986). Table 3 indicates skin activities as a percentage of liver activities. All enzyme activities in the skin generally appear to be less than 10% of the specific activities found in the liver. However, when the entire mass of skin is considered, it may indeed prove to be a significant site of foreign compound metabolism, especially where dermal exposure occurs over a wide area as may happen with occupational exposure, environmental agents and perfumed products.

TABLE 3 - Specific activities of enzymes in the skin

Enzyme	Activity (pmol/min/mg protein) as percentage of that in liver
Aryl hydrocarbon hydroxylase	2.7
Ethoxycoumarin dealkylase	0.5-7
Ethoxyresorufin dealkylase	4
Aniline hydroxylase	6
Aldrin epoxygenase	0.4
Epoxide hydratase	6
Glutathione S-transferase	15

From Finnen, 1987; Mukhtar and Bickers, 1981; Rettie et al., 1986; Lilienblum et al., 1986.

Effect of inducers

In general, inducers and inhibitors have the same effect in skin as in other organs although the magnitude of the response may be different (Mukhtar et al., 1982). The xenobiotic metabolizing enzymes of skin may be induced by various chemicals applied topically. 3-Methylcholanthrene and other polycyclic aromatic hydrocarbons cause a 7-10 fold increase in the metabolism of benzo(a)pyrene, ethoxyresorufin and ethoxycoumarin (Merk et al., 1984; Mukhtar and Bickers, 1981; Moloney et al., 1982b; Das et al., 1986). Steroids increase aldrin and ethoxycoumarin metabolism 5-8 fold (Finnen, 1987). Aroclor 1254 has caused a 10-15 fold increase in aryl hydrocarbon hydroxylase together with a 3-fold increase in UDPGT activity (Mukhtar and Bickers, 1981; Lilienblum et al., 1986). TCDD induces the metabolism of benzo(a)pyrene with a shift towards more water soluble conjugates (Kao et al., 1984).

Effects of inhibitors

There is also a significant response of skin metabolizing enzymes to the application of inhibitors. Imidazoles such as ketoconazole decrease ethoxyresorufin metabolism and AHH activity. Metyrapone decreases the metabolism of ethoxyresorufin and aldrin, and plant phenols such as ellagic acid and chlorogenic acid reduce benzo(a)pyrene metabolism (Merk et al., 1987; Mukhtar et al., 1984; Bickers et al., 1986; Rettie et al., 1986). Inhibitors of epoxide hydrolase reduce the formation of benzo(a)pyrene-4,5-dihydrodiol from benzo(a)pyrene (Del Tito et al., 1984), and SKF 525A, carbon monoxide and α-naphthoflavone inhibit ethoxycoumarin metabolism (Finnen and Shuster, 1985).

Other factors which may affect metabolism by the skin include animal species (Kao et al., 1985), anatomical site of application (Don et al., 1987), dose applied (Del Tito et al., 1984), and disease (Bickers et al., 1984).

SUMMARY

This paper has briefly described the methodology which may be employed to obtain data on the absorption and metabolism of xenobiotics by skin, factors affecting the absorption of chemicals through skin, and the contribution of metabolism in skin to the disposition of topically applied xenobiotics.

In conclusion, we may now consider the 'life-cycle' of a xenobiotic in the skin. Briefly, after a compound is applied or comes in contact with the skin it may be removed to some extent by evaporation, sweating and by washing or abrasion of the skin surface. The chemical may also be degraded by bacteria on the skin such as Staphylococcus epidermidis. There may also be absorption of the compound into the skin, the extent of which depends on the physicochemical properties of the drug molecule, the dose applied (concentration and surface area), the region of application, whether the skin is occluded, hydration, temperature, vehicle, animal species and skin damage. The chemical diffuses into the lipid-rich stratum corneum where it may be either sequestered (resulting in the establishment of a reservoir), or pass through into the epidermis. Here the chemical may bind to certain tissue constituents or be metabolized and the parent compound and/or metabolites may then diffuse into the dermis where further metabolism may occur. Finally the chemical will enter the blood stream to be distributed to distant sites of the body. These processes are illustrated in Figure 1.

Despite the fact that the skin is an important route of exposure of the body to environmental xenobiotics of all kinds, our knowledge of its barrier properties and metabolic activities is still very limited. The development of new methodologies and their application to the evaluation of problem compounds, is therefore of considerable importance.

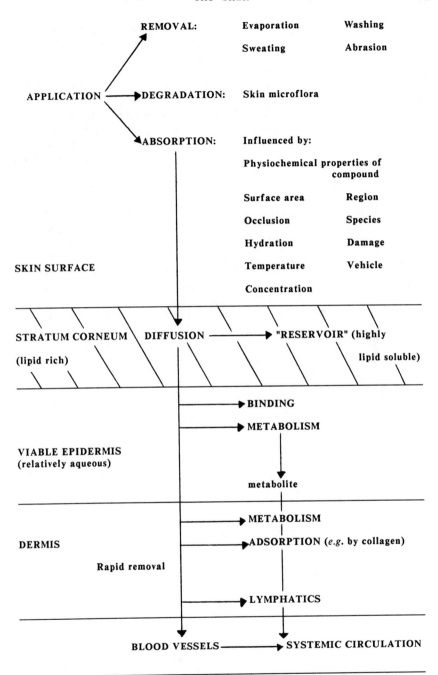

Figure 1. The life cycle of a xenobiotic in the skin

Acknowledgements. We acknowledge financial support from the Research Institute for Fragrance Materials towards original research contained in this paper.

REFERENCES

Akin, F.J. and Norred, W.P., 1976, Factors affecting measurement of aryl hydrocarbon hydroxylase activity in mouse skin. Journal of Investigative Dermatology, 67, 709-712.

Bartek, M.J., LaBudde, J.A. and Maibach, H.I., 1972, Skin permeability in vivo: comparison in rat, rabbit, pig and man. Journal of Investigative Dermatology, 58, 114-123.

Behl, C.R., Flynn, G.L., Kurihara, T., Harper, N., Smith, W., Higuchi, W.I., Ho, N.F.H. and Pierson, C.L., 1980, Hydration and percutaneous absorption: I. Influence of hydration on alcohol permeation through hairless mouse skin. Journal of Investigative Dermatology, 75, 346-352.

Bickers, D.R., Dutta-Choudhury, T. and Mukhtar, H., 1982, Epidermis: A site of drug metabolism in neonatal rat skin. Molecular Pharmacology, 21, 239-247.

Bickers, D.R., Mukhtar, H., Dutta-Choudhury, T., Marcelo, C.L. and Voorhees, J.J., 1984, Aryl hydrocarbon hydroxylase, epoxide hydrolase and benzo(a)pyrene metabolism in human epidermis: comparative studies in normal subjects and patients with psoriasis. Journal of Investigative Dermatology, 83, 51-56.

Bickers, D.R., Das, M. and Mukhtar, H., 1986, Pharmacological modification of epidermal detoxification systems. British Journal of Dermatology, 115, 9-16.

Bronaugh, R.L. and Franz, T.J., 1986, Vehicle effects on percutaneous absorption: in vivo and in vitro comparisons with human skin. British Journal of Dermatology, 115, 1-11.

Bronaugh, R.L. and Stewart, R.F., 1984, Methods for in vitro percutaneous absorption studies IV: The flow-through diffusion cell. Journal of Pharmaceutical Science, 74, 64-67.

Bronaugh, R.L., Stewart, R.F., Congdon, R.R. and Giles, A.L., 1982, Methods for in vitro percutaneous absorption studies I. Comparison with in vivo results. Toxicology and Applied Pharmacology, 62, 474-480.

Bronaugh, R.L., Stewart, R.F., Wester, R.C., Bucks, D., Maibach, H.I. and Anderson, J., 1985, Comparison of percutaneous absorption of fragrances by humans and monkeys. Food and Chemical Toxicology, 23, 111-114.

Bronaugh, R.L., Weingarten, D.P. and Lowe, N.J., 1986, Differential rates of percutaneous absorption through the eczematous and normal skin of a monkey. Journal of Investigative Dermatology, 87, 451-453.

Chidgey, M.A.J., Kennedy, J.F. and Caldwell, J., 1987, Studies on benzyl acetate III. The percutaneous absorption and disposition of [methylene-^{14}C]benzyl acetate in the rat. Food and Chemical Toxicology, 25, 521-525.

Coomes, M.W., Sparks, R.W. and Fouts, J.R., 1984, Oxidation of 7-ethoxycoumarin and conjugation of umbelliferone by intact viable epidermal cells from the hairless mouse. Journal of Investigative Dermatology, 82, 598-601.

Das, M., Asokan, P., Don, P.S.C., Krueger, G.G., Bickers, D.R. and Mukhtar, H., 1986, Carcinogen metabolism in human skin grafted onto athymic nude mice. A model system for the study of human skin carcinogenesis. Biochemical and Biophysical Research Communications, 138, 33-39.

Del Tito, B.J., Mukhtar, H. and Bickers, D.R., 1984, In vivo metabolism of topically applied benzo(a)pyrene-4,5-oxide in neonatal rat skin. Journal of Investigative Dermatology, 82, 378-380.

Denyer, S.P., Guy, R.H., Hadgraft, J. and Hugo, W.B. 1985, The microbial degradation of topically applied drugs. International Journal of Pharmaceutics, 26, 89-97.

Don, P.S.C., Mukhtar, H., Zaim, M.T. and Bickers, D.R., 1987, Regional differences in cutaneous carcinogen metabolism. Journal of Investigative Dermatology, 88, 485.

Dupuis, D., Rougier, A., Roguet, R., Lotte, C. and Kavopissis, G. 1984, In vivo relationship between horny layer reservoir effect and percutaneous absorption in human and rat. Journal of Investigative Dermatology, 82, 353-356.

Feldmann, R.J. and Maibach, H.I., 1967, Regional variation in percutaneous penetration of ^{14}C-cortisol in man. Journal of Investigative Dermatology, 48, 181-183.

Feldmann, R.J. and Maibach, H.I., 1970, Absorption of some organic compounds through the skin in man. Journal of Investigative Dermatology, 54, 399-404.

Finnen, M.J., 1987, Skin metabolism by oxidation and conjugation. In Pharmacology and the Skin Vol. 1, edited by B. Shroot and H. Schaefer, (Basel:Karger) pp. 163-169.

Finnen, M.J. and Shuster, S., 1985, Phase 1 and phase 2 metabolism in isolated epidermal cells from adult hairless mice and in whole human hair follicles. Biochemical Pharmacology, 34, 3571-3575.

Franz, T.J., 1975, Percutaneous absorption. On the relevance of in vitro data. Journal of Investigative Dermatology, 64, 190-195.

Golden, G.M., Guzak, D.B., Kennedy, A.H., McNeill, S.C., Wakshull, E. and Harm, C.S., 1987, Drug metabolism and distribution in human skin: in vivo and in vitro results. Journal of Investigative Dermatology, 88, 491.

Grissom, R.E., Brownie, C. and Guthrie, F.E., 1987, In vivo and in vitro dermal penetration of lipophilic and hydrophilic pesticides in mice. Bulletin of Environmental Contamination and Toxicology, 38, 917-924.

Guy, R.H., Hadgraft, J. and Bucks, D.A., 1987, Transdermal drug delivery and cutaneous metabolism. Xenobiotica, 17, 325-343.

Hotchkiss, S.A., Fraser, S., Miller, J.M. and Caldwell, J., 1988a, The percutaneous absorption of benzyl acetate through rat skin in vitro. British Journal of Pharmacology, in press.

Hotchkiss, S.A., Fraser, S., Miller, J.M. and Caldwell, J., 1988b, Percutaneous absorption of topically applied benzyl acetate in vitro: effect of vehicle on skin penetration. Human Toxicology, in press.

Hukkelhoven, M.W., Dykstra, A.C. and Vermonken, A.J., 1983, Human hair follicles and cultured hair follicle keratinocytes as indicators for individual differences in carcinogen metabolism. Archives of Toxicology, 53, 265-274.

Idson, B., 1975, Percutaneous absorption. Journal of Pharmaceutical Science, 64, 901-924.

Idson, B., 1983, Vehicle effects in percutaneous absorption. Drug Metabolism Reviews, 14, 207-222.

Jimbo, Y., 1983, Penetration of fragrance compounds through human epidermis. Journal of Dermatology, 10, 229-239.

Jimbo, Y., Ishihara, M., Osamura, H., Takano, M. and Ohara, M., 1983, Influence of vehicles on penetration through human epidermis of benzyl alcohol, isoeugenol and methyl isoeugenol. Journal of Dermatology, 10, 241-250.

Kao, J., Hall, J., Shugart, L.R. and Holland, J.M., 1984, An in vitro approach to studying cutaneous metabolism and disposition of topically applied xenobiotics. Toxicology and Applied Pharmacology, 75, 289-298.

Kao, J., Patterson, F.K. and Hall, J., 1985, Skin penetration and metabolism of topically applied chemicals in six mammalian species, including man: in vitro study with benzo(a)pyrene and testosterone. Toxicology and Applied Pharmacology, 81, 502-516.

Kao, J., Hall, J. and Helman, E., 1988, In vitro percutaneous absorption in mouse skin: influence of skin appendages. Toxicology and Applied Pharmacology, 94, 93-103.

Krueger, G.G., Wojciechowski, Z.J., Burton, S.A., Gilhar, A., Huether, S.E., Leonard, L.G., Rohr, U.D., Petelenz, T.J., Higuchi, W.I. and Pershing, L.K., 1985, The development of a rat/human skin flap served by a defined and accessible vasculature on a congenitally athymic (nude) rat. Fundamental and Applied Toxicology, 5, S112-S121.

Lilienblum, W., Irmscher, G., Fusenig, N.E. and Bock, K.W., 1986, Induction of UDP-glucuronyltransferase and aryl-hydrocarbon hydroxylase activity in mouse skin and in normal and transformed skin cells in culture. Biochemical Pharmacology, 35, 1517-1520.

Mackee, G.M., Sulzberger, M.B., Hermann, F. and Bare, R.L., 1945, Histologic studies on percutaneous penetration with special reference to the effects of vehicles. Journal of Investigative Dermatology, 6, 43-61.

Maibach, H.I., Feldmann, R.J., Milby, T.H. and Serat, W.F., 1971, Regional variation in percutaneous penetration in man. Archives of Environmental Health, 23, 208-211.

Merk, H.F., Rumpf, M., Bolsen, K., Wirth, G. and Goerz, G., 1984, Inducibility of AHH-activity in human hair follicles by topical application of liquor carbonis detergens. British Journal of Pharmacology, 111, 279-284.

Merk, H.F., Kaufmann, I., Vöpel, F., Röwert, H. and Jungiger, H., 1987, Influence of imidazoles on cutaneous cytochrome P450 activity after topical and systemic application. In Pharmacology and the Skin Vol. 1, edited by B. Shroot and H. Schaefer (Basel:Karger) pp. 184-189.

Merk, H.F., Mukhtar, H., Kaufmann, I., Das, M. and Bickers, D.R., 1987, Human hair follicle benzo(a)pyrene and benzo(a)pyrene 7,8-diol metabolism: effect of exposure to a coal tar-containing shampoo. Journal of Investigative Dermatology, 88, 71-76.

Moloney, S.T., Fromson, J.M. and Bridges, J.W., 1982a, Cytochrome P450 dependent deethylase activity in rat and hairless mouse skin microsomes. Biochemical Pharmacology, 31, 4011-4018.

Moloney, S.T., Fromson, J.M. and Bridges, J.W., 1982b, The metabolism of 7-ethoxycoumarin and 7-hydroxycoumarin by rat and hairless mouse skin strips. Biochemical Pharmacology, 31, 4005-4009.

Mukhtar, H. and Bickers, D.R., 1981, Comparative activity of the mixed function oxidases, epoxide hydratase, and glutathione S-transferase in liver and skin of the neo-natal rat. Drug Metabolism and Disposition, 9, 311-314.

Mukhtar, H., Link, C.M., Kushner, D.M. and Bickers, D.R., 1982, Effect of topical application of defined constituents of coal tar on skin and liver aryl hydrocarbon hydroxylase and ethoxycoumarin O-dealkylase. Toxicology and Applied Pharmacology, 64, 541-549.

Mukhtar, H., Das, M., Del Tito, B.J. and Bickers, D.R., 1984, Epidermal benzo(a)pyrene metabolism and DNA-binding in Balb/c mice: inhibition by ellagic acid. Xenobiotica, 14, 527-531.

Noonan, P.K. and Wester, R.C., 1983, Cutaneous biotransformation: some pharmacological and toxicological implications. In Dermatoxicology, edited by F.N. Marzulli and H.I. Maibach, 2nd ed., (New York: Hemisphere) pp. 71-90.

Pohl, R.J., Coomes, M.W., Sparts, R.W. and Fouts, J.R., 1984, 7-Ethoxycoumarin O-deethylation activity in viable basal and differentiated keratinocytes isolated from the skin of the hairless mouse. Drug Metabolism and Disposition, 12, 25-34.

Rettie, A.E., Williams, F.M. and Rawlins, M.D., 1986, Substrate specificity of the mouse skin mixed function oxidase system. Xenobiotica, 16, 205-211.

Riley, R.T. and Kemppainen, B.W., 1985, Effect of serum-parathion interactions on cutaneous penetration of parathion in vitro. Food and Chemical Toxicology, 23, 67-71.

Riviere, J.E., Bowman, K.F. and Monteiro-Riviere, N.A., 1986, The isolated perfused porcine skin flap (IPPSF) 1. A novel in vitro model for percutaneous absorption and cutaneous toxicology studies. Fundamental and Applied Toxicology, 7, 444-453.

Rougier, A., Depuis, D., Lotte, C. and Roguet, R., 1985, The measurement of the stratum corneum reservoir. A predictive method for in vivo percutaneous absorption studies. Journal of Investigative Dermatology, 84, 66-68.

Rougier, A., Lotte, C. and Depuis, D., 1987, An original predictive method for in vivo percutaneous absorption studies. Journal of the Society of Cosmetic Chemists, 38, 397-417.

Scheuplein, R.J. and Blank, I.H., 1971, Permeability of the skin. Physiological Reviews, 51, 702-747.

Scheuplein, R.J. and Bronaugh, R.L., 1983, Percutaneous absorption. In Biochemistry and Physiology of the Skin Vol. 2, edited by L.A. Goldsmith (New York:Oxford University Press), pp. 1255-1295.

Shah, V.P., 1987, Migration of drugs across the skin after oral administration: Griseofulvin. In *Pharmacology and the Skin Vol. 1*, edited by B. Shroot and H. Schaefer (Basel:Karger), pp. 41-49.

Smith, J.G., 1988, Paraquat poisoning by skin absorption: a review. Human Toxicology, 7, 15-19.

Smith, L.H. and Holland, J.M., 1981, Interaction between benzo(a)pyrene and mouse skin in organ culture. Toxicology, 21, 47-57.

Solomon, A.E. and Lowe, N.J., 1979, Percutaneous absorption in experimental epidermal disease. British Journal of Dermatology, 100, 717-722.

Täuber, U., 1982, Metabolism of drugs on and in the skin. In *Dermal and Transdermal Absorption*, edited by R. Brandau and B.H. Lippold (Stuttgart:Wissenschaftliche Verlagsgesellschaft), pp. 133-151.

Täuber, U. and Rost, K.L., 1987, Esterase activity of the skin including species variations. In *Pharmacology and the Skin Vol. 1*, edited by B. Shroot and H. Schaefer (Basel:Karger), pp. 170-183.

Thompson, S. and Slaga, T.J., 1976, Mouse epidermal aryl hydrocarbon hydroxylase. Journal of Investigative Dermatology, 66, 108-111.

Wester, R.C. and Maibach, H.I., 1976, Relationship of topical dose and percutaneous absorption in rhesus monkey and man. Journal of Investigative Dermatology, 67, 518-520.

Wester, R.C. and Noonan, P.K., 1980, Relevance of animal models for percutaneous absorption. International Journal of Pharmacy, 7, 99-110.

Wester, R.C., Noonan, P.K., Smeach, S. and Kosobud, L., 1983, Pharmacokinetics and bioavailability of intravenous and topical nitroglycerin in the rhesus monkey. Estimate of percutaneous first-pass metabolism. Journal of Pharmaceutical Science, 72, 745-748.

Wester, R.C., Maibach, H.I., Bucks, D.A.W. and Aufrere, M.B., 1984, *In vivo* percutaneous absorption of paraquat from hand, leg and forearm of humans, Journal of Toxicology and Environmental Health, 14, 759-762.

BIOTRANSFORMATION AND BIOACTIVATION OF XENOBIOTICS BY THE KIDNEY

M. W. Anders

Department of Pharmacology, University of Rochester,
601 Elmwood Avenue, Rochester, NY 14642, USA

INTRODUCTION
The kidney is a major site of drug and chemical biotransformation and bioactivation, in addition to being an excretory organ for polar drugs and chemicals and their metabolites. Moreover, studies conducted, largely in the last decade, have revealed that xenobiotic-induced renal damage and, perhaps, nephrocarcinogenicity are often the consequence of renal bioactivation of nephrotoxic chemicals. Thus, studies on renal xenobiotic biotransformation and bioactivation are warranted because they may help elucidate mechanisms of xenobiotic-induced nephrotoxicity. The objective of this review is to report new findings in renal xenobiotic biotransformation. Moreover, because most toxic xenobiotics require metabolism for the expression of toxicity (Anders, 1985), the presence of xenobiotic metabolizing enzymes in renal tissue raises the possibility that the nephrotoxicity and nephrocarcinogenicity of drugs and chemicals may be attributable to organ-selective bioactivation. Hence, selected examples of the renal bioactivation of nephrotoxic chemicals will be presented. Reviews about renal drug metabolism and books about extrahepatic drug metabolism, which include chapters on the kidney, have appeared (Anders, 1980; Gram, 1980; Rush and Hook, 1982; Rydström et al., 1983). Also, several reviews about renal bioactivation of xenobiotics have appeared (Bach et al., 1982; Gram et al., 1986; Rush et al., 1984; Walker and Duggin, 1988).

RENAL XENOBIOTIC BIOTRANSFORMATION AND BIOACTIVATION
Oxidative and oxygenation reactions
Cytochrome P-450-dependent oxidations and oxygenations. It is now well established that cytochromes P-450 are present

in renal cortical microsomal fractions (Anders, 1980), and cytochrome P-450 isozymes of renal cortex have been purified and characterized (Yoshimoto et al., 1986). The enzymology of extrahepatic, including renal, cytochromes P-450 has recently been reviewed (Masters et al., 1987; Schwab and Johnson, 1987).

The localization of renal cytochomes P-450 in the nephron has been studied with immunohistochemical techniques (Baron et al., 1983). These studies showed that antibodies to rat hepatic cytochrome P-450$_{MC-B}$ and P-450$_{PB-B}$ stained the proximal and distal convoluted tubules and the collecting ducts of rat kidney; antibodies to cytochrome P-450$_{MC-B}$ also stained the thick loops of Henle, whereas antibodies to cytochrome P-450$_{PB-B}$ reacted with the thin loops of Henle. Cytochrome P-450$_{PCN-E}$ was detected only in the collecting ducts.

The responsiveness of rat renal cytochromes P-450 to inducing agents is an area of continuing interest. As reported earlier (Anders, 1980), the most potent inducing agents for renal cytochromes P-450 are 3-methylcholanthrene (3-MC), which induces a low-spin cytochrome P-450 (P-448L) (Degawa et al., 1987), and 2,3,7,8-tetrachlorodibenzo-p-dioxin (TCDD), which induces cytochrome P-448$_{MC}$ (Goldstein and Linko, 1984). Polybrominated biphenyls (PBBs) are also potent inducers of renal aryl hydrocarbon hydroxylase and ethoxyresorufin O-deethylase activities (McCormack et al., 1979). 3-MC induces the formation of forms of cytochrome P-450 in rabbit kidney cortex with high specific activity for benzo(a)pyrene hydroxylase and ω- and (ω-1)-fatty acid hydroxylation (Ogita et al., 1982). Treatment of rats or mice with β-naphthoflavone and isosafrole, but not with 2,4,2',4'-tetrachlorobiphenyl, stimulated ethoxycoumarin and ethoxyresorufin O-deethylase activities in renal microsomes from male and female mice (Hook et al, 1982); these studies confirm earlier observations that the kidney is refractory to induction by inducers of the phenobarbital class, but is responsive to inducers of the polycyclic aromatic hydrocarbon class. cis-Platinum, which reduces hepatic cytochrome P-450 concentrations, is a potent inducer of renal cytochromes P-450 (Maines, 1986).

Finally, the important role of renal mitochondrial cytochrome P-450 in the metabolism of 25-hydroxycholecalciferol to 1α,25-dihydroxycholecalciferol merits note (Jefcoate, 1986).

Cytochrome P-450-dependent renal bioactivation. Renal cytochromes P-450 have been implicated in the bioactivation of many compounds. The nephrotoxicity of chloroform, which is dependent on its bioactivation in the kidney, provides a relevant example. Chloroform is nephrotoxic in many species of mammals (Smith, 1986), although the susceptibility to chloroform-induced renal damage differs among species. The in vivo nephrotoxicity of chloroform is characterized by necrosis of the proximal tubules, which is reflected in the observed proteinuria, glucosuria, and elevated blood urea nitrogen concentrations and in decreased organic anion and cation transport.

Mice are very susceptible to chloroform-induced nephrotoxicity (rabbits, rats and dogs being less susceptible), and male mice are much more susceptible than female mice. The susceptibility of male mice to chloroform-induced toxicity is decreased by castration, and the effect of castration is reversed by testosterone administration; the susceptibility of female mice to chloroform is increased by giving testosterone. The activity of cytochromes P-450 is higher in male mice than in female mice, and parallel effects of hormone administration on renal cytochrome P-450 concentrations and chloroform-induced toxicity are observed, indicating a central role for renal cytochromes P-450 in the bioactivation of chloroform.

Further studies showed a close correlation between chloroform-induced nephrotoxicity and the renal metabolism of chloroform. The metabolism of ^{14}C-chloroform to ^{14}C-carbon dioxide and the covalent binding of ^{14}C-chloroform-derived metabolites to renal macromolecules is greater in male mice than in female mice. Moreover, the in vitro metabolism of chloroform is inhibited by carbon monoxide, indicating a role for cytochromes P-450 in the renal bioactivation of chloroform. Phosgene is the predominant electrophilic metabolite in the liver and is thought to be responsible for the hepatotoxicity of chloroform. The inclusion of glutathione in incubation mixtures containing male kidney microsomal fractions increased the formation of water-soluble metabolites of ^{14}C-chloroform, presumably due to the formation of diglutathionyl dithiocarbonate formed by the reaction of glutathione with phosgene. Subsequent studies confirmed that phosgene is a metabolite of chloroform.

Hence, the renal bioactivation of chloroform can be summarized by the reactions shown in Figure 1. Chloroform (1) is metabolized by renal cytochromes P-450 to the oxygenated intermediate trichloromethanol (2), which loses hydrochloric acid to form phosgene (3). The hydrolysis of phosgene yields carbon dioxide (4), a known metabolite of chloroform, whereas reaction of phosgene with glutathione yields diglutathionyl dithiocarbonate (5), also a known

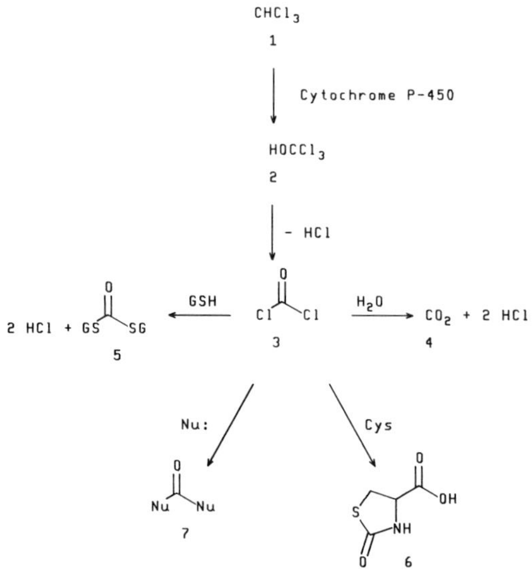

Figure 1. Renal bioactivation of chloroform.
1 = chloroform, 2 = trichloromethanol, 3 = phosgene,
4 = carbon dioxide, 5 = diglutathionyl dithiocarbonate,
6 = 2-oxothiazolidine-4-carboxylic acid, 7 = phosgene-
derived adduct formed with tissue nucleophiles (Nu:),
Cys = cysteine.

metabolite of chloroform. Phosgene can be trapped by reaction with cysteine to yield 2-oxothiazolidine-4-carboxylic acid (6). Finally, phosgene may react with sulphur-, nitrogen-, or oxygen-centered tissue nucleophiles to form covalently bound species (7), which have not been characterized but are thought to play a central role in the cytotoxicity of chloroform.

Also, a role of renal cytochromes P-450 in the bioactivation of mutagenic compounds has been reported. Degawa et al. (1988) described the presence of a male-specific, renal cytochrome P-450 that catalyzes the metabolism of 3-methoxy-4-aminoazobenzene, but not a tryptophan-pyrolysate component or a glutamic acid-pyrolysate component, to a mutagenic product.

Flavoprotein monooxygenase-dependent oxidations and oxygenations. Microsomal flavoprotein-dependent monooxygenase activity is present in renal tissues of several species (Ziegler, 1980), although the activity is much less than that found in liver microsomes. The role of this enzyme in the renal biotransformation and bioactivation of nucleophilic N- and S-containing compounds has received little attention, but a recent report indicates that it is more important than cytochromes P-450 in catalyzing the oxidation of the insecticide phorate to phorate sulphoxide (Kinsler et al., 1988).

Prostaglandin endoperoxide synthase-dependent cooxidations.
The biosynthesis of prostaglandins involves the fatty acid cyclooxygenase-dependent oxygenation of a polysaturated fatty acid, such as arachidonic acid, to a hydroperoxy endoperoxide, prostaglandin G_2 (PGG_2), followed by prostaglandin hydroperoxidase-dependent reduction of PGG to a hydroxy endoperoxide, prostaglandin H_2 (PGH_2) (Needleman et al., 1986); a single enzyme catalyzes both reactions, and the collective catalytic activity of the enzyme is termed PGH synthase or PGH endoperoxide synthase. PGH synthase activity is very high in seminal vesicles and is also present in microsomal fractions from other tissues, including the kidney. Whereas the fatty acid cyclooxygenase activity of PGH synthase exhibits considerable substrate specificity for polyunsaturated fatty acids, the prostaglandin hydroperoxidase activity of PGH synthase is not specific for PGG and will accept several organic hydroperoxides as substrates; the hydroperoxidase is, however, specific for an iron-containing porphyrin (haem).
After discovery of the role of PGH synthase in prostaglandin biosynthesis, it was observed that the enzyme also catalyzes the cooxidation of xenobiotics structurally unrelated to arachidonic acid or to endoperoxide intermediates (Eling and Krauss, 1985; Eling et al., 1983; Marnett, 1981; Marnett and Eling, 1983), indicating that PGH synthase may play an important role in the bioactivation of xenobiotics. The reaction requires the intermediate formation of PGG_2, and cooxidation is not supported by PGH_2; inhibitors of PGG_2 synthesis block cooxidation. Such cooxidations include dehydrogenation, demethylation, epoxidation, sulphoxidation, N-oxidation, C-oxidation, and dioxygenation reactions. Given the widespread distribution of PGH synthase activity in the body and its catalysis of the oxidation of xenobiotics, it is not surprising that extrahepatic PGH synthase activity has been implicated in the biotransformation and bioactivation of xenobiotics in the

kidney (Davis et al., 1981; Spry et al., 1986; Zenser and Davis, 1984).

The mechanism of PGH synthase-catalyzed cooxidations has been elucidated (Eling and Krauss, 1985; Marnett and Eling, 1983). The role of fatty acid cyclooxygenase activity is to generate a hydroperoxide; the prostaglandin hydroperoxidase activity then catalyzes reactions in which the hydroperoxide is reduced and an electron donor is oxidized or reactions in which the hydroperoxide is reduced and an acceptor is oxygenated.

The resting state of PGH synthase contains haem iron in the +3 oxidation state, which can reduce the hydroperoxide to an alcohol and generate an iron-oxo complex analogous to horseradish peroxidase Compound I, which contains haem iron in the formal oxidation state of +5:

$$Fe^{III} + ROOH \longrightarrow Fe^{V}\text{-}O^- + ROH$$

"Compound I" may then undergo a one-electron reduction by the donor DH_2 to yield the radical $DH\cdot$ and an iron-oxo complex analogous to horseradish peroxidase Compound II, which contains an iron-oxo complex with iron in the +4 oxidation state:

$$Fe^{V}\text{-}O^- + DH_2 \longrightarrow Fe^{IV}\text{-}O^- + DH\cdot$$

Reaction of "Compound II" with DH_2 yields $DH\cdot$ and the resting enzyme:

$$Fe^{IV}\text{-}O^- + DH_2 \longrightarrow Fe^{III} + DH\cdot + H_2O$$

Alternatively, transfer of hydroperoxide-derived oxygen to acceptor molecules may occur:

$$Fe^{V}\text{-}O^- + A \longrightarrow Fe^{III} + AO$$

but this has been shown to occur only in the case of sulphoxidation (Egan et al., 1981); most PGH synthase-catalyzed oxygenation reactions appear to be the consequence of the reaction of peroxyl radicals, generated by the reaction of dioxygen with hydroperoxide-derived, carbon-centered radicals, with acceptor molecules (Marnett, 1987).

PGH synthase-dependent renal bioactivation. A role for PGH synthase-dependent bioactivation of nephrotoxic and nephrocarcinogenic xenobiotics has been established (Davis et al., 1981; Spry et al., 1986; Zenser and Davis, 1984; Zenser et al., 1983). Oxidative metabolism of drugs in the kidney may be catalyzed by cytochrome P-450-dependent monooxygenases and by PGH synthase. The intrarenal distribution of the enzyme systems differs: cytochromes P-450 are primarily localized in the renal cortex, whereas PGH synthase activity predominates in the inner medulla; both catalytic activities are present in the outer medulla (Zenser et al., 1979).

The nephrotoxicity of acetaminophen provides a relevant example of PGH synthase-dependent renal bioactivation (Walker and Duggin, 1988). Ingestion of overdoses of acetaminophen may lead to acute nephrotoxicity characterized by tubular necrosis, which is confined largely to the renal cortex. The location of the renal lesion indicates the involvement of cytochromes P-450, which presumably catalyze the two-electron oxidation of acetaminophen to N-acetyl-p-benzoquinoneimine, which reacts with tissue nucleophiles (Figure 2).

The chronic ingestion of acetaminophen is associated with the development of the so-called analgesic nephropathy, which is characterized by papillary necrosis and interstitial nephritis. The development of acetaminophen-induced chronic nephrotoxicity is associated with the PGH synthase-dependent bioactivation of acetaminophen in the renal medulla, which is, as noted above, rich in PGH synthase activity. The PGH synthase-dependent metabolism of acetaminophen involves a one-electron oxidation of acetaminophen to the phenoxyl radical, which may undergo a second PGH synthase-catalyzed one-electron oxidation to yield N-acetyl-p-benzoquinoneimine (Figure 2) (West et al., 1984). The quinoneimine readily alkylates tissue nucleophiles, including glutathione. Although its reaction with glutathione is a detoxification, the concentration of glutathione is low in renal medullary tissue; hence, depletion of glutathione by metabolites of acetaminophen may render critical tissue macromolecules susceptible to alkylation by reactive intermediates of acetaminophen.

The PGH synthase-catalyzed bioactivation of several renal and bladder carcinogens has been described (Davis et al., 1981; Zenser and Davis, 1984; Zenser et al., 1983). For example, the urinary tract carcinogens benzidine, N-[4-(5-nitro-2-furyl)-2-thiazolyl]formamide, 2-amino-4-(5-nitro-2-furyl)thiazole, and 3-hydroxymethyl-1-{[3-(5-nitro-2-furyl)allylidene]amino}hydantoin are all metabolized by PGH synthase.

Reductive reactions
As noted previously (Anders, 1980), cytochrome P-450-dependent reductions of polyhalogenated alkanes, pyridine

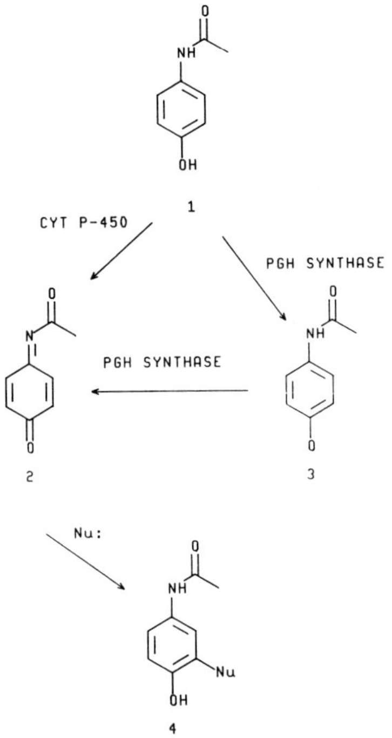

Figure 2. Cytochrome P-450- and PGH synthase-dependent bioactivation of acetaminophen. 1 = acetaminophen, 2 = N-acetyl-p-benzoquinoneimine, 3 = acetaminophen phenoxyl radical, 4 = acetaminophen-derived adduct formed with tissue nucleophiles (Nu:).

nucleotide-dependent reductions of aldehydes and ketones, and glutathione-dependent reductions of α-haloketones are catalyzed by renal enzymes.

Microsomal nitroreductase activity, with p-nitrobenzoic acid and N-[4-(5-nitro-2-furyl)-2-thiazolyl]formamide as substrates, is also present in rabbit renal cortex and inner and outer medulla (Zenser et al., 1981); renal nitroreductases may be involved in 5-nitrofuran-induced nephrotoxicity.

Sulindac, a nonsteroidal antiinflammatory agent, is a prodrug that is reduced to sulindac sulphide, which is pharmacologically active, and the major sites of reduction in vivo are the liver and kidney (Duggan et al., 1980). Subsequent studies showed that renal sulphoxide reduction is catalyzed by a novel, thioredoxin- and NADPH-dependent enzyme system (Anders et al., 1981). Recent studies revealed the presence of two thioredoxin-dependent methyl sulphoxide reductases in rat kidney cytosol, which were purified to near homogeneity (Fukazawa et al., 1987).

Hydrolytic reactions

Esterases and amidases, which are widely distributed in the body, are also present in renal tissue (Anders, 1980). Renal amidase activity may play a role in the nephrotoxicity of acetaminophen by catalyzing its hydrolysis to p-aminophenol, which undergoes PGH synthase-dependent bioactivation (Newton et al., 1985). Microsomal epoxide hydrase activity is present in the kidney (Anders, 1980).

The glutathione-dependent bioactivation of nephrotoxic halogenated alkenes requires metabolism of glutathione S-conjugates to the corresponding cysteine S-conjugates (Elfarra and Anders, 1984). This reaction may be catalyzed by renal γ-glutamyltransferase, aminopeptidase M, and cysteinylglycine dipeptidase, whose activities are high in renal tissue (Tate, 1980).

Conjugation reactions

There is substantial glucuronide and sulphate conjugate formation in the kidney (Anders, 1980; Aitio and Marniemi, 1980; Powell and Roy, 1980). The development of the "Specific Activity Difference Ratio," which allows quantification of conjugates formed in the kidney and excreted in the urine, is an important advance for investigating the role of the kidney in conjugate formation (Diamond and Quebbemann, 1981). Rush et al. (1983) used this method to study glucuronide and sulphate conjugate formation in the rat; they observed that the nephrogenic fraction of urinary p-nitrophenyl glucuronide was twice as large in female rats (22%) as in male rats (11%), whereas no sex difference in the nephrogenic fraction of p-nitrophenyl sulphate was observed.

Renal glutathione conjugate formation has been reviewed (Anders, 1980; Chasseaud, 1980). Recent studies show that renal cytosol contains glutathione S-transferases 1-1, 1-2, 2-2, 3-3, 7-7, 3-6, and 4-6 (Guthenberg et al., 1985; Trakshel and Maines, 1988). In rabbit kidney, both cytosolic and microsomal glutathione S-transferase activities are present in cortex, outer medulla, and inner medulla, with 1-chloro-2,4-dinitrobenzene as the substrate (Mohandas et al., 1984); the highest activities were present in the cortex, and the activity varied with the substrate.

Pyridoxal phosphate-dependent reactions
Cysteine conjugate β-lyase. The recent elucidation of the role of cysteine conjugate β-lyase (β-lyase) in the bioactivation of nephrotoxic cysteine S-conjugates has focused attention on the role of pyridoxal phosphate (PLP)-dependent enzymes in xenobiotic biotransformation and bioactivation (Elfarra and Anders, 1984). β-Lyase, which catalyzes transamination and β-elimination reactions of cysteine S-conjugates, has been purified from rat kidney cytosol (Stevens et al., 1986) and mitochondria (Stevens et al., 1988), from beef kidney cytosol (Ricci et al., 1986; Lash and Anders, 1988), and from human kidney cytosol (Nelson et al., 1988). The rat renal cytosolic β-lyase is identical with renal glutamine transaminase K (Stevens et al., 1986). Rat renal mitochondrial β-lyase activity is present in both the outer membrane and the matrix (Lash et al., 1986; Elfarra et al., 1987; Stevens et al., 1988).

Cysteine conjugate β-lyase-dependent bioactivation. A multistep pathway, which involves renal β-lyase activity, for the bioactivation of nephrotoxic haloalkenes has recently been elucidated and several reviews describing its mechanism have appeared (Anders et al., 1988; Dekant et al., 1988a, 1988b; Elfarra and Anders, 1984; Lash and Anders, 1986; Lock, 1987).

The first step is the glutathione S-transferase-catalyzed formation of S-haloalkyl- (from fluoroalkenes) or S-haloalkenyl- (from chloroalkenes) glutathione conjugates, which are metabolized to the corresponding cysteine S-conjugates by γ-glutamyltransferase and by cysteinylglycine dipeptidase and aminopeptidase M. The cysteine S-conjugates are then metabolized by renal β-lyase to ammonia, pyruvate, and unstable thiols that may give rise to reactive, electrophilic intermediates. The target organ selectivity of nephrotoxic cysteine S-conjugates is determined by two factors: (i) the active, probenecid-sensitive uptake of cysteine S-conjugates by renal proximal tubular cells and (ii) the presence of high activities of β-lyase in the target cells.

The glutathione-dependent bioactivation of the nephrocarcinogen hexachloro-1,3-butadiene is illustrated in Figure 3. The reaction of hexachloro-1,3-butadiene with glutathione is catalyzed by the hepatic glutathione S-transferases, the microsomal transferase being a much better catalyst than the cytosolic form. The glutathione conjugate S-(pentachlorobutadienyl)glutathione is excreted in the bile

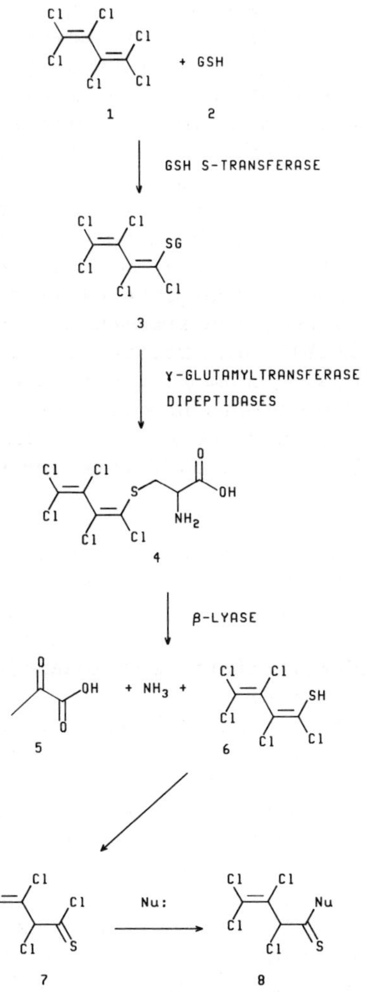

Figure 3. Cysteine conjugate β-lyase-dependent bioactivation of hexachloro-1,3-butadiene. 1 = hexachloro-1,3-butadiene, 2 = glutathione, 3 = S-(pentachlorobutadienyl)glutathione, 4 = S-(pentachlorobutadienyl)-L-cysteine, 5 = pyruvate, 6 = 1,2,3,4,4-pentachlorobutadienyl thiol, 7 = 2,3,4,4-tetrachlorothionobutenoyl chloride, 8 = hexachloro-1,3-butadiene-derived adduct formed with tissue nucleophiles (Nu:).

and is metabolized to the cysteine S-conjugate S-(pentachlorobutadienyl)-L-cysteine in the intestine or in the kidney. After uptake by the renal proximal tubular cells, S-(pentachlorobutadienyl)-L-cysteine is metabolized by β-lyase to ammonia, pyruvate, and, presumably, 1,2,3,4,4-pentachlorobutadienyl thiol; the thiol may tautomerize to form the thioacylating agent 2,3,4,4-tetrachlorothionobutenoyl chloride, which reacts with tissue nucleophiles (Dekant et al., 1988c). Renal mitochondria are the primary intracellular targets of cytotoxic cysteine S-conjugates, including S-(pentachlorobutadienyl)-L-cysteine, and renal mitochondrial β-lyase may play an important role in the cytotoxicity of these conjugates. The pathway described above may account for the observed nephrotoxicity of hexachloro-1,3-butadiene (Dekant et al., 1988a, b). Similar pathways have been described for the bioactivation of chlorotrifluoroethylene, tetrafluoroethylene, 1,1-difluoro-2,2-dichloroethylene, hexafluoropropene, tetrachloroethylene, and trichloroethylene.

REFERENCES

Aitio, A. and Marniemi, J., 1980, Extrahepatic glucuronide conjugation. In *Extrahepatic Metabolism of Drugs and Other Foreign Compounds*, edited by T.E. Gram (New York: SP Medical and Scientific Books), pp. 365-387.

Anders, M.W., 1980, Metabolism of drugs by the kidney. Kidney International, 18, 636-647.

Anders, M.W., editor, 1985, *Bioactivation of Foreign Compounds* (Orlando: Academic Press).

Anders, M.W., Ratnayake, J.H., Hanna, P.E. and Fuchs, J.A., 1981, Thioredoxin-dependent sulfoxide reduction by rat renal cytosol. Drug Metabolism and Disposition, 9, 307-310.

Anders, M.W., Lash, L., Dekant, W., Elfarra, A.A. and Dohn, D.R., 1988, Biosynthesis and biotransformation of glutathione S-conjugates to toxic metabolites. CRC Critical Reviews in Toxicology, 18, 311-341.

Bach, P.H., Bonner, F.W., Bridges, J.W. and Lock, E.A., editors, 1982, *Nephrotoxicity-Assessment and Pathogenesis* (Chichester: Wiley).

Baron, J., Kawabata, T.T., Redick, J.A., Knapp, S.A., Wick, D.G., Wallace, R.B., Jakoby, W.B. and Guengerich, F.P., 1983, Localization of carcinogen-metabolizing enzymes in human and animal tissues. In *Extrahepatic Drug Metabolism and Chemical Carcinogenesis*, edited by J. Rydström, J. Montelius, and M. Bengtsson (Amsterdam: Elsevier), pp. 73-88.

Chasseaud, L.F., 1980, Extrahepatic conjugation with glutathione. In Extrahepatic Metabolism of Drugs and Other Foreign Compounds, edited by T. E. Gram (New York: SP Medical and Scientific Books), pp. 427-452.

Davis, B.B., Mattammal, M.B. and Zenser, T.V., 1981, Renal metabolism of drugs and xenobiotics. Nephron, 27, 187-196.

Degawa, M., Yamada, H., Hishinuma, T., Masuko, T. and Hashimoto, Y., 1987, Organ selective induction of cytochrome P-448 isozymes in the rat by 2-methoxy-4-aminoazobenzene and 3-methylcholanthrene. Journal of Biochemistry, 101, 1437-1445.

Degawa, M., Namiki, M., Miura, S.-I., Ueno, H. and Hashimoto, Y., 1988, A male-specific renal cytochrome P-450 isozyme(s) responsible for mutagenic activation of 3-methoxy-4-aminoazobenzene in mice. Biochemical and Biophysical Research Communications, 152, 843-848.

Dekant, W., Lash, L.H. and Anders, M.W., 1988a, Fate of glutathione conjugates and bioactivation of cysteine S-conjugates by cysteine conjugate β-lyase. In Glutathione Conjugation: Its Mechanism and Biological Significance, edited by H. Sies and B. Ketterer (San Diego, Academic Press), in press.

Dekant, W., Vamvakas, S. and Anders, M. W., 1988b, Bioactivation of nephrotoxic haloalkenes by glutathione conjugation: Formation of toxic and mutagenic intermediates by cysteine conjugate β-lyase. Drug Metabolism Reviews, in press.

Dekant, W., Berthold, K., Vamvakas, S. and Henschler, D., 1988c, Thioacylating agents as ultimate intermediates in the β-lyase catalysed metabolism of S-(pentachlorobutadienyl)-L-cysteine. Chemico-Biological Interactions, in press.

Diamond, G.L. and Quebbemann, A.J., 1981, In vivo quantification of renal sulfate and glucuronide conjugation in the chicken. Drug Metabolism and Disposition, 9, 402-409.

Duggan, D.E., Hooke, K.F. and Hwang, S.S., 1980, Kinetics of the tissue distributions of sulindac and metabolites. Drug Metabolism and Disposition, 8, 241-246.

Egan, R.W., Gale, P.H., Baptista, E.M., Kennicott, K.L., VandenHeuvel, W. J. A., Walker, R. W., Fagerness, P. E. and Kuehl, F. A., Jr., 1981, Oxidation reactions by prostaglandin cycloxygenase-hydroperoxidase. Journal of Biological Chemistry, 256, 7352-7361.

Elfarra, A.A. and Anders, M.W., 1984, Renal processing of glutathione conjugates: Role in nephrotoxicity. Biochemical Pharmacology, 33, 3729-3732.

Elfarra, A. A., Lash, L. H. and Anders, M. W., 1987, α-Keto-acids stimulate renal cysteine conjugate β-lyase activity and potentiate the cytotoxicity of S-(1,2-dichlorovinyl)-L-cysteine. Molecular Pharmacology, 31, 208-212.

Eling, T. E. and Krauss, R. S., 1985, Arachidonic acid-dependent metabolism of chemical carcinogens and toxicants. In Arachidonic Acid Metabolism and Tumor Initiation, edited by L. J. Marnett (New York: Martinus Nijhoff), pp. 83-124.

Eling, T., Boyd, J., Reed, G., Mason, R. and Sivarajah, K., 1983, Xenobiotic metabolism by prostaglandin endoperoxide synthetase. Drug Metabolism Reviews, 14, 1023-1053.

Fukazawa, H., Tomisawa, H., Ichihara, S. and Tateishi, M., 1987, Purification and properties of methyl sulfoxide reductases from rat kidney. Archives of Biochemistry and Biophysics, 256, 480-489.

Goldstein, J. A. and Linko, P., 1984, Differential induction of two 2,3,7,8-tetrachlorodibenzo-p-dioxin-inducible forms of cytochrome P-450 in extrahepatic versus hepatic tissue. Molecular Pharmacology, 25, 185-191.

Gram, T.E., editor, 1980, Extrahepatic Metabolism of Drugs and Other Foreign Compounds (New York: SP Medical & Scientific Books).

Gram, T.E., Okine, L.K. and Gram, R.A., 1986, The metabolism of xenobiotics by certain extrahepatic organs and its relation to toxicity. Annual Review of Pharmacology and Toxicology, 26, 259-291.

Guthenberg, C., Jensson, H., Nyström, Österlund, E., Tahir, M.K. and Mannervik, B., 1985, Isoenzymes of glutathione transferase in rat kidney cytosol. Biochemical Journal, 230, 609-615.

Hook, J.B., Elcombe, C.R., Rose, M.S. and Lock, E.A., 1982, Characterization of the effects of known hepatic monooxygenase inducers on male and female rat and mouse kidneys. Life Sciences, 31, 1077-1084.

Jefcoate, C.R., 1986, Cytochrome P-450 enzymes in sterol biosynthesis and metabolism. In Cytochrome P-450. Structure, - Mechanism, and Biochemistry, edited by P.R. Ortiz de Montellano (New York: Plenum), pp. 387-428.

Kinsler, S., Levi, P.E. and Hodgson, E., 1988, Hepatic and extrahepatic microsomal oxidation of phorate by the cytochrome P-450 and FAD-containing monooxygenase systems in the mouse. Pesticide Biochemistry and Physiology, 31, in press.

Lash, L.H. and Anders, M.W., 1986, Bioactivation and cytotoxicity of nephrotoxic amino acids and glutathione S-conjugates. Comments on Toxicology, 1, 87-106.

Lash, L.H. and Anders, M.W., 1988, S-Conjugate metabolism by purified bovine kidney cytosolic cysteine conjugate β-lyase. FASEB Journal, 2, A1059.

Lash, L.H., Elfarra, A.A. and Anders, M.W., 1986, Renal cysteine conjugate β-lyase: Bioactivation of nephrotoxic cysteine S-conjugates in mitochondrial outer membrane. Journal of Biological Chemistry, 261, 5930-5935.

Lock, E.A., 1987, The nephrotoxicity of haloalkane and haloalkene glutathione conjugates. In Selectivity and Molecular Mechanisms of Toxicity, edited by F. DeMatteis and E. A. Lock (New York: Macmillan Press), 59-83.

Maines, M.D., 1986, Differential effect of cis-platinum (cis-diamminedichloroplatinum) on regulation of liver and kidney haem and haemoprotein metabolism. Biochemical Journal, 237, 713-721.

Marnett, L.J., 1981, Polycyclic aromatic hydrocarbon oxidation during prostaglandin biosynthesis. Life Sciences, 29, 531-546.

Marnett, L.J., 1987, Peroxyl free radicals: Potential mediators of tumor initiation and promotion. Carcinogenesis, 8, 1365-1373.

Marnett, L.J. and Eling, T.E., 1983, Cooxidation during prostaglandin biosynthesis: A pathway for the metabolic activation of xenobiotics. Reviews in Biochemical Toxicology, 5, 135-172.

Masters, B.S.S., Muerhoff, A.S. and Okita, R.T., 1987, Enzymology of extrahepatic cytochromes P-450. In Mammalian Cytochromes P-450, edited by F. P. Guengerich (Boca Raton: CRC Press), Vol. I, pp. 108-133.

McCormack, K.M., Cagen, S.Z., Rickert, D.E., Gibson, J.E. and Dent, J.G., 1979, Stimulation of hepatic and renal mixed-function oxidase in developing rats by polybrominated biphenyls. Drug Metabolism and Disposition, 7, 252-259.

Mohandas, J., Marshall, J.J., Duggin, G.G., Horvath, J.S. and Tiller, D.J., 1984, Differential distribution of glutathione and glutathione-related enzymes in rabbit kidney. Biochemical Pharmacology, 33, 1801-1807.

Needleman, P., Turk, J., Jakschik, B., Morrison, A. and Lefkowith, J.B., 1986, Arachidonic acid metabolism. Annual Review of Biochemistry, 55, 69-102.

Nelson, R., Van Dyke, R., Powis, G., Lash, L. and Anders, M., 1988, Characterization of human liver and kidney cysteine conjugate β-lyase activity. FASEB Journal, 2, A1355.

Newton, J.F., Kuo, C-H., DeShone, G.M., Hoefle, D., Bernstein, J. and Hook, J.B., 1985, The role of p-aminophenol in acetaminophen-induced nephrotoxicity: Effect of bis(p-nitrophenyl)phosphate on acetaminophen and p-aminophenol nephrotoxicity and metabolism in Fischer 344 rats. Toxicology and Applied Pharmacology, 81, 416-430.

Ogita, K., Kusunose, Ichihara, K. and Kusunose, M., 1982, Multiple forms of cytochrome P-450 in kidney cortex microsomes of rabbits treated with 3-methylcholanthrene. Journal of Biochemistry, 92, 921-928.

Powell, G.M. and Roy, A.B., 1980, Sulphate conjugation. In Extrahepatic Metabolism of Drugs and Other Foreign Compounds, edited by T. E. Gram (New York: SP Medical and Scientific Books), pp. 389-425.

Ricci, G., Nardini, M., Federici, G. and Cavallini, D., 1986, The transamination of L-cystathionine, L-cystine and related compounds by a bovine kidney transaminase. European Journal of Biochemistry, 157, 57-63.

Rush, G.F. and Hook, J.B., 1982, Renal drug metabolism and nephrotoxicity. In Nephrotoxicity. Assessment and Pathogenesis, edited by P.H. Bach, F.W. Bonner, J.W. Bridges and E. A. Lock (New York: Wiley), pp. 237-245.

Rush, G.F., Newton, J.F. and Hook, J.B., 1983, Sex differences in the excretion of glucuronide conjugates: The role of intrarenal glucuronidation. Journal of Pharmacology and Experimental Therapeutics, 227, 658-662.

Rush, G.F., Smith, J.H., Newton, J.F. and Hook, J.B., 1984, Chemically induced nephrotoxicity: Role of metabolic activation. CRC Critical Reviews in Toxicology, 13, 99-160.

Rydström, J., Montelius, J. and Bengtsson, M., editors, 1983, Extrahepatic Drug Metabolism and Chemical Carcinogenesis (Amsterdam: Elsevier).

Schwab, G.E. and Johnson, E.F., 1987, Enzymology of rabbit cytochromes P-450. In Mammalian Cytochromes P-450, edited by F. P. Guengerich (Boca Raton: CRC Press), Vol. I, pp. 56-107.

Smith, J.H., 1986, Role of renal metabolism in chloroform nephrotoxicity. Comments on Toxicology, 1, 125-144.

Spry, L.A., Zenser, T.V. and Davis, B.B., 1986, Bioactivation of xenobiotics by prostaglandin H synthase in the kidney: Implications for therapy. Comments on Toxicology, 1, 109-123.

Stevens, J.L., Robbins, J.D. and Byrd, R.A., 1986, A purified cysteine conjugate β-lyase from rat kidney cytosol. Journal of Biological Chemistry, 261, 15529-15537.

Stevens, J.L., Ayoubi, N. and Robbins, J.D., 1988, The role of mitochondrial matrix enzymes in the metabolism and toxicity of cysteine conjugates. Journal of Biological Chemistry, 263, 3395-3401.

Tate, S.S., 1980, Enzymes of mercapturic acid formation. In Enzymatic Basis of Detoxication, edited by W. B. Jakoby (New York: Academic Press), pp. 95-120.

Trakshel, G.M. and Maines, M.D., 1988, Characterization of glutathione S-transferases in rat kidney. Biochemical Journal, 252, 127-136.

Walker, R.J. and Duggin, G.G., 1988, Drug nephrotoxicity. Annual Review of Pharmacology and Toxicology, 28, 331-345.

West, P.R., Harman, L.S., Josephy, P.D. and Mason, R.P., 1984, Acetaminophen: enzymatic formation of a transient phenoxyl free radical. Biochemical Pharmacology, 33, 2933-2936.

Yoshimoto, M., Kusunose, E., Yamamoto, S., Maekawa, M. and Kusunose, M., 1986, Purification and characterization of 2 forms of cytochrome P-450 from rat-kidney cortex microsomes. Biochemistry International, 13, 749-755.

Zenser, T.V. and Davis, B.B., 1984, Enzyme systems involved in the formation of reactive metabolites in the renal medulla: Cooxidation via prostaglandin H synthase. Fundamental and Applied Toxicology, 4, 922-929.

Zenser, T.V., Mattammal, M.B. and Davis, B.B., 1979, Demonstration of separate pathways for the metabolism of organic compounds in rabbit kidney. Journal of Pharmacology and Experimental Therapeutics, 208, 418-421.

Zenser, T.V., Mattammal, M.B., Palmier, M.O. and Davis, B.B., 1981, Microsomal nitroreductase activity of rabbit kidney and bladder: Implications in 5-nitrofuran-induced toxicity. Journal of Pharmacology and Experimental Therapeutics, 219, 735-740.

Zenser, T.V., Cohen, S.M., Mattammal, M.B., Wise, R.W., Rapp, N.S. and Davis, B.B., 1983, Prostaglandin hydroperoxidase-catalyzed activation of certain N-substituted aryl renal and bladder carcinogens. Environmental Health Perspectives, 49, 33-41.

Ziegler, D.M., 1980, Microsomal flavin-containing monooxygenase: Oxygenation of nucleophilic nitrogen and sulfur compounds. In: Enzymatic Basis of Detoxication, edited by W. B. Jakoby (New York: Academic Press), pp. 201-227.

BIOACTIVATION AND DETOXICATION OF ORGANOPHOSPHORUS
INSECTICIDES IN RAT BRAIN

Janice E. Chambers, Carol S. Forsyth and
Howard W. Chambers

Departments of Biological Sciences and Entomology,
Mississippi State University,
Mississippi State, Mississippi 39762,
USA

INTRODUCTION
  The primary action of the organophosphorus (OP)
insecticides is the inhibition of acetylcholinesterase
(AChE) which leads to accumulation of the neurotransmitter
acetylcholine and thus hyperactivity within the cholinergic
pathways. In a case of an acute poisoning, this leads to
tremors, convulsions, respiratory collapse and death. Death
is usually the result of respiratory failure resulting from
four actions: bronchoconstriction, excess mucus accumulation
in the respiratory tract, paralysis of the respiratory
muscles, and inhibition of the respiratory control centre in
the medulla oblongata. The last of these actions implies
that active anticholinesterase molecules must be present at
target sites in the brain.

ORGANOPHOSPHORUS INSECTICIDE METABOLISM AND BIOCHEMICAL ACTION
  Many of the OP insecticides are applied in the field as
phosphorothionates, which are relatively weak
anticholinesterases. The I_{50} values for the inhibition of
AChE of various rat brain regions by the three OP
insecticides parathion, methyl parathion and EPN are 10^{-5} -
10^{-4}M (Table 1) (Forsyth and Chambers, 1988). In contrast,
the corresponding I_{50} values for the active metabolites (the
phosphate or oxon forms) paraoxon, methyl paraoxon, and EPN-
oxon are 10^{-8} - 10^{-7}M, an increase in potency of three
orders of magnitude. The oxons are clearly the compounds of
concern from a toxicological perspective.
  The oxons readily phosphorylate serine residues in
proteins, releasing the leaving group. The phosphorylated
product is frequently quite persistent, lasting for several
days. The critical serine residue involved in OP
insecticide intoxication is that at the catalytic site of

TABLE 1 - I_{50} values for three phosphorothionate insecticides and their oxon analogs for rat brain acetylcholinesterase[1]

Compound	I_{50}, M[2]	
	Female	Male
Phosphorothionates		
Parathion	$4.35 \pm 0.13 (\times 10^{-5})$	$4.00 \pm 0.53 (\times 10^{-5})$
Methyl parathion	$7.25 \pm 0.30 (\times 10^{-4})$	$6.46 \pm 0.25 (\times 10^{-4})$
EPN	$2.73 \pm 0.08 (\times 10^{-5})$	$2.40 \pm 0.06 (\times 10^{-5})$
Oxons		
Paraoxon	$2.62 \pm 0.18 (\times 10^{-8})$	$2.37 \pm 0.10 (\times 10^{-8})$
Methyl paraoxon	$1.01 \pm 0.14 (\times 10^{-7})$	$0.94 \pm 0.08 (\times 10^{-7})$
EPN-oxon	$3.55 \pm 0.13 (\times 10^{-8})$	$3.30 \pm 0.12 (\times 10^{-8})$

[1] I_{50} calculated by linear regression of log-logit transformations of per cent inhibition data resulting from 15 min pre-incubations of rat brain homogenate with inhibitor. Brain homogenate was 12% cerebral cortex, 42% cerebellum, 10% corpus striatum and 36% medulla oblongata/pons, as a representative sample of critical brain regions. Acetylcholinesterase was assayed using the technique of Ellman et al. (1961).

[2] Mean ± S.E.M., 3-4 replications.

AChE which normally becomes transiently acetylated during the hydrolysis of its substrate acetylcholine.

However, there are other serine containing enzymes in the organism which could also serve as targets for the oxons (for example, serine esterases such as tissue B-esterases, serum B-esterases, serum butyrylcholinesterase and erythrocyte AChE). The B-esterases of both plasma and liver are considerably more sensitive to inhibition by paraoxon and EPN-oxon than is the target AChE in the brain (Table 2). Additionally, the liver B-esterases are more sensitive to inhibition by paraoxon and EPN-oxon than are the plasma B-esterases. In contrast, methyl paraoxon is a more potent inhibitor of AChE than of the B-esterases and of the plasma B-esterases than of the liver B-esterases. This suggests

TABLE 2 - Ratios of I_{50} values for three oxon analogues of phosphorothionate insecticides to target (acetylcholinesterase, AChE) and non-target (B-esterases, B-est) esterases in rat tissues[1]

Compound	Brain AChE/ plasma B-est	Brain AChE/ liver B-est	Plasma B-est/ liver B-est
Paraoxon	3.8	16.7	4.4
Methyl paraoxon	0.9	0.3	0.3
EPN-oxon	6.3	38.9	6.2

[1] I_{50} values resulted from linear regression of log-logit transformations of per cent inhibition data resulting from 15 min pre-incubations of rat brain or liver homogenate or diluted rat plasma with inhibitor. Acetylcholinesterase way assayed using the technique of Ellman et al. (1961). B-Esterases were assayed using 4-nitrophenyl valerate as substrate. Regressions were the result of 3 or more replications.

that the B-esterases could serve to protect the target from inhibition in the case of paraoxon or EPN-oxon intoxication but would afford little, if any, protection against methyl paraoxon. This idea is supported by the fact that the addition of plasma or a liver homogenate appreciably reduces the amount of brain AChE inhibition elicited by paraoxon or EPN-oxon, but not methyl paraoxon (Table 3). In addition, the oxon can be degraded by liver or plasma A-esterases, further reducing the amount of oxon reaching the brain targets.

This idea of a limited amount of oxon being available in the blood to enter the brain is also supported by the work of De Schryver et al. (1987) on methyl paraoxon toxicokinetics in the dog. The liver readily extracted methyl paraoxon from the blood following an intravenous administration. By extrapolation, the liver must contain factors or metabolic systems which would sequester or inactivate the oxon formed in that organ before it can migrate into the blood. In addition, these researchers found that the volume of distribution was greater than the total body water, which implied that tissue binding of the

TABLE 3 - Effects of plasma or liver homogenate on rat
brain acetylcholinesterase (AChE) inhibition
by organophosphates[1]

Compound	Per cent inhibition of AChE			
	Conc., nM	Brain	Plasma + brain	Liver + brain
Paraoxon	31.6	55	45	7
Methyl paraoxon	316.0	78	82	73
EPN-oxon	31.6	51	41	10

[1] Inhibitors were incubated in buffer, diluted plasma (12.5 µl/ml) or liver homogenate (12.5 mg/ml) for 15 min prior to addition of brain homogenate. Per cent inhibition was determined after a 2nd 15-min incubation. AChE was assayed using the technique of Ellman et al. (1961).

methyl paraoxon contributed to the distribution pattern. Also, the plasma clearance was higher than the plasma blood flow through the liver, suggesting that the blood had some detoxication capacity.

Thus, although the liver is typically the most active tissue metabolically in the organism and would be expected to generate the activated metabolite most rapidly, these alternate targets could readily sequester the hepatically-generated oxon and prevent it from reaching the target AChE in the brain. Little of the oxon produced in the liver may be able to exit the liver until the hepatic B-esterases are substantially inhibited. Once in the blood, three blood enzymes (serum butyrylcholinesterase, serum B-esterases and erythrocyte AChE) could all be inhibited, and, in addition, sites on serum albumin could also be phosphorylated non-specifically. All of these phosphorylations are essentially "detoxications", in that the oxon is destroyed at a slow rate during the phosphorylation reaction. Thus, because of these non-target phosphorylations as well as the A-esterase-mediated detoxication, it seems rather unlikely that appreciable amounts of hepatically-generated oxon would arrive intact at the brain target sites. Additionally phosphorothionates in the animal resulting from dermal or pulmonary exposures (more likely routes in occupational exposures) are not subject to first pass metabolism in the liver; therefore appreciable quantities of these unaltered compounds could circulate directly to the brain. So even though the brain is not very active in xenobiotic metabolism, if the brain has the ability to activate the phosphorothionate that circulates to it, then production of the activated metabolite in close proximity to the target AChE may be extremely important in intoxication. The importance of extrahepatic activation of the phosphorothionates was suggested two decades ago (Neal, 1967), but this phenomenon has not been extensively investigated. Because some oxon exited the liver following parathion infusion but not after chlorpyrifos infusion, Sultatos et al. (1985) concluded that extrahepatic activation alone was important in chlorpyrifos poisoning whereas both hepatic and extrahepatic activation were involved with parathion. If the brain can also detoxify the oxon, then it may be able to partially protect itself. Both activation and detoxication routes need to be understood to assess the significance of brain metabolism during phosphorothionate exposures.

The usual route of phosphorothionate activation is through the process of oxidative desulphuration catalyzed by cytochrome P-450-dependent monooxygenases which require NADPH and molecular oxygen (Kulkarni and Hodgson, 1984; Neal and Halpert, 1982; Neal, 1980). In this reaction, an oxygen atom is added to the phosphorothionate to form a phosphooxythirane intermediate which can be converted to either of two products: if S is released, the oxon is formed (the activation reaction of interest here) and if the leaving group is released, a dialkyl thiophosphate detoxication product is formed (a dearylation reaction which was not studied here). The latter degradation reaction is specific for the phosphorothionate (and not the oxon) (Nakatsugawa and Dahm, 1967). These reactions are inhibited by carbon monoxide and are inducible in liver by phenobarbitone, 3,4-benzo(a)pyrene or 3-methylcholanthrene pretreatment (Murphy and DuBois, 1958; Neal, 1967).

The oxons are substrates for a group of Ca^{++}-dependent carboxylesterases termed A-esterases or phosphotriesterases. These enzymes hydrolyze the oxons and serve as the major route of detoxication of the activated metabolites (Sultatos and Murphy, 1983b; Butler et al., 1985).

MONOOXYGENASE ACTIVITY IN BRAIN

There is a body of knowledge accumulating on the existence of cytochrome P-450 and xenobiotic metabolizing enzyme activities in the brain. The brain has been shown to metabolize several xenobiotics: aminopyrine (Marietta et al., 1979), morphine (Fishman et al., 1976), benzo(a)pyrene (Cohn et al., 1977) and 7,12-dimethylbenzanthracene (Juchau et al., 1979). Qato and Maines (1985) reported cytochrome P-450, NADPH-cytochrome c reductase, benzo(a)pyrene hydroxylase and 7-ethoxycoumarin deethylase activities in rat brain microsomes (Table 4). Aryl hydrocarbon hydroxylase, using naphthalene as a substrate, has also been reported for rat brain microsomes (Mesnil et al., 1985). However, others have found higher levels of cytochrome P-450 in both synaptic and non-synaptic mitochondria than in microsomes from brains of rats, guinea pigs, rabbits and swine (Walther et al., 1986). These researchers found equivalent amounts of cytochrome P-450 in brain non-synaptic and synaptic mitochondria (48-49 pmol per mg protein) whereas brain microsomes possessed less than half as much (about 18 pmol per mg protein) (Table 4).

TABLE 4 - Cytochrome P-450 concentration and monooxygenase activities in rat brain

Enzyme	Mitochondria	Microsomes	Reference
Cytochrome P-450 (pmol/mg protein)	48.0±1.2	17.7±4.9	Walther et al., 1986.
		5.1±0.9	Qato and Maines, 1985.
NADPH cytochrome c reductase (nmol/mg protein/min)	0.13±0.05	16.6±2.4	Walther et al., 1986.
		133.7±9.5	Qato and Maines, 1985.
Benzo(a)pyrene hydroxylase (pmol/mg/h)		0.9±0.1	Qato and Maines, 1985.
7-Ethoxycoumarin deethylase (pmol/mg/h)		53.3±6.7	Qato and Maines, 1985.
7-Ethoxyresorufin O-deethylase (pmol/mg protein/h)	8.2±1.2	18.1±2.3	Walther et al., 1987.

A similar distribution between mitochondria and microsomes was also found in the brains from guinea pigs, rabbits and swine with from 2.7- to 4.1-fold more cytochrome P-450 observed in non-synaptic mitochondria than in microsomes. The cytochrome P-450 was located primarily in the mitochondrial inner membrane; when the outer membrane was stripped off, the cytochrome P-450 concentration in the remaining membrane was about 2.5-fold higher (129 pmol per mg protein). However the concentrations of cytochrome P-450 in brain were very low compared to those in liver (580 pmol per mg protein) with 12- and 33-fold higher contents in liver than in brain mitochondria and microsomes, respectively. The absorbance maxima for the mitochondrial and microsomal cytochrome P-450's were 448 and 450 nm, respectively. In contrast, the specific activity of NADPH cytochrome c reductase was about 5-fold higher in brain microsomes than mitochondria (16.6±2.35 and 3.27±0.35 nmol per min per mg protein, respectively). The authors speculated that the brain monooxygenase activity, particularly that in the mitochondria, was associated with the brain's endocrine function, notably steroid hormone metabolism. The specific activity of ethoxyresorufin O-deethylase was about 2-fold higher in brain microsomes than in mitochondria and liver specific activity was about 39-87 fold higher than that observed in either brain fraction (Table 4). Both NADPH and NADH served as cofactors; however the activity was greatest when both cofactors were present, compared to either cofactor alone. As expected, NADPH alone yielded greater activity than NADH alone (Walther et al., 1987). This double cofactor dependence has also been reported for brain 7-ethoxycoumarin O-deethylase activity (Srivastava et al., 1983). We are now accumulating data on the presence of NADPH cytochrome c reductase and aminopyrine N-demethylase in both mitochondrial and microsomal fractions from rat brain.

7-Ethoxyresorufin O-deethylase activity in brain mitochondria and microsomes was induced by 2- and 1.6-fold, respectively, by pretreatment with 3-methylcholanthrene, while liver activity was induced 75-fold by the same treatment (Walther et al., 1987). The authors suggested that this low capacity for induction serves to protect the brain from environmentally-elicited enhancement by toxicity of xenobiotics.

Kapitulnik et al. (1987) also located cytochrome P-450 in the rat brain (immunohistochemically) and found it most concentrated in the basal ganglia, septum and hypothalamus. The only immunoreactivity observed in brain was with antibodies raised to rat liver cytochrome P450c, the isozyme induced by 3-methylcholanthrene treatment. There was no brain reactivity to antibodies raised against rat liver cytochrome P-450's induced by phenobarbitone, isosafrole or pregnenolone-16α-carbonitrile.

Thus the cytochrome P-450 present in brain is of a limited type, similar to the cytochrome P450c of rat liver and the cytochrome P-450scc of adrenal gland (Le Goascogne et al., 1987). It has catalytic properties similar to those of the cytochrome P-450's of the liver in terms of cofactor requirements, inhibitory profiles and potential substrates. The enzymes located in the mitochondria are at least as important as, if not more important than, the enzymes located in the microsomes. Finally, the enzymes have a limited capacity to be induced by 3-methylcholanthrene.

ORGANOPHOSPHORUS INSECTICIDE METABOLISM IN BRAIN

Our primary interest is in the disposition and metabolism of organophosphorus insecticides. Because of the above stated evidence that there are very sensitive liver and blood non-target esterases which could sequester the hepatically-generated oxons before they have an opportunity to reach the target AChE in the brain, we have developed the hypothesis that the phosphorothionate activation activity in the brain, even though probably very low compared to other tissues, is significant in the formation of the oxon responsible for the inhibition of brain AChE during phosphorothionate intoxication. Our work has concentrated on the monooxygenase-mediated activation activity, as well as the oxon degradation activity to assess the overall importance of brain metabolism in the poisoning phenomenon. The emphasis of the investigations reported here is a study of the desulphuration of three phosphorothionate insecticides which have wide current and/or recent usage (parathion, EPN and methyl parathion) and the A-esterase mediated detoxication of their oxons (paraoxon, EPN-oxon and methyl paraoxon, respectively) analogues. The reactions are illustrated in Figure 1.

To confirm initially that the target brain AChE was indeed more sensitive to the oxons and to determine if there were regional differences in sensitivity, inhibition studies were run on brain AChE from major important regions of the brain from both sexes. As an example, the inhibition data for methyl paraoxon and methyl parathion in different brain regions are given in Table 5. As expected, methyl paraoxon was a more potent inhibitor by almost three orders of magnitude than was methyl parathion.

No outstanding differences in sensitivity among the four brain regions were observed in either sex. The AChE from

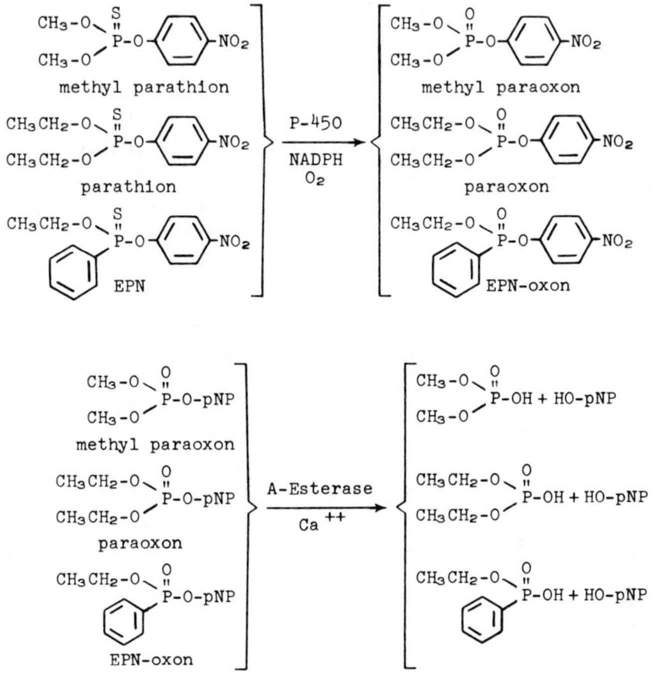

Figure 1. Monooxygenase-mediated phosphorothionate activation reactions and A-esterase-mediated oxon hydrolysis reactions. (pNP = p-nitrophenyl)

males tended to be slightly more sensitive than that from females; however the differences were small, and would not be expected to contribute substantially, if at all, to sex differences in toxicity. The same pattern was also observed with parathion/paraoxon and EPN/EPN-oxon (Forsyth and Chambers, 1988), except that I_{50}'s for the phosphorothionates and oxons were in the range of 10^{-5}M and 10^{-8}M, respectively (about one order of magnitude more potent than methyl parathion/methyl paraoxon, Table 1). Also no outstanding differences in sensitivities among brain regions were observed, and AChE from males was slightly more sensitive than from females, but probably not enough to contribute substantially to sex differences in toxicity. Thus, the oxons are the compounds of toxicological concern to all major brain regions.

TABLE 5 - I_{50} values for methyl parathion and methyl paraoxon for rat brain acetylcholinesterase from several brain regions[1]

Compound	Region	I_{50}, M[2]	
		Female	Male
Methyl parathion (X 10^{-4}M)	Cerebral cortex	7.34±0.12	6.31±0.09
	Corpus striatum	6.66±0.10	6.58±0.44
	Cerebellum	7.60±0.36	6.92±0.18
	Medulla/pons	6.93±0.32	6.22±0.05
Methyl paraoxon (X 10^{-7}M)	Cerebral cortex	1.10±0.02	0.90±0.01
	Corpus striatum	0.82±0.17	0.93±0.03
	Cerebellum	1.25±0.09	1.08±0.02
	Medulla/pons	1.07±0.05	0.97±0.02

[1]I_{50} calculation method as described in Table 1.
[2]Mean ± S.E.M. (N=3)

Parathion activation in brain microsomes was originally reported by Norman and Neal (1976). Metabolite detection in those studies was by thin layer chromatography of radiolabelled compounds. Researchers more recently have utilized HPLC techniques for metabolite detection (Sultatos and Murphy, 1983a; Wallace and Dargan, 1987). Both of these techniques rely on free metabolite being available for extraction. However, the premise behind all of our experiments is that the oxons are so reactive that they will phosphorylate non-target esterases before they have an opportunity to be transported intact to the brain. By extrapolation, there may well be sites for phosphorylation available in the homogenate or subcellular fraction (such as AChE in brain) which could react with the oxon as it is formed and thus bias the results from an in vitro assay downwards with respect to the amount of oxon produced. To counteract this possibility, we have chosen to assay for the oxon based on its biological activity by adding to the incubation mixture an exogenous source of AChE (a bovine brain suspension) to serve as a trap for the oxon as it is formed (Forsyth and Chambers, 1988). This method, by quantifying the resultant inhibition to the bovine AChE, has the potential of assaying far more of the oxon formed, than a method which quantifies only the metabolite remaining intact at the end of the incubation period. In fact, preliminary studies in our laboratory indicated that substantially less inhibition of the exogenous AChE was observed when it was added after the incubation period than when it was present during the incubation period.

A low but significant amount of phosphorothionate activation was observed with all three substrates, [about 0.2-1.8 nmoles oxon formed per min per g for both mitochondria (P_2 pellet) and microsomes for each sex, Figure 2].

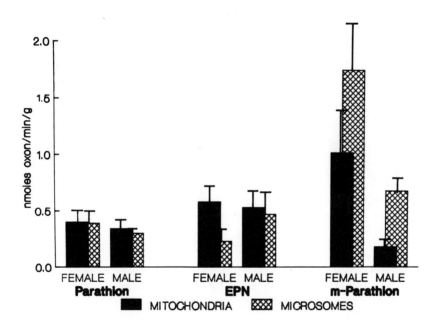

Figure 2. Oxon production from parathion, EPN and methyl parathion (m-Parathion) by rat brain mitochondrial [17,000g(15 min) pellet of 5000g(10 min) supernatant] and microsomal fractions during a 5 min incubation in the presence of O_2 and an NADPH-generating system. Oxon estimated by the amount of inhibition resulting to an exogenous source of acetylcholinesterase (bovine brain) present concurrently during incubation. Acetylcholinesterase was determined by the method of Ellman et al. (1961). Data are expressed as nmoles oxon produced per min per g, mean ± S.E.M.

Activities are expressed in terms of wet weight rather than protein so that more meaningful comparisons between subcellular fractions can be made. Although activities of female brain mitochondria and microsomes with methyl parathion appeared higher than other activities, they were not significantly higher (P> 0.05) than the corresponding male activities. There were no significant (P> 0.05) differences between sexes or between subcellular fractions for any compound. Thus, the frequently observed sex difference in monooxygenase activity in liver (Figure 3) is not present with this activity in brain.

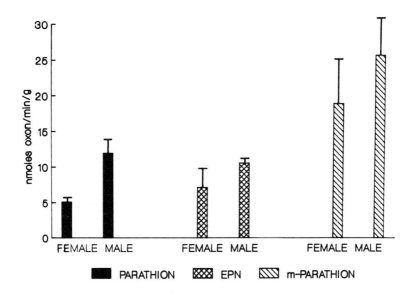

Figure 3. Oxon production by rat liver microsomes. Conditions are as described in Figure 2, except for a 2 min incubation

As expected, liver microsomal activities were appreciably higher than brain activities (5-26 nmoles oxon per min per g). When the activities of brain mitochondria and microsomes are combined to estimate the overall activation capacity in the brain, there is a 6-19-fold greater activity in liver than brain for parathion per unit weight, 9-11-fold for EPN and 7-30-fold for methyl parathion (with the higher number in the range for the male because of the higher male hepatic activities). However, the disparity between liver and brain metabolic capacity is even greater if the relative size of the two organs is taken into account. Although liver activation was similar for both parathion and EPN, livers activated higher amounts of methyl parathion (Figure 3). The smaller methoxy groups apparently make methyl parathion a more suitable substrate than parathion and EPN which contain ethoxy groups. Preliminary results have suggested that brain activity is not induced by either phenobarbitone or β-naphthoflavone pretreatment, even though liver activity is substantially induced by both.

The amount of A-esterase activity was estimated spectrophotometrically by the amount of p-nitrophenol produced by the hydrolysis of the oxons. This reaction was performed in the presence of 0.1mM eserine, a carbamate AChE inhibitor (i.e., to inhibit the release of p-nitrophenol by phosphorylation of AChE by oxons). There was considerably less degradation activity in the brain than the liver, and the patterns of degradation by the two organs were different. While the brain degraded the most EPN-oxon and the least paraoxon (Figure 4), the liver degraded the most methyl paraoxon and the least EPN-oxon (Figure 5). The degradation rate of all oxons tested was not significantly ($P > 0.05$) affected by animal sex.

The overall degradation activities (1-11 nmoles per min per g for brain and 31-134 for liver) were considerably higher than the overall (mitochondrial plus microsomal) activation activities (0.6-2.8 nmoles per min per g for brain and 5-26 for liver). Ratios of degradation to activation activities were greater than 1 in all cases and ranged up to 13 (Figure 6). This implies that all the oxon formed in either tissue would be immediately degraded, and therefore no oxon accumulation could occur. However, in fact this is not true since AChE inhibition does occur in vivo following phosphorothionate intoxication. Kinetic studies in liver have indicated that the Km's for oxon hydrolysis (about 180 μM) are considerably higher than those for phosphorothionate activation (about 10-20 μM) (Sultatos and Murphy, 1983 a and b; Wallace and Dargan, 1987). Because the A-esterases will not function efficiently at the low substrate concentrations provided by the monooxygenases, accumulation of low levels of oxons in vivo can be predicted. The ratio of degradation to activation in the brain for parathion/paraoxon is quite low suggesting that paraoxon might be especially likely to accumulate in the brain. Since brain AChE can be inhibited at considerably lower concentrations than either of these Km values (I_{50}'s of 0.02-0.1 μM, Table 1), it is reasonable to conclude that the generated oxon would phosphorylate AChE far more readily than it would be degraded by the A-esterases.

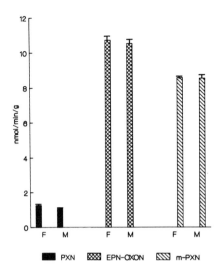
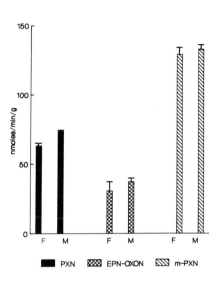

Figure 4. Oxon degradation by rat brain homogenates in the presence of Ca^{++} and Mg^{++} (1mM each) and 0.1mM eserine during a 45 min incubation. Data are expressed as nmoles p-nitrophenol produced per min per g, mean $\pm$ S.E.M. (PXN = paraoxon, m-PXN = methyl paraoxon)

Figure 5. Oxon degradation by rat liver homogenates. Conditions were as described in Figure 4, expect for a 20 min incubation.

SUMMARY

In summary, the brain displays xenobiotic metabolizing enzyme activity. Cytochrome P-450 (present in both microsomes and mitochondria, in the brain) is capable of metabolizing a variety of substrates, including benzo(a)pyrene, naphthalene, aminopyrine, 7-ethoxycoumarin, 7-ethoxyresorufin and the phosphorothionate insecticides parathion, methyl parathion and EPN. The brain cytochrome P-450 appears to resemble the adrenal cytochrome P-450scc involved in steroidogenesis. The brain cytochrome P-450 has a very limited capacity to be induced compared to the liver cytochromes. With respect to the phosphorothionate insecticides, the activation activity in brain is low but its close proximity to target AChE may make it highly significant in phosphorothionate poisoning. Although A-

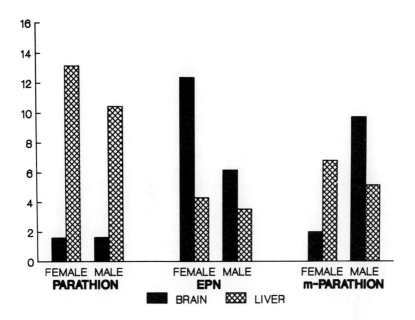

Figure 6. Ratios of oxon degradation to phosphorothionate activation activities in rat brain and liver for parathion/paraoxon, EPN/EPN-oxon, and methyl parathion (m-PARATHION)/methyl paraoxon

esterase activity is higher than activation activity, the higher Km values for A-esterases may limit their ability to effectively degrade the oxons before they have an opportunity to inhibit the target AChE. Thus, the brain may well have the capacity to intoxicate itself during phosphorothionate exposures.

Acknowledgements. This research was supported by National Institutes of Health grant ES04394.

REFERENCES
Butler, E.G., Eckerson, H.W. and LaDu, B.N., 1985, Paraoxon hydrolysis vs. covalent binding in the elimination of paraoxon in the rabbit. Drug Metabolism and Disposition, 13, 640-645.

Chambers, J.E. and Yarbrough, J.D., 1973, Organophosphate degradation by insecticide-resistant and susceptible populations of mosquitofish (Gambusia affinis). Pesticide Biochemistry and Physiology, 3, 312-316.

Cohn, J.A., Alvares, A.P. and Kappas, A., 1977, On the occurrence of cytochrome P-450 and aryl hydrocarbon hydroxylase activity in the rat brain. Journal of Experimental Medicine, 145, 1606-1611.

De Schryver, E., De Reu, L., Belpaire, F. and Willems, J., 1987, Toxicokinetics of methyl paraoxon in the dog. Archives of Toxicology, 59, 319-322.

Ellman, G.L., Courtney, K.D., Andres, V., Jr. and Featherstone, R.M., 1961, A new and rapid colorimetric determination of acetylcholinesterase activity. Biochemical Pharmacology, 7, 88-95.

Fishman, J., Hahn, E.F. and Norton, B.I., 1976, N-Demethylation of morphine in rat brain is localized in sites of high opiate receptor content. Nature, 261, 64-65.

Forsyth, C.S. and Chambers, J.E., 1988, Activation and degradation of the phosphorothionate insecticides parathion and EPN by rat brain. Biochemical Pharmacology, accepted.

Juchau, M.R., Di Giovanni, J., Namkung, M.J. and Jones, A.H., 1979, A comparison of the capacity of fetal and adult liver, lung, and brain to convert polycyclic aromatic hydrocarbons to mutagenic and cytotoxic metabolites in mice and rats. Toxicology and Applied Pharmacology, 49, 171-178.

Kulkarni, A.P. and Hodgson, E., 1984, The metabolism of insecticides: the role of monooxygenase enzymes. Annual Review of Pharmacology and Toxicology, 24, 19-42.

Le Goascogne, C., Robel, P., Gouezou, M., Sananes, N., Baulieu, E.E. and Waterman, M., 1987, Neurosteroids: Cytochrome P-450 in rat brain. Science, 237, 1212-1215.

Levi, P.E. and Hodgson, E., 1985, Oxidation of pesticides by purified cytochrome P-450 isozymes from mouse liver. Toxicology Letters, 24, 221-228.

Marietta, M.P., Vesell, E.S., Hartman, R.D., Weisz, J. and Dvorchik, B.H., 1979, Characterization of cytochrome P-450-dependent aminopyrine N-demethylase in rat brain: comparison with hepatic aminopyrine N-demethylation. Journal of Pharmacology and Experimental Therapeutics, 208, 271-279.

Mesnil, M., Testa, B. and Jenner, P., 1985, Aryl hydrocarbon hydroxylase in rat brain microsomes. Biochemical Pharmacology, 34, 435-436.

Murphy, S.D. and DuBois, K.P., 1958, The influence of various factors on the enzymatic conversion of organic thiophosphates to anticholinesterase agents. Journal of Pharmacology, 124, 194-202.

Nakatsugawa, T. and Dahm, P.A., 1967, Microsomal metabolism of parathion. Biochemical Pharmacology, 16, 25-38.

Neal, R.A., 1967, Studies on the metabolism of diethyl 4-nitrophenyl phosphorothionate (parathion) in vitro. Biochemical Journal, 103, 183-191.

Neal, R.A., 1980, Microsomal metabolism of thiono-sulfur compounds: mechanisms and toxicological significance. In Reviews in Biochemical Toxicology, 2nd edition, edited by E. Hodgson, J.R. Bend and R.M. Philpot (New York: Elsevier North Holland), pp. 131-171.

Neal, R.A. and Halpert, J., 1982, Toxicity of thiono-sulfur compounds. Annual Review of Pharmacology and Toxicology, 22, 321-339.

Norman, B.J. and Neal, R.A., 1976, Examination of the metabolism in vitro of parathion (diethyl p-nitrophenyl phosphorothionate) by rat lung and brain. Biochemical Pharmacology, 25, 37-45.

Srivastava, S.P., Seth, P.K. and Mukhtar, H., 1983, 7-Ethoxycoumarin O-deethylase activity in rat brain microsomes. Biochemical Pharmacology, 32, 3657-3660.

Sultatos, L.G., Minor, L.D. and Murphy, S.D., 1985, Metabolic activation of phosphorothioate pesticides: role of the liver. Journal of Pharmacology and Experimental Therapeutics, 232, 624-628.

Sultatos, L.G. and Murphy, S.D., 1983a, Kinetic analyses of the microsomal biotransformation of the phosphorothioate insecticides chlorpyrifos and parathion. Fundamental and Applied Toxicology, 3, 16-21.

Sultatos, L.G. and Murphy, S.D., 1983b, Hepatic microsomal detoxification of the organophosphates paraoxon and chlorpyrifos oxon in the mouse. Drug Metabolism and Disposition, 11, 232-238.

Wallace, K.B. and Dargan, J.E., 1987, Intrinsic metabolic clearance of parathion and paraoxon by livers from fish and rodents. Toxicology and Applied Pharmacology, 90, 235-242.

Walther, B., Ghersi-Egea, J.F., Minn, A. and Siest, G., 1986, Subcellular distribution of cytochrome P-450 in the brain. Brain Research, 375, 338-344.

Walther, B., Ghersi-Egea, J.F., Jayyosi, Z., Minn, A. and Siest, G., 1987, Ethoxyresorufin O-deethylase activity in rat brain subcellular fractions. Neuroscience Letters, 76, 58-62.

2. Intermediary Xenobiotic Metabolites and their Significance

METABOLITES RESULTING FROM OXIDATIVE AND REDUCTIVE PROCESSES

Patricia E. Levi and Ernest Hodgson

Toxicology Program, Box 7633,
North Carolina State University,
Raleigh, NC 27695, USA

INTRODUCTION

Metabolism by a wide array of enzymes that are capable of using xenobiotics as substrates is the principal means by which the toxicity of exogenous chemicals is modified in vivo, being either increased or decreased. Most xenobiotics that enter the body are lipophilic, a property that enables them to penetrate lipid membranes and to be transported by lipoproteins in the blood. Xenobiotic metabolism generally consists of two phases. In phase one, a polar reactive group is introduced into the molecule, making the metabolite a suitable substrate for phase two conjugation reactions with endogenous cofactors such as sugars and amino acids. Such conjugation products are exceedingly water-soluble and are readily excreted. Although the overall process is generally a detoxication sequence, reactive intermediates may be formed that are much more toxic than the parent compound. It is, however, usually a sequence that increases water solubility and hence decreases the half-life in vivo. Formation of reactive intermediates occurs more frequently with phase one oxygenations since the products are often potent electrophiles capable of reacting with nucleophiles within the cells, within proteins, RNA, and DNA. These covalent bonds with cellular macromolecules are thought to be the molecular lesions initiating chemically induced toxicity or carcinogenicity.

Phase one reactions include microsomal monooxygenations, such as those catalyzed by the cytochrome P-450 monooxygenase system (P-450) and the flavin-containing monooxygenase (FMO); peroxidase-type oxidations, such as co-oxidation by prostaglandin synthetase (PGS); cytosolic oxidations, e.g. xanthine and aldehyde oxidases (often referred to as molybdenum hydroxylases); and reduction reactions, many of which are catalyzed by P-450. Table 1 summarizes some of the more important enzymes and substrates involved.

The vast majority of drugs and other xenobiotics are initially metabolized via the P-450 pathway; in addition, the FMO also makes a significant contribution to xenobiotic metabolism due to its ability to carry out a wide variety of nitrogen and sulphur oxidations. Both of these systems will be discussed in more detail later, particularly the relative involvement of the two enzymes in the metabolism of common substrates.

COOXIDATION BY PROSTAGLANDIN SYNTHETASE

The most important peroxidase reaction involved in the metabolic oxidation of xenobiotics is probably that catalyzed by prostaglandin synthetase (PGS) (Marnett and Eling, 1983). PGS catalyzes the oxygenation of poly-unsaturated fatty acids to hydroxy endoperoxides, with the preferential substrate in vivo being arachidonic acid (AA). PGS catalyzes two activities; the fatty acid cyclooxygenase activity which brings about the oxygenation of polyunsaturated fatty acids, such as arachidonic acid, to form a hydroperoxy endoperoxide, prostaglandin G (PGG); this compound is then reduced to a hydroperoxy endoperoxide termed prostaglandin H (PGH). A number of xenobiotics may be cooxidized during this second hydroperoxidase reaction.

The enzyme, which is membrane bound, is present in virtually all mammalian tissues, but is especially concentrated in seminal vesicles, a rich source which has been used extensively in experimental studies. PGS is also high in platelets, lungs, skin, kidney medulla, endothelial cells, and embryonic tissue. A wide variety of compounds can undergo oxidation during PGH biosynthesis; the types of cooxidations catalyzed include dehydrogenation, demethylation, epoxidation, sulphoxidation, N-oxidation, C-hydroxylation and dioxygenation. Some of the more important compounds which can be cooxidized are listed in Table 1.

It is apparent that many xenobiotics can be metabolized to reactive metabolites by PGS. The bladder carcinogens FANFT (N-[4-(5-nitro-2-furyl)-2-thiazolyl]-formamide) and ANFT (2-amino-4-(5-nitro-2-furyl)-thiazole), and benzidine are converted to metabolites that covalently bind to proteins and DNA. Acetaminophen (paracetamol), a widely used analgesic, can produce hepatic and renal toxicity. The inner medullary region of the kidney, which has little or no cytochrome P-450 activity, but is rich in PGS activity, is the region most susceptible to damage by acetaminophen. The metabolism of polycyclic aromatic hydrocarbons(PAH) has

TABLE 1 - Summary of some important oxidation and reduction reactions of xenobiotics.

ENZYMES	EXAMPLES
Cytochrome P-450 (P-450)	
Epoxidation and hydroxylation	Aldrin, benzo(a)pyrene, aflatoxin, nicotine, bromobenzene
N-Dealkylation	Ethylmorphine, atrazine, dimethyl-nitrocarbamate, dimethylaniline
O-Dealkylation	p-Nitroanisole, chlorfenvinphos
S-Dealkylation	Methylmercaptan
S-Oxidation	Thiobenzamide, phorate, endosulfan, methiocarb, chlorpromazine
N-Oxidation	2-Acetylaminofluorene
P-Oxidation	Diethylphenylphosphine
Desulphuration	Parathion, fonofos, carbon disulphide
Dehalogenation	CCl_4, $CHCl_3$
Flavin-containing monooxygenase (FMO)	
N-Oxygenation	Nicotine, dimethylaniline, imipramine
S-Oxygenation	Thiobenzamide, phorate, thiourea
P-Oxygenation	Diethylphenylphosphine
Desulphuration	Fonofos
Prostaglandin synthetase (PGS)-cooxidation	
Dehydrogenation	Acetaminophen, benzidine, DES, epinephrine
N-Demethylation	Dimethylaniline, benzphetamine, aminocarb
Hydroxylation	B(a)P, 2-aminofluorene, phenyl-butazone
Epoxidation	7,8-dihydrobenzo(a)pyrene
Sulphoxidation	Methylphenylsulphide
Oxidations	FANFT, ANFT, bilirubin
Molybdenum hydroxylases (aldehyde oxidase, xanthine oxidase)	
Oxidations	Purines, pteridine, methotrexate, quinolones, 6-deoxycyclovir
Reductions	Aromatic nitrocompounds, azo dyes, nitrosoamines, N-oxides, sulphoxides
Alcohol dehydrogenase	
Oxidations	Methanol, ethanol, isopropanol, glycols, glycol ethers (2-butoxy-ethanol)
Aldehyde dehydrogenase	
Oxidations	Aldehydes resulting from alcohol and glycol oxidations

been studied extensively, and a number of compounds have been shown to be activated by PGS to metabolites which bind to protein and DNA. Compounds activated by PGS to intermediates which are mutagenic in Salmonella tests include benzo(a)pyrene-7,8-diol, benzo(a)anthracene-3,4-diol, and 7,12-dimethylbenzanthracene. Several aromatic amines, 2-aminofluorene, benzidine and 2-naphthylamine, are also activated to mutagens by PGS; however, 2-acetylaminofluorene and 1-naphthylamine are not activated by PGS.

Recent studies by Wells and coworkers (Wells and Nagai, 1988; Wells et al., 1988), indicate that the anticonvulsant drugs phenytoin (Dilantin) and trimethadione are activated by PGS, probably in embryonic tissue, which is low in P-450 but high in PGS, to a reactive free radical which may be the metabolite responsible, in part, for the teratogenic potential of these drugs. Phenytoin teratogenicity in mice was reduced by pretreatment with the cyclooxygenase inhibitor acetylsalicylic acid, the antioxidant caffeic acid, and the free radical spin trapping agent PBN (α-phenyl-N-t-butylnitrone). On the other hand, pretreatment of mice with the tumor promoter TPA, an inducer of PGS activity, increased the teratogenicity of phenytoin.

It is now increasingly evident that PGS can serve as an alternate enzyme for xenobiotic metabolism, particularly in tissues with low monooxygenase activity. Consequently, the cooxidation of drugs, chemical carcinogens, and other xenobiotics by PGS is an important area of study necessary in order to increase our understanding of the mechanisms underlying toxicity. In addition this enzyme system, like the monooxygenases, may be useful for screening chemicals for mutagenic and teratogenic potential.

MOLYBDENUM HYDROXYLASES

There are a number of molybdenum-containing enzymes, but those which are important in carbon oxidation are aldehyde oxidase (AO) and xanthine oxidase (XO), also referred to as molybdenum hydroxylases. Both enzymes catalyze the oxidation of a wide range of aldehydes and N-heterocycles (Beedham, 1988). The name aldehyde oxidase is somewhat misleading, however, since oxidation of N-heteroaromatics is more significant. The differences in substrate specificities between monooxygenases and molybdenum hydroxylases is partly based on chemistry since with monooxygenases, the mechanism involves an electrophilic attack on the carbon, whereas the hydroxylases catalyze nucleophilic addition at an unsaturated carbon, which can be represented as an attack by the hydroxyl ion.

$$RH + OH^- \longrightarrow ROH + 2e^- + H^+$$

Although molecular oxygen may be involved in the overall oxidation process, the oxygen atom incorporated into the product is ultimately derived from water, whereas with monooxygenations, it derives from molecular oxygen. In N-heteroaromatic ring compounds, the most electropositive carbon in a heteroaromatic ring is usually adjacent to a ring N atom and this is the normal position of nucleophilic molybdenum hydroxylase attack. In contrast, microsomal oxidations would tend to occur distant from the N, probably in an adjacent carbocyclic ring. Uncharged carbocyclic compounds, such as naphthalene, are not substrates for these enzymes, but as the number of N atoms increases, so does the affinity for the enzyme. Thus the bases purine and pteridine are oxidized exclusively by AO and XO, although the oxidations do not necessarily occur at the same positions.

These enzymes are found predominantly in the cytosol and occur in most tissues. However, the highest levels are found in those tissues most often exposed to ingested foreign compounds; AO is most prevalent in the liver, whereas the highest levels of XO occur in the small intestine, milk and mammary gland. Both enzymes react with endogenous substrates; XO is involved in the final stages of purine catabolism and AO catalyzes the oxidation of vitamins such as pyridoxal and N-methylnicotinamide. Aldehyde oxidase and xanthine oxidase also catalyze the oxidation of a wide range of xenobiotics including aldehydes, uncharged bases, quaternary N-heterocycles and carbocyclics. In addition to their oxidative role, these enzymes may be involved in reductive pathways in vivo, and they are known to catalyze the in vitro reduction of aromatic nitro compounds, azo dyes, nitrosamines, N-oxides, and sulphoxides. Table 1 lists some of their known substrates. Since these enzymes metabolize such a wide range of xenobiotics, including many which are also substrates for the monooxygenases, the role of the molybdenum hydroxylases in detoxication can no longer be discounted. In addition, they may also be important in the production as well as deactivation of toxic metabolites via reductive pathways.

ALCOHOL AND ALDEHYDE DEHYDROGENASE

Alcohol and aldehyde dehydrogenases are probably best known for their sequential activation of alcohols, such as methanol and isopropanol, to toxic metabolites. Although methanol toxicity may occur after skin absorption or inhalation, the main route of uptake is ingestion, either inadvertantly or deliberately, as an adulterant in alcoholic beverages. Methanol is converted, principally in the liver, by alcohol dehydrogenase to formaldehyde and then to formic acid by aldehyde dehydrogenase. The local production of formaldehyde in the retina is thought to be responsible for the production of retinal edema and blindness characteristic of methanol poisoning. For this reason exposure to methanol vapors in industrial and laboratory settings is to be avoided or minimized as much as possible.

Ethylene glycol (CH_2OHCH_2OH) ingestion also results in serious and dramatic poisoning. As with methanol, ethylene glycol has been ingested in beverages mixed with ethanol. Ethylene glycol itself is relatively non-toxic, but it is metabolically activated, primarily in the liver, to aldehydes, glycolate, oxalate, and lactate. Survivors of the acute toxicity phase may experience renal failure caused by deposition of calcium oxalate crystals in the renal tubules. The activation of ethylene glycol depends on initial oxidation by alcohol dehydrogenase, and as in methanol poisoning, ethanol is administered to act as a preferential substrate for the enzyme. During subsequent haemodialysis, ethanol is commonly added to the dialysate to maintain plasma levels of ethanol.

Recent studies (Ghanayem et al., 1987a, 1987b) have established the role of alcohol and aldehyde dehydrogenase in the toxicity of 2-butoxyethanol (BE). This glycol ether ($CH_3CH_2CH_2CH_2OCH_2CH_2OH$) has the solvent properties of both alcohols and ethers and thus is used extensively as a cleaning agent and a general solvent, with many of the products containing BE intended for household use. The major target of acute BE toxicity is the haematopoietic system; exposure of rats to a single dose of BE results in acute haemolytic anaemia as shown by a dose-dependent reduction in the number of circulating red blood cells and a decrease in haemoglobin concentration. In addition there is an increase in the concentration of free haemoglobin in the plasma and urine (haemoglobinuria) of rats treated with BE. Liver and kidney alterations have also been observed in BE treated rats, but these effects are thought to be secondary to the haematotoxic events. Pretreatment of rats with the alcohol dehydrogenase inhibitor pyrazole or the aldehyde dehydrogenase inhibitor cyanamide protected rats against BE-induced haematotoxicity as well inhibiting the formation of the major urinary metabolite butoxyacetic acid (BAA). In addition, inhibition of BE metabolism to butoxyacetic acid was accompanied by increased formation of BE-glucuronide and BE-sulphate (detoxication reactions). Thus activation of BE via alcohol and aldehyde dehydrogenase is a prerequisite for the development of BE-induced haematotoxicity which can be attributed to formation of the reactive metabolite butoxyacetic acid. The understanding of these mechanisms of toxicity suggests a potential treatment of acute exposure to BE or other glycol ethers by using alcohol dehydrogenase inhibitors to block the formation of toxic metabolites.

FLAVIN CONTAINING MONOOXYGENASE

The flavin-containing monooxygenase (FMO), like the cytochrome P-450 monooxygenase system (P-450) is located in the endoplasmic reticulum and is involved in the monooxygenation of a wide variety of xenobiotics. Although originally characterized as an amine oxidase, it is now recognized that the FMO, like P-450, can catalyze oxygenation of a wide variety of xenobiotic nitrogen, sulphur, and phosphorous containing xenobiotics (Hodgson and Levi, 1988; Ziegler, 1980), but unlike P-450 cannot catalyze carbon hydroxylations. Table 2 summarizes some of the similarities and differences between the FMO and P-450.

TABLE 2 - Comparison of cytochrome P-450 (P-450) and the the flavin-containing monooxygenase (FMO)

FEATURE	FMO	P-450
Cofactors	NADPH, O_2	NADPH, O_2, reductase
Location	Microsomes	Microsomes
Inducers	None	PB, PAH'S, EtOH, PCN
Inhibitors	None	CO, SKF-525A, PBO
Isozymes	Few	Many
Substrates	N,S,P, compounds	N,S,P,C, compounds
Reactions	Oxygenation	Oxygenation, epoxidation Reduction, dealkylation

The FMO was first purified to homogeneity from pig liver microsomes (Zeigler and Poulsen, 1978), and in our laboratory, we have purified the FMO from both pig liver and mouse liver (Sabourin et al., 1984; Sabourin and Hodgson,

1984). Subsequently our group (Tynes et al., 1985) as well as Williams et al. (1984) purified an FMO from rabbit lung which was shown to be catalytically and immunologically distinct from the liver enzyme. The mouse and rabbit lung FMOs have a unique ability for N-oxidation of the primary aliphatic amine, n-octylamine, a chemical commonly included in microsomal incubations to inhibit P-450. In the mouse lung, this compound not only serves as a substrate, but is also a positive effector of metabolism. The mouse and rabbit lung enzymes have a higher pH optimum, near 9.8, compared to the liver form which has a peak near pH 8.8. Using antibodies raised in goats, Ouchterlony immunodiffusion analysis indicated that the liver and lung proteins were immunochemically dissimilar (Tynes and Hodgson, 1985). It has now become evident that there are several FMO enzymes with overlapping substrate specificities, and it is likely that the relative proportions of these isozymes vary in different tissues within and between species (Tynes and Philpot, 1987).

Since a wide variety of drugs and other foreign compounds are metabolized by both P-450 and FMO systems (Table 1), it is of interest to delineate the relative contributions of these two enzymes. Changes in the activities of these enzymes either through environmental, physiological, or genetic factors will ultimately effect the efficacy and/or toxicity of drugs and other xenobiotics.

RELATIVE IMPORTANCE OF FMO AND P-450 IN MICROSOMAL OXIDATIONS

Often the same substrate is metabolized by both P-450 and FMO; this is especially prevalent with many N and S containing pesticides. In order to study the relative contributions of these two enzymes with common substrates, we developed methods to measure each separately in the same microsomal preparation. The most useful of these techniques is inhibition of P-450 activity by using an antibody to NADPH cytochrome P-450 reductase thus permitting measurement of FMO activity alone. A second procedure is heat treatment of microsomal preparations ($50^{\circ}C$ for 1 minute) to inactivate the FMO, thus permitting determination of P-450 activity, which is unchanged by the heat treatment (Tynes and Hodgson, 1983).

The relative contributions of the two enzyme systems with thiobenzamide as substrate are summarized in Table 3. Note the higher FMO activity in female than in male mouse liver as well as the difference in relative contribution in lung and liver for the mouse but not the rat. Similar studies have been carried out using the insecticide, phorate, examining the relative contribution of the two

TABLE 3 - Relative contributions of FMO and P-450 to the microsomal oxidation of thiobenzamide[1]

SPECIES	SEX	TISSUE	RELATIVE ACTIVITY (%)	
			P-450	FMO
Mouse	M	Liver	50	50
Mouse-Pb[2]	M	Liver	65	35
Mouse	F	Liver	25	75
Mouse	M	Lung	20	80
Rat	M	Liver	35	65
Rat	M	Lung	40	60

[1]Modified from Tynes and Hodgson, 1983, Biochemical Pharmacology, 32, 3419.

[2]Mice pretreated with phenobarbitone (Pb).

systems in the production of phorate sulphoxide in different tissues with special emphasis on how the level of enzyme activity is affected by in vivo exposure of animals to both inducers and inhibitors of monooxygenase activity (Table 4) (Kinsler et al., 1988a, 1988b). In the livers of untreated animals both enzymes have important roles in the sulphoxidation of phorate (60 percent by P-450 and 40 percent by FMO); by contrast in the kidney and lung, while the overall activity is low compared to liver, the relative contribution by FMO is significantly higher. These levels of activity, however, are easily disturbed by compounds which alter the level of P-450. Thus pretreatment of the mice with phenobarbitone significantly increases, not only the rate of phorate oxidation, but the proportion which is metabolized by the P-450 pathway. Such alterations may assume toxicological importance when the products from the two enzymes differ, and particularly when one metabolite is more toxic or more pharmacologically active than others.

COMPLEX METABOLIC PATHWAYS

Purified enzymes can be useful in elucidating the reactions and products involved in complex metabolic pathways, especailly when more than one enzyme is involved and the oxidative pathways involve both detoxication and activation. Purified FMO and P-450 isozymes were used to examine in detail the separate oxidative pathways (Figure 1)

TABLE 4 - Relative contributions of P-450 and FMO to the microsomal oxidation of phorate in mouse.

TISSUE	SEX	PRODUCT FORMATION[1]				
		CONTROL	+AR[2]	+HEAT	%FMO	%P-450
Liver	M	12.7	2.8	9.3	21.7	78.3
Liver	F	14.4	3.7	11.7	24.0	76.1
Lung	M	3.3	1.9	--	59.1	41.3
Lung	F	5.7	3.1	--	54.0	46.0
Kidney	M	1.6	1.2	--	72.0	28.1
Kidney	F	2.0	1.8	--	90.0	10.0
Liver-Pb[3]	M	69.7	10.1	59.6	14.3	85.5
Liver-PBO[4]						
2 hr	M	11.1	--	6.5	41.4	58.6
36 hr	M	19.4	--	16.3	16.0	84.0

[1]nmols phorate sulphoxide/mg protein/min. Values within 10% ±SD of means, n=3.

[2]Antibody to P-450 reductase

[3]Phenobarbitone-treated mice

[4]Piperonyl butoxide treated mice
  Kinsler et al., 1988a, 1988b

of phorate metabolism (Levi and Hodgson, 1988).
  Both P-450 and FMO catalyze the initial sulphoxidation of thioether-containing organophosphate insecticides such as phorate and disulfoton to form the sulphoxide. Subsequent oxidation reactions, however, such as formation of the sulphone and oxidative desulphuration to the corresponding oxons are catalyzed entirely by P-450. Although both the FMO and P-450 produce phorate sulphoxide, the products are stereochemically different (Table 5). The FMO catalyzed the formation of (-) phorate sulphoxide while two of the P-450 isozymes (P-450 B2, a major constitutive form, and P-450 PB, the principal form induced by phenobarbitone) yielded (+) phorate sulphoxide. The other three P-450 isozymes examined gave racemic mixtures. Thus in vivo, the net optical activity of sulphoxidation is a function of such factors as the presence of activators, inhibitors, and inducers, as well as sex and organ.

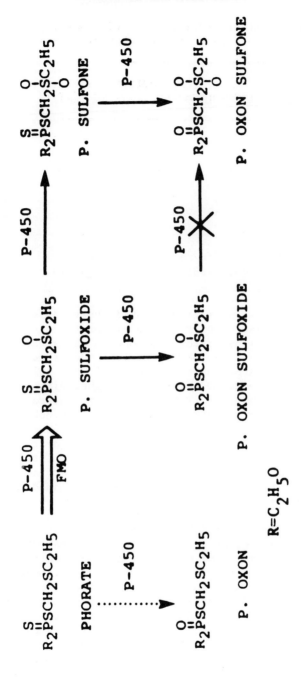

Figure 1. Oxidation of phorate showing metabolites produced by P-450 and FMO

TABLE 5 - Oxidation of phorate to phorate sulphoxide
by purified FMO and P-450 isozymes

ENZYME	PHORATE SULPHOXIDE OPTICAL ACTIVITY	Km (μM)	Vmax[1]
P-450 A1	none	111	31
P-450 B1	none	125	50
P-450 B2	(+)	100	111
P-450 B3	none	275	50
P-450 PB	(+)	67	91
FMO	(-)	32	80

[1]nmol sulphoxide formed/nmol enzyme/min.
Levi and Hodgson, 1988, Xenobiotica 18, 29.

Both (+) and (-) phorate sulphoxide serve as substrates for further oxidation by P-450 to either the oxon (an activation reaction) or the sulphone (a detoxication pathway). However, (+) phorate sulphoxide is not only the preferred substrate, but the percent of oxon sulphoxide relative to the percent sulphone is higher with the (+) sulphoxide as substrate (Table 6). It is interesting to note that the isozyme of P-450 induced by phenobarbitone (P-450 PB) which most rapidly catalyzed the formation of (+) phorate sulphoxide also produced the highest percent of oxon sulphoxide of any of the P-450s. It is interesting to speculate that environmental or physiological factors which increase the level of this isozyme in vivo could potentially enhance the toxicity of this compound.

FORMATION OF MULTIPLE OXIDATIVE METABOLITES

Frequently different sites on the same xenobiotic molecule are attacked by different enzymes, as is the case with the drug thioridizine. Thioridazine is a phenothiazine neuroleptic that is extensively metabolized after administration. S-Oxidation is the predominant route of metabolism in man (Hale and Poklis, 1985; Kilts et al., 1982), producing the 2-sulphoxide, the 2-sulphone, and the 5-sulphoxide. The 2-sulphoxide and the 2-sulphone have greater antipsychotic activity than the parent compound (Kilts et al., 1981, 1982) while the ring sulphoxides are

TABLE 6 - Oxidation of (+) and (-) phorate sulphoxide by purified cytochrome P-450 isozymes.

ENZYME	(+) PHORATE SULPHOXIDE			(-) PHORATE SULPHOXIDE		
	SULPHONE[1]	OXON[1]	%OXON	SULPHONE[1]	OXON	%OXON
P-450 PB	4.8	5.3	52.5	3.9	2.4	38.1
P-450 A1	1.4	0.5	26.3	0.6	0.1	14.3
P-450 B1	1.2	0.4	25.0	0.6	0.2	25.0
P-450 B2	2.3	2.0	46.5	1.3	0.9	40.9
P-450 B3	0.9	0.6	40.0	0.6	0.2	25.0

[1]nmol product formed/30 min/0.4 nmol P-450. Values represent the average of replicates which were within 10%.
Levi and Hodgson, 1988, Xenobiotica 18, 29.

largely responsible for the cardiotoxic side effects of thioridazine (Hale and Poklis, 1986). Northioridazine, the demethylation product, is formed in significant quantities in rats, but not in man. Additional minor metabolites include ring hydroxylations and combinations of the above sulphoxides, sulphones, demethylations, and phenols. The enzymes responsible for the various routes of oxidative metabolism have not been determined, but preliminary results (Lembke, unpublished results) indicate the involvement of both P-450 and FMO in the oxidation ot thioridazine. Incubations with partially purified mouse FMO and thioridazine in which NADPH consumption was monitored gave a Km of 8.8 μM and Vmax of 352 nmol/min/unit FMO, indicating that thioridazine is, in fact, a good substrate for the FMO.

The effect of heat pretreatment to eliminate FMO activity in microsomal oxidations is shown in Table 7. Heat pretreatment results in a decrease in the amounts of the 2-sulphoxide and northioridazine (an N-demethylation reaction), indicating that these reactions may be mediated in part by the FMO. On the other hand the amount of the 5-sulphoxide and 5-sulphone produced is increased in incubations with heat treated microsomes. Such differential routes of metabolism may assume pharmacoloical significance in the case of activation of a prodrug or when one of the metabolites is more toxic.

TABLE 7 - Oxidation of thioridazine by rabbit liver microsomes.

PRODUCT[1]	UNTREATED	HEAT TREATED
Thioridazine	4.65^2	2.28^2
Northioridazine	0.29	0.04
Sulforidazine	ND[3]	0.03
Mesoridazine	0.96	0.49
Thioridazine-5-sulphoxide	0.61	0.94

[1] nmol product/mg protein/min

[2] Values represent disappearance of substrate

[3] No detected response

Lembke, Mailman, and Hodgson; unpublished results.

XENOBIOTICS AS SUBSTRATES, INHIBITORS, AND INDUCERS OF THE SAME ENZYME

Occasionally a compound may be a substrate, an inhibitor, and an inducer of the same enzyme; this type of situation is illustrated with the methylenedioxyphenyl compounds (MDP). As a chemical class, MDP compounds are widespread in the environment, and include carcinogens (eg. isosafrole), food constituents (eg. myristicin), and insecticide synergists (eg. piperonyl butoxide).

The activity of MDP compounds as insecticide synergists and drug potentiators is due to their ability to inhibit P-450 monooxygenase reactions that are critical in the primary metabolism of insecticides and other xenobiotics. MDP compounds owe this inhibitory effect to their ability to form a metabolite complex at the active site of the P-450 molecule (Figure 2) (Wilkinson et al., 1984). The most probable mechanism appears to be an attack on the methylene carbon, followed by elimination of water to yield a carbene. The highly reactive carbene either reacts with the haem iron to form a P-450-inhibitory complex or breaks down to yield the catechol. The formation of this stable MDP metabolite-cytochrome P-450 complex is responsible for the appearance of a type III double Soret optical difference spectrum and for inhibition of monooxygenase activity.

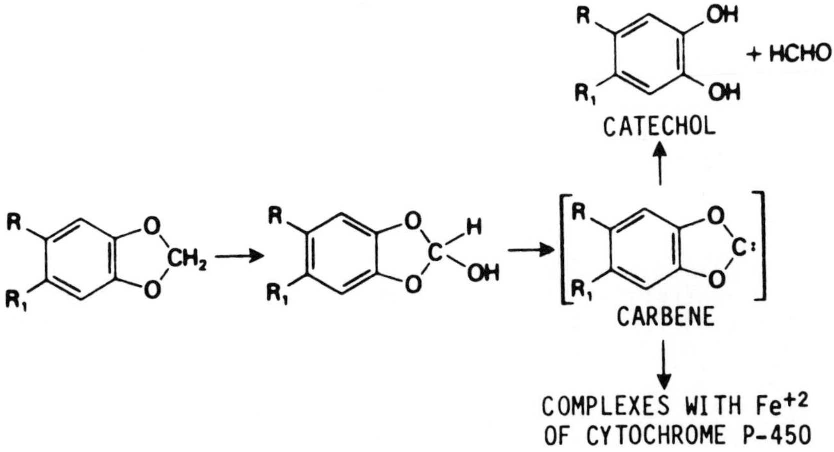

Figure 2. Proposed activation for methylenedioxyphenyl compounds by P-450

Methylenedioxyphenyl compounds have a biphasic effect on P-450 levels and P-450 related monooxygenase activities, namely inhibition followed by induction. Neither the mechanism for MDP induction of P-450 nor the pattern of isozyme induction is well known, and recently we have focused on these events. We have examined the structure-activity relationship (QSAR) between MDP compounds which differed in either phenyl ring substitutents or substitution on the methylene bridge and the ability of these compounds to induce specific P-450 isozymes as well as their ability to inhibit <u>in vitro</u> monooxygenase activities. In order to determine if induction patterns are modulated by interactions with the Ah receptor, two strains of mice C57/BL6 (Ah positive) and DBA (Ah negative) were used. Previous studies in our laboratory (Cook and Hodgson, 1983) confirmed that substitution on the methylene carbon (methyl and dimethyl butylbenzodioxole) abolished the ability of the MDP compound to form the type III complex, to inhibit monooxygenase activities, and to induce P-450. The dimethyl benzodioxole did, however, inhibit the activities of cytochrome P-450 reductase and epoxide hydrolase (Cook and Hodgson, 1984).

TABLE 8 - Effect of in vivo administration of MDP compounds on in vitro microsomal monooxygenase activities

TREATMENT	ETHYL-MORPHINE[1]	ACET-ANILIDE[1]	BENZO(a)-PYRENE[2]	ETHOXY-RESORUFIN[1]
Control	3.07	2.88	0.99	0.10
3-MC	3.18	12.79*	7.46*	0.88*
PB	7.61*	4.88*	2.78*	0.30*
Safrole	5.10*	6.13*	1.02	0.29*
Isosafrole	5.22*	7.40*	0.95	0.48*
Dihydrosafrole	3.10	2.99	0.50*	0.22*
n-Butyl-BD	5.09*	3.44	1.09	0.21*
t-Butyl-BD	6.23*	4.35*	0.96	0.21*
Methyl-BD	3.00	3.27	0.70	0.11
Nitro-BD	3.21	3.15	0.74	0.12
Bromo-BD	3.89	3.58	0.79	0.10

(BD=benzodioxole)

*Significantly different ($p<0.05$) from control values.
Values within 10% ± SD of means (n=4)

[1] nmol product/mg protein/min

[2] nmol substrate consumed/mg protein/min
Lewandowski et al., 1988

The effects of substitution on the phenyl ring on monooxygenase induction are summarized in Table 8. The propyl and butyl substituted compounds were effective inducers of P-450 associated enzyme activities while the methyl, bromo, and nitro substituted compounds were very poor inducers. As can be seen from examining the inductive pattern of monooxygenase activities, there was a significant increase in ethylmorphine N-demethylase by safrole, isosafrole, n-butyl-BD, and t-butyl-BD, indicating a PB type induction. There was, however, no increase over control of benzo(a)pyrene hydroxylase activity, as would be expected with 3-MC type induction. On the other hand, there is an increase in both ethoxyresorufin O-deethylase activity and acetanilide hydroxylase activity, both of which are associated with 3-MC induction, suggesting 3-MC type induction. Evaluating MDP induction on the basis of enzyme activities is complicated by the fact that both induction and inhibition (due to formation of the MDP-metabolite complex) are occurring simultaneously. Since a significant portion of the induced isozyme is complexed, and there is no reliable method of removing the complex in vitro, the level of enzyme activity represents only the uncomplexed isozyme, and this level varies both with P-450 isozyme and MDP compound.

The use of specific antibodies to individual P-450 isozymes provides an additional tool for assessing the induction pattern of individual isozymes. Western blots and immunoquantitation using antibodies to the P-450s induced by PB and 3-MC showed that, in fact, both isozymes were induced by the MDP compounds (Table 9), and that there is no significant difference in the level of 3-MC isozyme induction in either Ah positive (C-57) or Ah negative (DBA) mice.

TABLE 9 - Quantitation of MDP induction of P-450 isozymes from Western blots

TREATMENT	P-450 PB[1]	P-450 3-MC[1]
Control	ND[2]	ND
3-MC	ND	4.14
PB	1.28	ND
Safrole	0.38	0.85
Isosafrole	0.61	1.88
Dihydrosafrole	0.11	0.88
n-Butyl-BD	0.35	2.52
t-Butyl-BD	0.48	0.85
Methyl-BD	ND	ND
Nitro-BD	ND	ND
Bromo-BD	ND	ND
(BD=benzodioxole)		

[1]pmol P-450/5µg microsomal protein

[2]No P-450 detected
Lewandowski et al., 1988.

Additional studies with purified isozymes will be essential, however, in order to elucidate the ability of the MDP compounds to complex with and inhibit specific P-450 isozymes so as to assess more accurately induced monooxygenase activities.

REFERENCES

Beedham, C., 1988, Molybdenum hydroxylases. In Metabolism of Xenobiotics, edited by J.W. Gorrod, H. Oelschlager, and J. Caldwell (London: Taylor and Francis), pp. 51-58.

Cook, J.C. and Hodgson, E., 1983, Induction of cytochrome P-450 by methylenedioxyphenyl compounds: importance of the methylene carbon. Toxicology and Applied Pharmacology, 68, 131-139.

Cook, J.D., and Hodgson, E., 1984, 2,2-Dimethyl-5-t-butyl-1,3-benzodioxole: an unusual inducer of microsomal enzymes. Biochemical Pharmacology, 33, 3941-3946.

Ghanayem, B.I., Burka, L.T., and Matthews, H.B., 1987a, Metabolic basis of ethylene glycol monobutyl ether (2-butoxyethanol) toxicity: role of alcohol and aldehyde dehydrogenases. Journal of Pharmacology and Experimental Therapeutics, 242, 222-231.

Ghanayem, B.I., Burka, L.T., Sanders, J.M., and Matthews, H.B., 1987b, Metabolism and disposition of ethylene glycol monobutyl ether (2-butoxyethanol) in rats. Drug Metabolism and Disposition, 15, 478-484.

Hale, P.W. Jr. and Poklis, A., 1985, Thioridazine-5-sulfoxide diastereoisomers in serum and urine from rat and man following chronic thioridazine administration. Journal of Analytical Toxicology, 9, 179-201.

Hale, P.W. Jr. and Poklis, A., 1986, Cardiotoxicity of thioridazine and two stereoisomeric forms of thioridazine-5-sulfoxide in the isolated perfused rat heart. Toxicology and applied Pharmacology, 86, 44-55.

Hodgson, E. and Levi, P.E., 1988, The flavin-containing monooxygenase as a sulfur oxidase. In Metabolism of Xenobiotics, edited by J.W. Gorrod, H. Oelschlager, and J. Caldwell (London: Taylor and Francis), pp. 81-88.

Kilts, C.D., Mailman, R.B., Hodgson, E., and Breese, G.R., 1981, Simultaneous determination of thioridazine and its sulfoxidized metabolites by HPLC: use in clinical and preclinical metabolic studies. Federation Proceedings, 40, 283.

Kilts, C.C., Patrick, K.S., Breese, G.R., and Mailman, R.B., 1982, Simultaneous determination of thioridazine and its S-oxidized and N-demethylated metabolites using High Performance Liquid Chromatography on radially compressed silica. Journal of Chromatography, 231, 377-391.

Kinsler, S., Levi, P.E., and Hodgson, E., 1988a, Hepatic and extrahepatic microsomal oxidation of phorate by the cytochrome P-450 and FAD-containing monooxygenase systems in the mouse. Pesticide Biochemistry and Physiology, 31, in press.

Kinsler, S., Levi, P.E., and Hodgson, E., 1988b, Effects of pretreatment with xenobiotics on the relative contributions of the cytochrome P-450 monooxygenase and flavin-containing monooxygenase systems in the microsomal oxidation of phorate in the mouse. In progress.

Levi, P.E., and Hodgson, E., 1988, Stereospecificity in the oxidation of phorate and phorate sulphoxide by purified FAD-containing mono-oxygenase and cytochrome P-450 isozymes. Xenobiotica, 18, 29-39.

Lewandowski, M., Chui, Y.C., Levi, P.E., and Hodgson, E., 1988, Induction of hepatic cytochrome P-450 monooxygenase activities by methylenedioxyphenyl compounds in mice. In progress.

Marnett, L.J. and Eling, T.E., 1983, Cooxidation during prostaglandin biosynthesis: a pathway for the metabolic activation of xenobiotics. In Reviews in Biochemical Toxicology 5, edited by E. Hodgson, J.R. bend, and R.M. Philpot (New York: Elsevier Science), pp. 135-172.

Sabourin, P.J., and Hodgson, E., 1984, Characterization of the purified microsomal FAD-containing monooxygenase from mouse and pig liver. Chemical and Biological Interactions, 51, 125-139.

Sabourin, P.J., Smyser, B.P., and Hodgson, E., 1984, Purification of the flavin-containing monooxygenase from mouse and pig liver microsomes. International Journal of Biochemistry, 16, 713-720.

Tynes, R.E. and Hodgson, E., 1983, Oxidation of thiobenzamide by the FAD-containing and cytochrome P-450-dependent monooxygenases of liver and lung microsomes. Biochemical Pharmacology, 32, 3419-3428.

Tynes, R.E., and Hodgson, E., 1985, Catalytic activity and substrate specificity of the flavin-containing monooxygenase in microsomal systems: characterization of the hepatic, pulmonary, and renal enzymes of the mouse, rabbit, and rat. Archives of Biochemistry and Biophysics, 240, 77-93.

Tynes, R.E., amd Philpot, R.M., 1987, Tissue- and species-dependent expression of multiple forms of mammalian microsomal flavin-containing monooxygenase. Molecular Pharmacology, 31, 569-574.

Tynes, R.E., Sabourin, P.J., and Hodgson, E., 1985, Identification of distinct hepatic and pulmonary forms of microsomal flavin-containing monooxygenase in the mouse and rabbit. Biochemical and Biophysical Research Communications, 126, 1069-1075.

Wells, P.G. and Nagai, M.K., 1988, Chemical teratogenesis IV. Inhibition of trimethadione teratogenicity by the cyclooxygenase inhibitor acetylsalicylic acid: a unifying hypothesis for the teratologic effects of hydantoins and structurally related anitconvulsants. Toxicology and Applied Pharmacology, in press.

Wells, P.G., Zubovits, J.T., and Wong, S.T., 1988, Chemical teratogenesis I. Modulation of phenytoin teratogenicity covalent binding by the cyclooxygenase inhibitor acetylsalicyclic acid. Toxicology and Applied Pharmacolo in press.

Wilkinson, C.F., Murray, M., Marcus, C.B., 1984, Interactions of methylenedioxyphenyl compounds with cytochrome P-450 and effects on microsomal oxidation. In Reviews in Biochemical Toxicology 6, edited by E. Hodgso J.R. Bend, and R.M. Philpot (New York: Elsevier Science) pp.27-64.

Williams, D.E., Ziegler, D.M., Hordin, D.J., Hale, S.E., an Masters, B.S.S., 1984, Rabbit lung flavin-containing monooxygenase is immunochemically and catalytically distinct from the liver enzyme. Biochemical and Biophysical Research Communications, 125, 116-122.

Ziegler, D.M., 1980, Microsomal flavin-containing monooxygenase: Oxygenation of nucleophilic nitrogen and sulfur compounds. In Enzymatic Basis of Detoxification, vol I, edited by W.B. Jakoby (New York: Academic Press), pp.201-227.

Ziegler, D.M. and Poulsen, L.L., 1978, Hepatic microsomal mixed-function amine oxidase. Methods in Enzymology, 52, 155.

BIOACTIVATION RESULTING FROM HYDROLYTIC PROCESSES

Walter C. Dauterman

North Carolina State University,
Raleigh, North Carolina 27695, USA

INTRODUCTION

A variety of xenobiotics which includes food additives, drugs, pesticides, chemical intermediates, etc. contain carboxyl esters, carboxyl amides or phosphoric anhydride bonds which are susceptible to hydrolytic cleavage. Hydrolases are proteins capable of catalyzing hydrolytic reactions of the following types:

$$RC(O)OR' + H_2O \longrightarrow RC(O)OH + HOR'$$

$$RC(O)NR_1R_2 + H_2O \longrightarrow RC(O)OH + HNR_1R_2$$

$$(RO)_2P(O)OX + H_2O \longrightarrow (RO)_2P(O)OH + HOX$$

Hydrolysis of xenobiotics can result in (i) detoxification, (ii) no change in toxicity or (iii) toxification (or bioactivation). In most instances the hydrolysis of carboxyl esters, carboxyl amides or phosphoric anhydride bonds in biologically active compounds results in detoxification, that is, in the loss of biological activity. Increased water solubility usually accompanies hydrolysis and this accelerates excretion. Hydrolysis is the only Phase I reaction that does not require energy expenditure by the organism (Hodgson and Dauterman, 1980). A number of hydrolases involved in xenobiotic metabolism and their properties have been described in recent reviews (Heymann, 1980, 1982; Dauterman, 1983, 1984).

Xenobiotic hydrolysis can lead to the formation of reactive metabolites, i.e. toxic acids, alcohols, or amines. These products can directly illicit a toxic response or may be further metabolized to reactive intermediates. Thus hydrolytic activity which initially results in detoxification may, after further metabolism, result in hydrolysis products that are toxic to the whole organism, or to a specific tissue or cell type. In this case, detoxification may be afforded for one target site while bioactivation is enabled at an entirely different target site. There are several instances of hydrolytically-mediated toxicity now known. A representative selection of these is given below.

EXAMPLES OF HYDROLYSIS THAT AFFORD BIOACTIVATION

Sodium fluoroacetate is a naturally occurring toxicant which is extremely toxic to mammals and has been utilized as a controversial control method for predators and rodents. Besides being toxic to mammals, fluoroacetate also shows insecticidal and miticidal activity. The main lesion of toxic action is the inhibition of aconitase by fluoro-citrate which is formed from fluoroacetate via the citric acid cycle. Fluoroacetate and MNFA (2-fluoro-$\underline{N}$-methyl-$\underline{N}$-(1-naphthyl)acetamide) are selectively toxic to mammalian species (Table 1).

TABLE 1 - Toxicity of fluoroacetate (FA) and MNFA to mammalian species

Species	Oral LD_{50} (mg/kg)	
	FA	MNFA
Guinea pig	0.5	2.0
Rat	4.7	120.0
Mouse	7.0	250.0

Noguchi et al. (1968)

The toxicity of MNFA appears to be due to the toxicity of fluoroacetate produced by enzymatic hydrolysis. The product of hydrolysis of MNFA in liver homogenates is $\underline{N}$-methyl-1-naphthylamine and fluoroacetate:

$$FCH_2C(O)N(CH_3)(C_{10}H_7) + H_2O \longrightarrow FCH_2C(O)OH + CH_3NHC_{10}H_7$$

A definite relationship between the relative toxicity and the ability of liver homogenates to hydrolyze MNFA is shown in Table 2 (Noguchi et al., 1968). Clearly hydrolysis of MNFA results in toxicant activation.

TABLE 2 - Hydrolysis of MNFA by liver homogenates

Species	μmoles hydrolyzed/gm liver/hr
Guinea pig	36.0
Rat	1.2
Mouse	0.8

Another example of activation via hydrolysis appears to involve N-acetyl-N-methylcarbamates (Miskus et al., 1968). The toxicity of mexacarbate and its N-acetylated analogue are presented in Table 3.

TABLE 3 - Biological activity of mexacarbate and acetylated mexacarbate

Compound	Spruce budworm Topical LD_{50} (μg/gm)	Mouse Oral LD_{50} (mg/kg)
Mexacarbate	1.4	30 - 50
Acetylated mexacarbate	2.4	>1000

Metabolism studies with acetylated mexacarbate in spruce budworm and mouse (Miskus et al., 1969) showed that the mouse cleaved the carbamic ester bond while the insect deacetylated the acetylated mexacarbate. Cleavage of the carbamic ester bond resulted in detoxication in the mouse while the spruce budworm removed the acetyl group and regenerated mexacarbate (Figure 1).

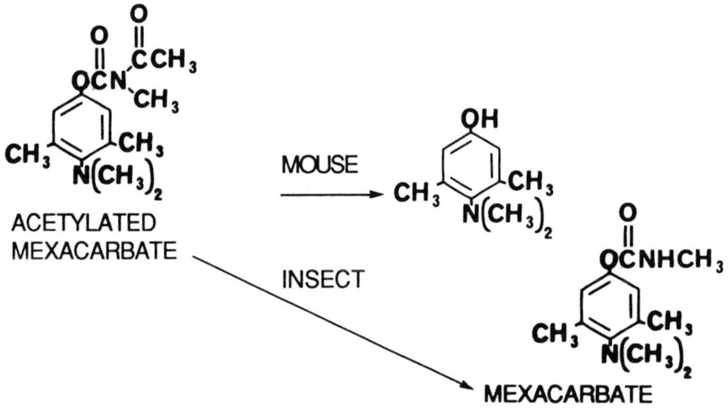

Figure 1. Major pathways of metabolism of acetylated mexacarbate

Fenazaflor, an acaricide, is hydrolyzed in vivo to 5,6-dichloro-2-trifluoromethyl-benzimidazole, a known uncoupler of oxidative phosphorylation (Corbett and Johnson, 1970):

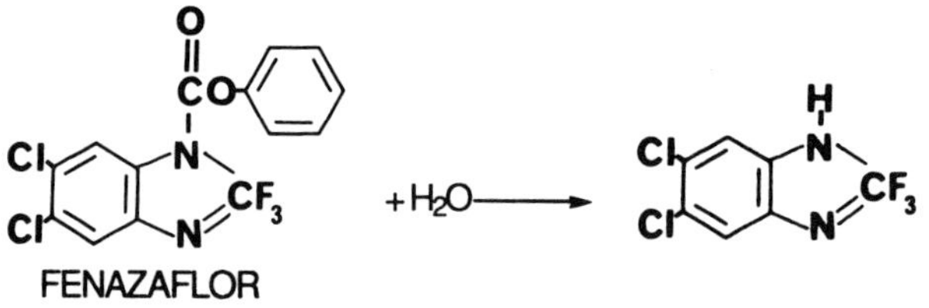

Butonate is hydrolyzed to trichlorfon by the action of esterases in insects and plants (Dedek, 1968). Trichlorfon then rearranges non-enzymatically and under physiological conditions forms dichlorvos. Each step in the reaction results in an increase in toxicity (Figure 2).

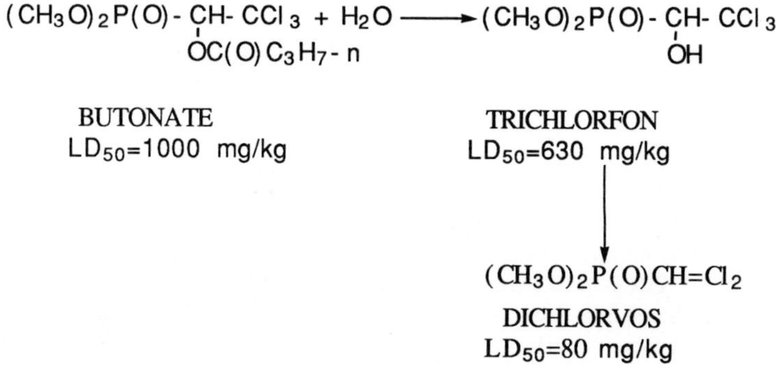

$$(CH_3O)_2P(O)-\underset{OC(O)C_3H_7\text{-}n}{CH\text{-}CCl_3} + H_2O \longrightarrow (CH_3O)_2P(O)-\underset{OH}{CH\text{-}CCl_3}$$

BUTONATE $\qquad\qquad\qquad$ TRICHLORFON
$LD_{50}=1000$ mg/kg $\qquad\quad$ $LD_{50}=630$ mg/kg

$$\downarrow$$

$$(CH_3O)_2P(O)CH=Cl_2$$
DICHLORVOS
$LD_{50}=80$ mg/kg

Figure 2. Hydrolytic bioactivation of butonate

Propanil (3,4-dichloropropionanilide) is a herbicide used on rice crops to kill weeds while not damaging rice plants. Rice plants contain an acylamidase that catalyzes the hydrolytic detoxification of propanil (Frear and Still, 1968). Similarly an acylamidase is present in the liver of rats, rabbits, mice and dogs which hydrolyzes propanil to 3,4-dichloroaniline and propionic acid. This enzyme, which is located in microsomes, is inhibited by various organophosphates and carbamates administered in vivo and in vitro:

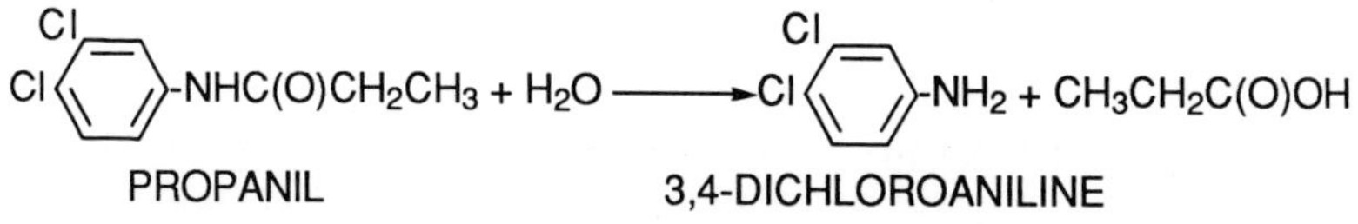

$$Cl_2C_6H_3\text{-}NHC(O)CH_2CH_3 + H_2O \longrightarrow Cl_2C_6H_3\text{-}NH_2 + CH_3CH_2C(O)OH$$

PROPANIL $\qquad\qquad\qquad\qquad\qquad$ 3,4-DICHLOROANILINE

Toxic signs seen after ip doses of 200 to 800 mg of propanil to mice include CNS depression, loss of righting reflex, cyanosis and death (Singleton and Murphy, 1973). It is well known that free aromatic amines are potent methaemoglobin forming agents in vivo. They are converted by microsomal N-hydroxylation to the corresponding aromatic hydroxylamines and aromatic nitroso derivatives in intact animals and by liver microsomes (Kiese, 1966):

$$R-\langle\bigcirc\rangle-NH_2 \xrightarrow{\text{OXIDATION}} R-\langle\bigcirc\rangle-NHOH$$

$$\downarrow$$

NITROSO DERIVATIVES

Singleton and Murphy (1973) were able to inhibit propanil-induced methaemoglobinemia by pretreating mice with 125 mg/kg of the esterase inhibitor TOCP (tri-o-cresyl phosphate) (Table 4). TOCP did not affect 3,4-dichloroaniline-induced methaemoglobinemia but apparently caused inhibition of the propanil-amidase activity. Propanil- and 3,4-dichloroaniline-induced methaemoglobinemia have similar time-response relationships.

TABLE 4 - Propanil and 3,4-dichloroaniline-induced methaemoglobin formation in control and TOCP-pretreated mice[1]

Propanil or DCA dose (mmol/kg)	Percent Methaemoglobin			
	Propanil-treated		DCA-treated	
	Control	+TOCP	Control	+TOCP
0.92	6.1	0.9	39.1	39.6
1.83	11.2	3.6	53.6	56.0

[1] 125 mg/kg TOCP; blood samples taken 1 hour after challenge with propanil and 30 min. after 3,4-dichloroaniline (DCA)

The hydrolysis of propanil to yield 3,4-dichloroaniline by liver homogenates was completely inhibited 18 hours after *in vivo* treatment of mice with 125 mg/kg TOCP. However, while the pretreatment with TOCP prevented cyanosis, it did not alter the ability of propanil to depress CNS activity.

Clearly, hydrolysis by an acylamidase results in the activation of propanil to 3,4-dichloroaniline, and leads to increased methaemoglobin formation in mice, whereas in rice plants the action of acylamidase is to destory the herbicidal activity of propanil.

In parallel findings, Heymann and Krisch (1969) reported that the carboxylesterase inhibitor bis [p-nitrophenyl phosphate] (BNPP), when administered simultaneously with phenacetin or acetanilide to rats, inhibited the formation of methaemoglobin.

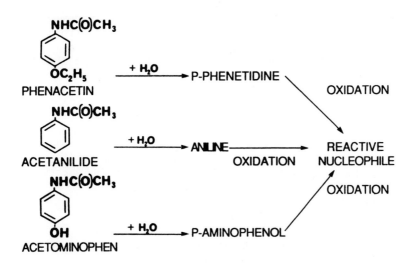

Figure 3. Hydrolysis in the toxicity of N-acetylated anilines

In the rat the amide bonds of both phenacetin and acetanilide are hydrolyzed by a microsomal carboxylesterase which can be inhibited stoichiometrically by organophorous compounds such as paraoxon or BNPP. Here hydrolysis is part of the activation process, leading to an intermediate which can be oxidized to the ultimate reactive metabolite (Figure 3).

Renal failure and hepatic necrosis have been reported with acute acetaminophen poisoning. A number of studies have indicated that metabolic activation of acetaminophen is a necessary requirement to manifest renal cortical necrosis (Nelson, 1982). Recently, it has been suggested that the formation of reactive intermediates from acetaminophen within the renal cortex requires prior deacetylation and then renal metabolism to an electrophilic intermediate (Newton et al., 1985) which binds to renal macromolecules. Hydrolysis of acetaminophen by cytosolic fractions of the renal cortex has been demonstrated (Newton et al., 1983). Histopathologic renal lesions produced by p-aminophenol were indistinguishable from the renal lesions formed following acetaminophen administration.

Several esters of allyl alcohol have been approved as food flavouring agents and are used to simulate fruit flavours in a variety of baked goods, candy, ice cream, etc. Allyl alcohol is a hepatotoxin which causes periportal necrosis. Certain esters of allyl alcohol also produce liver damage. The hepatotoxicity of allyl alcohol is associated with its oxidation to acrolein (Reid, 1972), which undergoes a reaction at the double bond with nucleophilic groups found in tissue macromolecules (Figure 4).

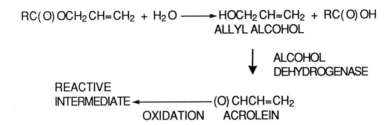

Figure 4. Bioactivation of allyl esters via hydrolysis to allyl alcohol

Inhibition of the hydrolysis of allyl acetate, allyl cinnamate, and allyl phenoxyacetate by pretreatment with 125 mg/kg TOCP or 10 mg/kg $\underline{S},\underline{S},\underline{S}$,-tributyl phosphorotrithioate (DEF), both carboxylesterase inhibitors, provided protection against hepatotoxicity (Silver and Murphy, 1978). These studies demonstrated that enzymatic hydrolysis was a necessary step in the activation of allyl esters to hepatotoxicity and that pyrazole, an inhibitor of alcohol dehydrogenase, prevented the oxidation of allyl alcohol and thus provided protection against hepatotoxicity.

Phthalate esters are used extensively as industrial solvents and plasticizers in the manufacture of a wide variety of plastics; di-(2-ethylhexyl)phthalate (DEHP) is the most commonly used phthalate in PVC formulation. DEHP has a low order of acute toxicity but exposure to the compound can induce a variety of changes in hepatic morphology and biochemical functions. The most notable effects involve the liver and reproductive organs. DEHP, following oral ingestion, is substantially hydrolyzed by lipases present in the pancreas, liver and intestinal mucosa to form the mono ester MEHP (Albro and Thomas, 1973):

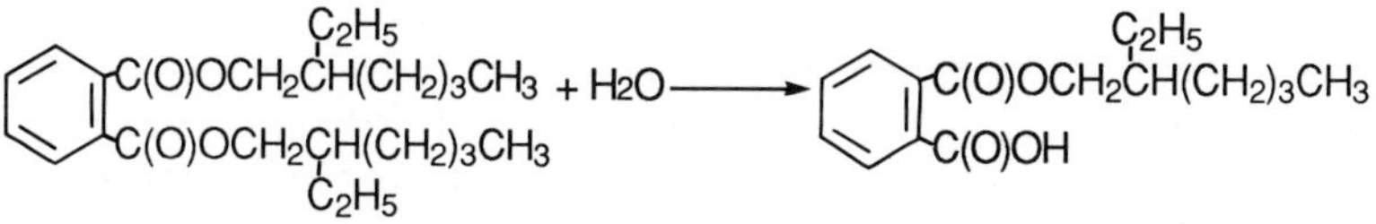

$$\text{C}_6\text{H}_4[\text{C(O)OCH}_2\text{CH(C}_2\text{H}_5)(\text{CH}_2)_3\text{CH}_3]_2 + \text{H}_2\text{O} \longrightarrow \text{C}_6\text{H}_4[\text{C(O)OCH}_2\text{CH(C}_2\text{H}_5)(\text{CH}_2)_3\text{CH}_3][\text{C(O)OH}]$$

DEHP
DI(2-ETHYLHEXYL)PHTHALATE

MEHP
MONO(2-ETHYLHEXYL)PHTHALATE

A pronounced dose route dependency for the de-esterification of DEHP has been observed. Lake and coworkers (1975) demonstrated that ingested DEHP in the rat caused characteristic biochemical and ultrastructural changes in the hepatic endoplasmic reticulum and mitochondria. The same results were reproducible with MEHP administration. The spectrum of biological activity of MEHP is remarkably similar to that of the diester suggesting that the monoester may contribute significantly to the activity of DEHP.

Miserotoxin (3-nitro-1-propyl β-D-glucopyranoside), a product of <u>Astragalus</u> species (leguminosae), is poisonous to cattle and sheep but relatively innocuous to rats. In cattle and sheep (and probably other ruminants) the glycoside is rapidly hydrolyzed by ruminal microorganisms (Majak and Clark, 1980) and the aglycone (3-nitropropanol) is released in the digestive tract (Figure 5). In rodents a low level of microbial hydrolysis in the gastrointestinal tract accounts for the low toxicity.

$$\underset{\text{MISEROTOXIN}}{\text{NO}_2(\text{CH}_2)_3\text{O-}\beta\text{-D-PYRANOSIDE}} + \text{H}_2\text{O} \xrightarrow{\text{RUMEN FLUID}} \text{NO}_2(\text{CH}_2)_3\text{OH}$$

$$\text{NO}_2(\text{CH}_2)_3\text{OH} \xrightarrow{\text{OXIDATION}} \text{NO}_2(\text{CH}_2)_2\text{C(O)OH} \longrightarrow \text{NITRITES}$$

Figure 5. Hydrolytic bioactivation of miserotoxin

The aglycone is then oxidized to nitropropionic acid and to nitrites; acute poisoning is accompanied by sublethal methaemoglobinemia (Muir et al., 1984). Here hydrolysis of the glucoside bond results in the release of the aglycone, with toxicity occurring after further metabolism.

CONCLUSIONS

It is quite obvious that hydrolysis of foreign compounds does not always lead to less active compounds. Some hydrolysis products are less toxic, some are of equal toxicity and others are more toxic than the parent compound. It is also clear that the further metabolism of the hydrolysis products (the acids, alcohols, amines) usually affords the most dramatic examples of hydrolytic bioactivation. A majority of these examples involve the hydrolysis of carboxyl amide bonds and then subsequent oxidation of the aromatic amines to form "active" metabolites. Similarly, whereas hydrolysis may decrease one type of biological activity, i.e. resulting in detoxification, the same hydrolytic reaction may form a product which causes a completely different but adverse biological response, in this case, activation.

In summary, hydrolysis cannot be assumed to result in detoxification of a foreign compound.

REFERENCES

Albro, P.W. and Thomas, R.O., 1973, Enzymatic hydrolysis of di-(2-ethylhexyl)phthalate by lipases. Biochimica et Biophysica Acta, 300, 380-390.

Corbett, J.R. and Johnson, E.R., 1970, Biochemical mode of action of the acaricide fenazaflor. Pesticide Science, 1, 120-123.

Dauterman, W.C., 1983, The role of hydrolases in insecticide metabolism and the toxicological significance of the metabolites. In Analytical and Toxicological Significance of Pesticide Metabolites, edited by J.E. Chambers and R.C. Honeycutt. Journal of Toxicology - Clinical Toxicology, 19, 623-635.

Dauterman, W.C., 1984, Environmental aspects of xenobiotic metabolism: hydrolysis. In Foreign Compound Metabolism, edited by J. Caldwell and G.D. Paulson (London: Taylor and Francis), pp. 161-170.

Dedek, W., 1968, Abbau and mickstande von 32P-butonat in Fruchten. Zeitschrift fur Naturforschung, 23b, 504-506.

Frear, D.S. and Still, G.G., 1968, The metabolism of 3,4-dichloropropionanilide in plants. Partial purification and properties of an aryl acylamidase from rice. Phytochemistry, 7, 913-920.

Heymann, E., 1980, Carboxylesterases and amidase, In *Enzymatic Basis of Detoxication*, Vol. II, edited by W.B. Jakoby (New York: Academic Press), pp. 291-323.

Heymann, E., 1982, Hydrolysis of carboxylic esters and amides. In *Metabolic Basis of Detoxication*, edited by W.B. Jakoby, John R. Bend and John Caldwell (New York: Academic Press), pp. 229-245.

Heymann, E. and Krisch, K., 1969, Inhibition of phenacetin- and acetanilide-induced methaemoglobinemia in the rat by the carboxylesterase inhibitor bis(p-nitrophenyl) phosphate. Biochemical Pharmacology, 16, 317-328.

Hodgson, E. and Dauterman, W.C., 1980, Metabolism of toxicants: phase one reactions. In *Introduction to Biochemical Toxicology*, edited by E. Hodgson and F.E. Guthrie (New York: Elsevier North-Holland), pp. 67-91.

Kiese, M., 1966, The biochemical production of ferrihemoglobin-forming derivatives from aromatic amines, and mechanisms of ferrihemoglobin formation. Pharmacological Reviews, 18, 1091-1161.

Lake, B.G., Gangolli, S.D., Grasso, P. and Lloyd, A.G., 1975, Studies on the hepatic effects of orally administered di(2-ethylhexyl) phthalate in the rat. Toxicology and Applied Pharmacology, 32, 355-367.

Majak, W. and Clark, L.J., 1980, Metabolism of aliphatic nitro compounds in bovine rumen fluids. Canadian Journal of Animal Science, 60, 319-325.

Miskus, R.P., Look, M., Andrews, T.L. and Lyon, R.L., 1968, Biological activity as an effect of structural changes in aryl N-methylcarbamates. Journal of Agriculture and Food Chemistry, 16, 605-607.

Miskus, R.P., Andrews, T.L. and Look, M., 1969, Metabolic pathways affecting toxicity of N-acetyl Zectran. Journal of Agricultural and Food Chemistry, 17, 842-844.

Muir, A.D., Majak, W., Pass, M.A. and Yost, G.S., 1984, Conversion of 3-nitropropanol (Miserotoxin aglycone) to 3-nitropropionic acid in cattle and sheep. Toxicology Letters, 20, 137-141.

Nelson, S.D., 1982, Metabolic activation and drug toxicity. Journal of Medicinal Chemistry, 25, 753-765.

Newton, J.F., Bailie, M.B. and Hook, J.B., 1983, Acetominphen nephrotoxicity in the rat: renal metabolic activation *in vitro*. Toxicology and Applied Pharmacology, 78, 433-444.

Newton, J.F., Pasino, D.A. and Hook, J.B., 1985, Acetaminophen nephrotoxicity in the rat: quantitation of renal metabolic activation *in vivo*. Toxicology and Applied Pharmacology, 78, 39-46.

Noguchi, T., Hashimoto, J. and Miyata, H., 1968, Studies of the biochemical lesions caused by a new fluorine pesticide, N-methyl-N-(1-naphthyl)monofluoroacetamide. Toxicology and Applied Pharmacology, 13, 189-198.

Reid, W.D., 1972, Mechanism of allyl alcohol induced hepatic necrosis. Experientia, 28, 1058-1061.

Silver, E.H. and Murphy, S.D., 1978, Effect of carboxylesterase inhibitors on the acute hepatotoxicity of esters of allyl alcohol. Toxicology and Applied Pharmacology, 45, 377-389.

Singleton, S.D. and Murphy, S.D., 1973, Propanil (3,4-dichloropropionanilide)-induced methemoglobin formation in mice in relation to acylamidase activity. Toxicology and Applied Pharmacology, 25, 20-29.

N-ACETYLTRANSFERASE POLYMORPHISM AND ARYLAMINE TOXICITY: RELATIONSHIP TO O-ACYLATION REACTIONS

Thomas J. Flammang[1], Glenn Talaska[1,2], David Z.J. Chu[2], Nicholas P. Lang[2], and Fred F. Kadlubar[1]

[1]National Center for Toxicological Research, Jefferson, AR, 72079, USA; [2]John L. McClellan Memorial Veterans Hospital, Little Rock, AR, 72205, USA

INTRODUCTION

N-Substituted aromatic compounds are ubiquitous environmental contaminants (Figure 1). They have been detected in urban air particulates (Schuetzle, 1983), tobacco smoke (Patriankos and Hoffman, 1979), complex dyes (Cerniglia et al., 1982), cooked foods (Sugimura and Sato, 1983), and in diesel exhaust (Wei and Shu, 1983). Furthermore, increased levels of aromatic amine haemoglobin adducts have been detected in the smoker vs nonsmoker (Bryant et al., 1987). Several of these compounds are known human and/or animal carcinogens. Thus, considerable resources are utilized to determine their mechanism of action. It is generally recognized that these compounds are converted by metabolic processes to electrophilic products that can react spontaneously with DNA forming a covalent carcinogen-DNA adduct (Kadlubar and Beland; 1985); an event believed to participate in the initiation of the neoplastic process. The term "metabolic activation" is used to describe the sum of the metabolic steps which convert a carcinogen to the final electrophile ("ultimate carcinogen") which reacts covalently with nucleic acids. This terminology will be used here, however, the following discussion is directed towards the final enzymatic step in the metabolic activation process.

In vitro experiments and animal models for carcinogenesis predict that several different enzymatic acylation reactions contribute to the detoxification and/or conversion of N-substituted aromatic carcinogens to reactive metabolites; these activities include N-acyltransferase, N,O- acyltransferase, O-acyltransferase and N-deacylase (Figure 2). Investigations into arylamine-induced carcinogenesis suggest that the contribution of the different acyl-

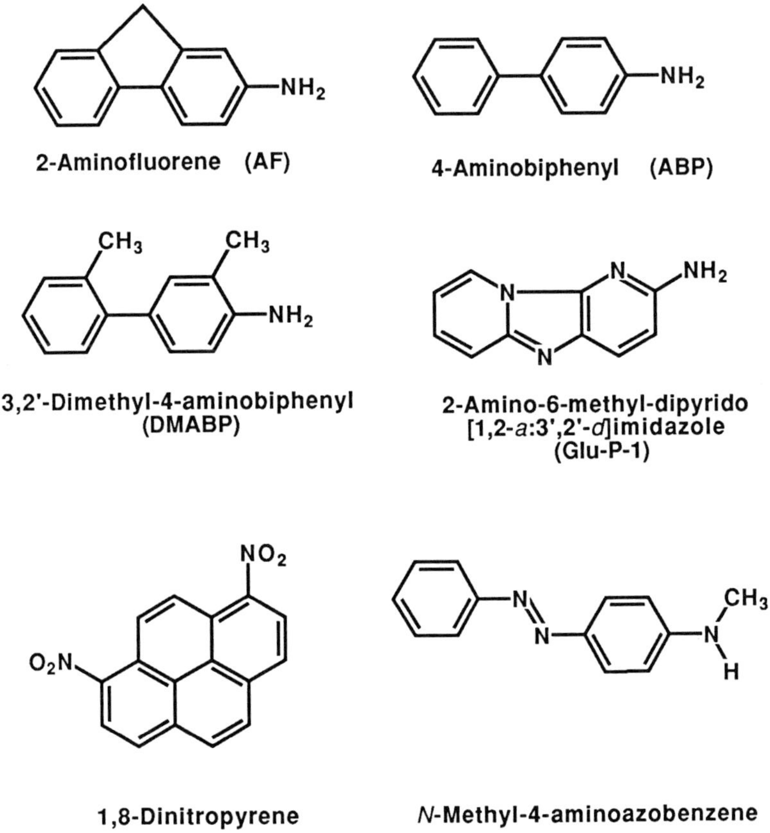

2-Aminofluorene (AF)

4-Aminobiphenyl (ABP)

3,2'-Dimethyl-4-aminobiphenyl (DMABP)

2-Amino-6-methyl-dipyrido [1,2-a:3',2'-d]imidazole (Glu-P-1)

1,8-Dinitropyrene

N-Methyl-4-aminoazobenzene

Figure 1. Structures of representative arylamine and nitroaromatic carcinogens

transferase activities towards the production of arylamine-DNA adducts may vary according to the target site. For the liver model, hepatic N-acetylation and N-oxidative reactions produce arylhydroxamic acids (Lotlikar and Luha, 1971; Lower and Bryan, 1973; Peters and Gutmann, 1955). The hydroxamic acids are subsequently converted by hepatic N,O-acyl-transferases (and sulphotransferases) to N-acetoxy-arylamines (and N-sulphonyloxy arylamides) which can react spontaneously with cellular DNA to produce arylamine-

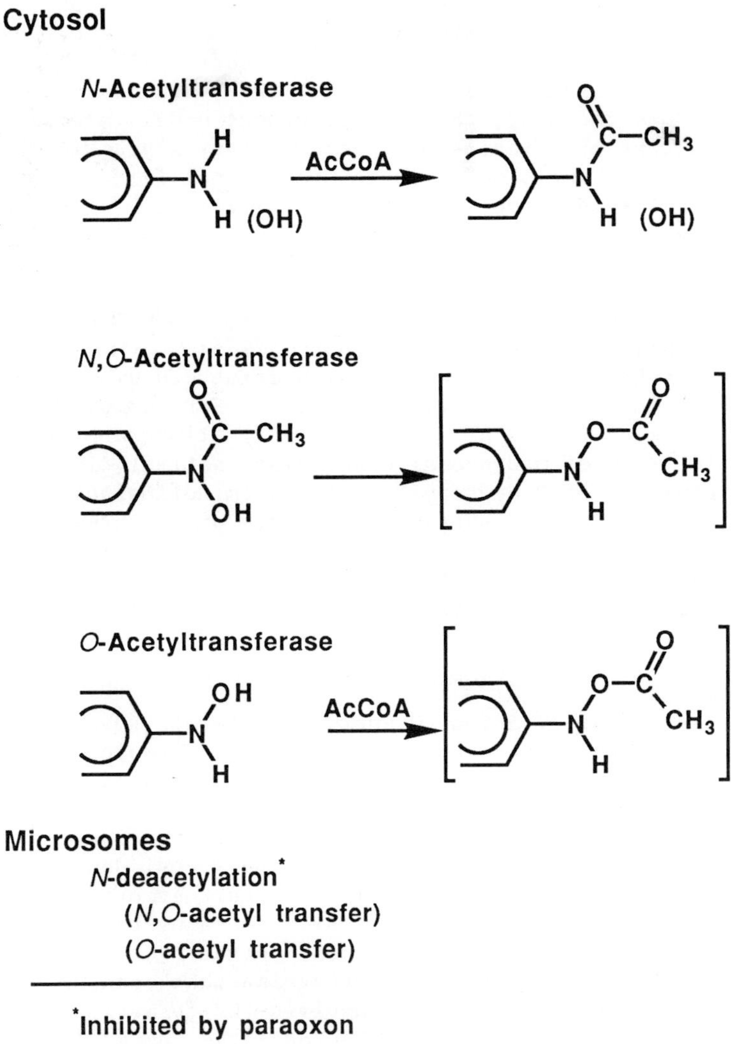

Cytosol

N-Acetyltransferase

N,O-Acetyltransferase

O-Acetyltransferase

Microsomes
N-deacetylation*
(N,O-acetyl transfer)
(O-acetyl transfer)

---

*Inhibited by paraoxon

Figure 2. Acetyltransferase reactions in mammalian tissues that are known to metabolically transform arylamine carcinogens and their metabolites

arylamide) adducts (Bartsch et al., 1972, 1973; King, 1974). In contrast, hepatic N-acetylation is considered a detoxification step for arylamine-induced bladder and colon

carcinogenesis. In the bladder model, non-acetylated
N-hydroxy-arylamines are transported via the urine or bile
to the bladder or colon lumen where they may react directly
with cellular DNA or be converted to electrophiles through
further metabolism (Kadlubar et al., 1977; Radomski et al.,
1977). More recently, the bladder cancer model in dogs
predicts that high levels of free N-hydroxy-arylamine
circulating in the blood lead to their renal filtration and
concentration in urine, resulting in high levels of
arylamine-DNA adducts in the bladder epithelium (Kadlubar
et al., 1988; Young and Kadlubar, 1982). DNA adducts of
several arylamines that are predicted by these reactions
have subsequently been detected in the cellular DNA of
treated-animals (Beland and Kadlubar, 1985).

N-Deacetylation reactions can also affect these models
(reviewed in King and Glowinski, 1983). For example, the
recent work of Lai et al. (1988) clearly illustrates the
essential role of microsomal deacetylase activity in the
tumorigenicity of N-hydroxy-N-acetyl-2-aminofluorene in the
infant mouse bioassay system. The focus of the present
paper is the recent advances in the study of the acetyl
coenzyme A-dependent reaction and its relationship to the
N,O-acetyltransferase activity and to the genetic
polymorphism for N-acetyltransferase.

O-ACYLATION

The terms O-acetylation, O-acetyltransferase (O-acetylase)
and N,O-acyltransferase (both intra- and intermolecular)
have been used variously to describe both cytosolic
enzymatic activity and enzymatic mechanism. Historically,
the usage of each term has depended on the source of the
activity being tested, the source of the acetyl donor, and
on the nature of the acetyl acceptor. Furthermore, the
microsomal fractions of the cell also contains the capacity
to catalyze each of these reactions. Regardless of these
qualifiers, it is important to recognize that each of these
enzymatic activities probably involves the formation of an
acylated-enzyme intermediate. The acylated-enzyme in turn
transfers the acyl group to an N-hydroxylated derivative
forming an O-acyloxy ester intermediate which will react
spontaneously with nucleophilic sites in DNA to form a
covalently bound DNA-adduct. The formation of a covalent
product with nucleic acid is the usual measure of metabolic
activation in the studies reported here. The typical assay
involves the incubation of the radiolabeled test compound
with DNA, RNA or guanosine, suitable co-factors or co-
substrates, and a source of enzyme (e.g. the cytosolic
subcellular fraction of the cell isolated by centrifugation)

at 37°C and physiological pH. Following termination of the incubation, the nucleic acid component is purified by multiple solvent extractions and the level of the covalently-bound radiolabeled carcinogen is then determined by liquid scintillation techniques.

Cytosolic arylhydroxamic acid-dependent acyltransferase

Liver cytosol from several species will catalyze the conversion of carcinogenic arylhydroxamic acids to reactive intermediates which bind covalently to macromolecular nucleophiles including RNA, DNA, and protein; the capacity for this activity in decreasing order is: rabbit > hamster > rat >> monkey,pig >> human >> guinea pig,mouse (King, 1974). The major product in the in vitro reaction with nucleic acids is the non-acetylated arylamine bound to carbon eight of guanine (DeBaun et al., 1968; King and Phillips, 1968; King and Phillips, 1969). The same major arylamine deoxyguanosine product is also recovered from the hepatic DNA of animals treated with either the hydroxamic acid or the parent arylamine (Beland et al., 1980). The mechanism for this reaction (Figure 3) is thought to involve the intramolecular transfer of the N-acyl group to the N-hydroxy position forming the unstable N-acyloxy-arylamine ester, thus the term N,O-acyltransferase (Bartsch et al., 1973; King, 1974). Although the acyloxy-intermediate has not been recovered from these reaction mixtures due to its presumed reactivity, several lines of evidence support both its existence and the formation of an acyl-enzyme intermediate as follows: 1) the level of aminofluorene adducts formed in reaction mixtures containing the parent hydroxamic acid and liver cytosol is increased upon addition of N-hydroxy-2-aminofluorene (Bartsch et al., 1972); 2) the formation of the arylamine adduct from N-hydroxy-N-acetyl-2-aminofluorene is decreased by addition of 2-aminofluorene with the concomitant production of N-acetyl-2-aminofluorene (King, 1974); 3) liver cytosol does not catalyze the formation of the 2-aminofluorene adducts from N-methoxy-N-acetyl-2-aminofluorene (King, 1974); and 4) N-hydroxy-2-aminofluorene forms covalent adducts in reaction mixtures fortified with cytosol and the arylhydroxamic acid, N-hydroxy-N-acetyl-4-aminobiphenyl, as the acetyl donor (Bartsch et al., 1973).

The extent of arylhydroxamic acid-dependent acyltransferase involvement in arylamine-induced carcinogenesis is not known. Clearly this activity is present in several target tissues (King and Glowinski, 1983). However, in the case of mammary gland tumorigenesis, a single injection of N-hydroxy-N-acetyl-2-aminofluorene to the mammary gland of female Sprague-Dawley rat results in mammary gland tumors while treatments with N-acetyl-2-aminofluorene or N-hydroxy-2-aminofluorene in the same assay were negative (Malejka-Giganti et al., 1973).

Cytosolic acetyl coenzyme A-dependent acetyltransferase

The early description of arylhydroxamic acid-dependent activity of rat cytosol by Bartsch et al. (1972) also observed that acetyl coenzyme A-fortified reaction mixtures catalyzed a low level of adduct formation from N-hydroxy-2-aminofluorene (<10% of the arylhydroxamic acid-dependent activity). Thus, the acetyl coenzyme A-dependent pathway appeared to be of minor importance to arylamine-induced carcinogenesis.

However, in the continuing work in our laboratory on the mechanisms of arylamine-induced carcinogenesis, we made a fortuitous observation with the potent rat colon carcinogen, 3,2'-dimethyl-4-aminobiphenyl which caused us to re-evaluate the importance of the acetyl coenzyme A-dependent pathway. Using either [^{3}H]3,2'-dimethyl-4-aminobiphenyl or its related [^{3}H]hydroxamic acid, we identified the non-acetylated C-8 and N^2-deoxyguanosine-substituted products in the hepatic and colon DNA of the treated rats (Flammang et al., 1985; Westra et al., 1985). Since no acetylated adducts were recovered, this suggested that the metabolic activation pathway for the compounds occurred through a common deacetylated metabolite, i.e. N-hydroxy-3,2'-dimethyl-4-aminobiphenyl, or through an identical electrophilic intermediate. Parallel in vitro studies demonstrated two unexpected results: 1) of all the possible known metabolic activation pathways, only the cytosol-mediated acetyl coenzyme A-dependent and the intermolecular arylhydroxamic acid-dependent pathways gave high levels of DNA adducts; 2) the levels of adducts using acetyl coenzyme A as co-substrate were comparable to the reaction using the hydroxamic acid, N-hydroxy-N-acetyl-4-aminobiphenyl as co-substrate; and 3) that the related hydroxamic acid, N-hydroxy-N-acetyl-3,2'-dimethyl-4-aminobiphenyl, was not metabolically activated to DNA-binding either by itself (intramolecular N,O-acyltransferase activity), or as the intermolecular acetyl donor (Table 1 and Flammang and Kadlubar, 1986; Flammang et al., 1985).

These surprising results suggested two important conclusions for arylamine induced carcinogenesis: 1) for certain arylamines, an endogenous acetyl donor was a suitable co-substrate for the enzymatic generation of the electrophile; and 2) the lack of intramolecular transfer was evidence for an exclusive formation of an O-acetoxy-

TABLE 1. Comparison of acetyl donors during rat liver cytosol-catalyzed activation of [^{3}H]N-hydroxy-3,2'-4-aminobiphenyl

Acetyl donor	DNA binding (pmol bound/mg DNA/15 min)[a]
none added (pH 4.6 buffer)	1154 ± 20
none added (pH 7.4 buffer)	5 ± 1
acetyl CoA (heat inactivated cytosol)	2 ± 1
acetyl CoA	267 ± 48
N-hydroxy-N-acetyl-4-aminobiphenyl	290 ± 22
N-hydroxy-N-acetyl-2-aminofluorene	137 ± 13
N-hydroxy-N-acetyl-3,2'-dimethyl-4-aminobiphenyl[b]	18 ± 1

[a] Assay conditions: 100 µM [^{3}H]N-hydroxy-3,2'-dimethyl-4-aminobiphenyl, 1 mM acetyl donor, 2 mg DNA/ml, 1 mg liver cytosol protein/ml, 1mM DTT, and 50 mM sodium pyrophosphate buffer (pH 7.4); n = 3 animals ± SD.

[b] Binding reactions containing [^{3}H]N-hydroxy-N-acetyl-3,2'-4-aminobiphenyl alone in the presence of liver cytosol and DNA were 3 ± 2 pmol bound/mg DNA.

arylamine without formation of an intermediate hydroxamic acid (Mechanism I in Figure 3). Each of these theories will be discussed separately.

The acetyl coenzyme A- and the arylhydroxamic acid-dependent binding activity of rat hepatic cytosol was repeated with several known carcinogenic N-hydroxy-arylamines (Table 2). Again the results indicated that both acetyl donors catalyzed each of the binding reactions to the same extent while the carcinogenic N-hydroxy-N-methyl-4-aminoazobenzene was not a substrate as published previously (Kadlubar et al., 1976). Since these data were in apparent conflict with Bartsch et al. (1972), their experiments were repeated and their results were confirmed (Flammang, 1985; Flammang and Kadlubar, 1986). Additional experiments, exchanging the differences in buffer components between the two studies, revealed that dithiothreitol (Cleland's reagent) increased the level of binding of the acetyl coenzyme A-dependent reaction and it had little

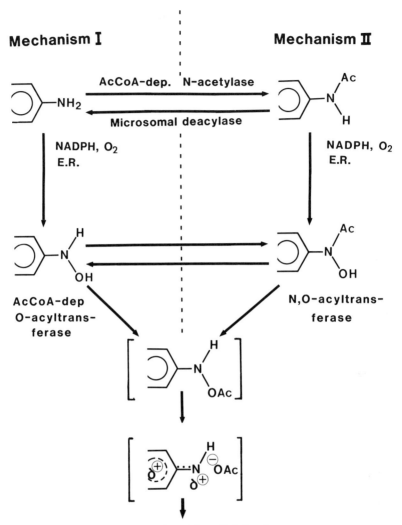

Figure 3. Proposed mechanisms for the metabolic conversion of arylamines and arylamides to electrophilic N-acetoxy-arylamine intermediates which may subsequently react with DNA. Mechanism I: the direct formation of the electrophilic intermediate via AcCoA-dependent O-acetyltransferase or arylhydroxamic acid-dependent transacetylase. Mechanism II: the formation of the electrophilic intermediate via arylhydroxamic acid-dependent N,O-acyltransferase

TABLE 2 - AcCoA-Dependent metabolic activation of [^{3}H]N-OH-arylamines by rat liver cytosol[a]

[^{3}H]Substrate	DNA Binding (pmol bound/mg DNA/15 min)		
	-AcCoA	+ AcCoA	+N-OH-AABP
N-OH-3,2'-diMe-4-aminobiphenyl	4 ± 3	275 ± 22	305 ± 12
N-OH-2-aminofluorene	20 ± 1	287 ± 1	325 ± 21
N-OH-4-aminobiphenyl	3 ± 1	228 ± 20	119 ± 2
N-OH-N'-acetylbenzidine	13 ± 2	78 ± 3	135 ± 15
N-OH-2-naphthylamine	11 ± 4	38 ± 6	18 ± 2
N-OH-N-methyl-4-aminoazobenzene	15[b]	25[b]	17[b]

[a] Assay conditions: 100 μM [^{3}H]N-OH-arylamine, 1 mM AcCoA, 2 mg DNA/ml, 1 mg liver cytosol protein/ml, 1 mM DTT, and 50 mM sodium pyrophosphate buffer (pH 7.4); n = 3 animals ± s.d. Incubations carried out with heat-denatured cytosol gave binding levels similar to those obtained without AcCoA. AcCoA = acetyl coenzyme A; N-OH = N-hydroxy; AABP = N-acetyl-4-aminobiphenyl.

[b] An average of two determinations at 50 and 100 μM [^{3}H]N-OH-arylamine.

effect on the levels in the hydroxamic acid dependent reaction (Flammang and Kadlubar, 1986). It is likely that dithiothreitol, a thiol-protecting reagent, stabilizes the enzyme since the reagent is required to stabilize rat liver N,O-acyltransferase during purification (King, 1974). Work in other laboratories confirmed these reactions and also extended the substrate specificity to two additional chemical classes, the highly mutagenic dinitropyrenes (Djuric et al., 1985) and the heterocyclic amines (Shinohara et al., 1986). In addition to acetyl coenzyme A, acetoacetyl coenzyme A and in particular propionyl coenzyme A supported the rat hepatic cytosol catalyzed binding of the N-hydroxy-arylamines to DNA; the additional endogenous coenzyme A derivatives including malonyl, succinyl, and oleolyl were not suitable co-substrates in the same reaction (Table 3 and Flammang and Kadlubar, 1986; Shinohara et al., 1986).

TABLE 3 - Metabolic activation of [^{3}H]N-OH-DMABP and [^{3}H]N-OH-AF by rat liver cytosol with various acyl donors[a]

	DNA Binding (pmol bound/mg DNA/15 min)	
Acyl	[^{3}H]N-OH-DMABP	[^{3}H]N-OH-AF
AcCoA	315 ± 15	321 ± 11
Acetoacetyl CoA	114 ± 40	42 ± 21
Propionyl CoA	122 ± 36	424 ± 79
Methylmalonyl CoA	<5[b]	69 ± 8
Malonyl, Succinyl, Palmitoyl, Oleoyl CoAs	<5[b]	<10[b]
Folinic Acid, L-Acetyl-carnitine	<5[b]	<10[b]
Acetyl phosphate	<5[b]	<10[b]
S-Acetyl-thiocholine, N-Acetyl-L-cysteine	<5[b]	<10[b]
N-Acetyl-2-aminofluorene	18 ± 10	22 ± 4
p-Nitrophenyl acetate	461 ± 28	721 ± 132

[a] Assay conditions: 100 μM [^{3}H]N-OH-arylamine, 1 mM acyl dono 1 mg rat cytosol protein/ml, 2 mg DNA/ml, 1 mM DTT, and 50 mM sodium pyrophosphate buffer (pH 7.4.); n = 3 animals ± s.d. Control incubations were carried out with heat-denatured cytosol which gave DNA binding levels that were 5-10 pmol bound/mg DNA/15 min. N-OH-DMABP = N-hydroxy-3,2'-dimethyl-4-aminobiphenyl; N-OH-AF = N-hydroxy-2-aminofluorene; AcCoA = acetyl coenzyme A.

[b] Judged as the limit of detection on duplicate assays.

The enzymatic acetyl coenzyme A-dependent DNA-binding of
N-hydroxy arylamines is widely distributed in several
tissues and species with varying susceptibility to
arylamine-induced tumorigenesis (Table 4 and Flammang and
Kadlubar, 1986; Shinohara et al., 1986). Similar to the
metabolic activation involving the cytosolic arylhydroxamic
acid-dependent acyltransferases (King, 1974; Flammang and
Kadlubar, 1986) the acetyl coenzyme A-dependent binding
levels were highest in the reactions containing cytosol from
rabbit, hamster and rat; binding was not catalyzed by dog
liver cytosol. The dog is known to be deficient in both N-
and N,O-acetyltransferase (King, 1974; Poirier et al.,
1963). In contrast to the arylhydroxamic acid-dependent
pathways (Table 4 and Flammang and Kadlubar, 1986), high
levels of acetyl coenzyme A-dependent binding were also
catalyzed by the hepatic cytosol of guinea pig, mouse and
human.

The mechanism for the acetyl coenzyme A-dependent
metabolic activation of the N-hydroxy-arylamines has not
been carefully studied. However, hplc analysis (Flammang
and Kadlubar, 1986) of the DNA recovered from the acetyl
coenzyme A-dependent reaction of N-hydroxy-3,2'-dimethyl-4-
aminobiphenyl showed only the same non-acetylated
deoxyguanosine adducts that were previously observed in vivo
(Westra et al., 1985) and in the synthetic reaction of N-
hydroxy-3,2'-dimethyl-4-aminobiphenyl with DNA in vitro
(Flammang et al., 1985). Clearly, the reaction did not
proceed through an N-acetyl-N-acetoxy intermediate which
should yield acetylated adducts (Kriek et al., 1967).
Historically, it could have been argued that the acetyl
coenzyme A-dependent mechanism involved the formation of the
hydroxamic intermediate (Mechanism II in Figure 3). The
evidence of Bartsch et al. (1972) demonstrated that the
hydroxamic acids are superior acetyl donors and the studies
of King (1974) demonstrated the high capacity of several
species to catalyze the intramolecular N,O-acyltransferase
reaction. However, the hydroxamic acid, N-hydroxy-N-acetyl-
3,2'-4-aminobiphenyl was not a suitable substrate for either
inter- or intramolecular N,O-acyltransferase reactions
tested (Table 1 and Table 4, respectively). This implies
that the acetyl coenzyme A-dependent reaction proceeds by
direct acylation of the N-hydroxy-arylamine as shown in
Figure 3, Mechanism I. Recent work with the hydroxamic acid
and the N-hydroxy-arylamine metabolite of the carcinogenic
food pyrolysis product, Glu-P-1, also demonstrates that the
preferred metabolic activation pathway is through the acetyl
coenzyme A mechanism (Shinohara et al., 1986). The reaction
mechanism also probably involves an acylated enzyme

TABLE 4 - Hepatic cytosol acetyltransferase-mediated binding of [^{3}H]N-OH-arylamines and [^{3}H]arylhydroxamic acids to DNA: A comparison of species and sexes[a]

Substrate/Species (n, sex)	DNA Binding (pmol bound/mg DNA/15min)		
N-OH-DMABP	+AcCoA[b]	+N-OH-AABP[b]	N-OH-ADMABP only[c]
Rabbit (6, male)	3102 ± 497	593 ± 193	<10[d]
Hamster (3, male)	1081 ± 36	623 ± 24	<10[d]
F344/N Nctr rat (6, male)	290 ± 30	341 ± 50	<10[d]
F344/N Nctr rat (3, female)	116 ± 40	247 ± 16	<10[d]
Human (male) No. 1	202[e]	41[e]	<10[d,e]
Human (male) No. 2	40[e]	<10[d,e]	<10[d,e]
Guinea pig (3, male)	160 ± 5	32 ± 5	<10[d]
B6C3F1/Nctr mouse (3, male)	58 ± 4	15 ± 4	<10[d]
B6C3F1/Nctr mouse (3, female)	50 ± 5	10d	<10[d]
Dog (2, male)	<10[d,e]	<10[d,e]	10[d,e]
N-OH-AF	+ AcCoA[b]	+N-OH+AABP[b]	N-OH-AAF only[c]
Rabbit (3, male)	5816 ± 1000	1182 ± 416	2728 ± 526
Hamster (3, male)	1382 ± 220	473 ± 57	734 ± 212
F344/N Nctr rat (4, male)	267 ± 14	285 ± 21	124 ± 14
F344/N Nctr rat (3, female)	165 ± 84	209 ± 60	119 ± 11
Human (male) No. 1	281[e]	44[e]	<20[d,e]
Human (male) No. 2	158[e]	<20[d,e]	<20[d,e]
Guinea pig (3, male)	1498 ± 307[f]	21 ± 3	79 ± 19
B6C3F1/Nctr mouse (3, male)	723 ± 81	<20[d]	<20[d]
B6C3F1/Nctr mouse (3, female)	458 ± 55	<20[d]	<20[d]
Dog (2, male)	<20[d,e]	<20[d,e]	<20[d,e]

Cont'd...

a   Control incubations without acyl donor or with heat-denatured cytosols showed binding levels that were <10% of the standard assay. AcCoA = acetyl coenzyme A; N-OH-DMABP = N-hydroxy-3,2'-dimethyl-4-aminobiphenyl; N-OH-ADMABP = N-acetyl-N-OH-DMABP; N-OH-AABP = N-hydroxy-N-acetyl-4-aminobiphenyl; N-OH-AAF = N-Hydroxy-N-acetyl-2-aminofluorene.
b   Assay conditions: 200 μM [^{3}H] N-OH-arylamine, 1 mM acyl donor, 2 mg DNA/ml, 1-2 mg cytosol protein/ml, 1 mM DTT, and 50 mM sodium pyrophosphate buffer (pH 7.4); n = number of animals ± s.d.
c   Same as above except 100 μM [3-H]arylhydroxamic acid was used and no exogenous acetyl donor was added.
d   Judged to be the limit of detection.
e   Average of duplicate assays which were within 10% of each other.
f   Binding was insensitive to 10^{-4}M paraoxon and 100 mM sodium fluoride.

---

intermediate as was discussed for the arylhydroxamic acid-dependent acyltransferases. The reaction kinetics of the acetyl coenzyme A-dependent binding reaction demonstrate a biphasic appearance of product (Figure 4), and regraphing the extrapolated y-intercepts of the biphasic curves, as described by Hartley and Kilby (1954), demonstrates the linear response (Figure 5).

The role of the acetyl coenzyme A-dependent metabolic activation of N-hydroxy-arylamines in carcinogenesis is not known. However, it is known that mutations are reduced and the adduct formation is decreased in cell-free extracts of the acetylase-deficient Salmonella typhimurium tester strain, TA98/1,8-DNP$_6$, as compared to the normal strain, TA98, for certain arylamine carcinogens (Saito et al., 1985); it is known that these strains do not contain the N,O-acyltransferase activity (Weeks et al., 1980). Similar to the arylhydroxamic acid-dependent activity, the acetyl coenzyme A-dependent reaction is widely distributed in several tissues and species that are susceptible to arylamine-induced carcinogenesis. The acetyl coenzyme A-dependent pathway may also be critical for activation of certain arylamine carcinogens to form DNA-adducts, e.g. 3,2'-dimethyl-4-aminobiphenyl, as they are poor substrates for the other known metabolic activation routes as discussed above. It is also possible that this reaction pathway is important to those species that are relatively deficient in the alternative arylhydroxamic acid-dependent acyltransferases, e.g. man, mouse, guinea pig.

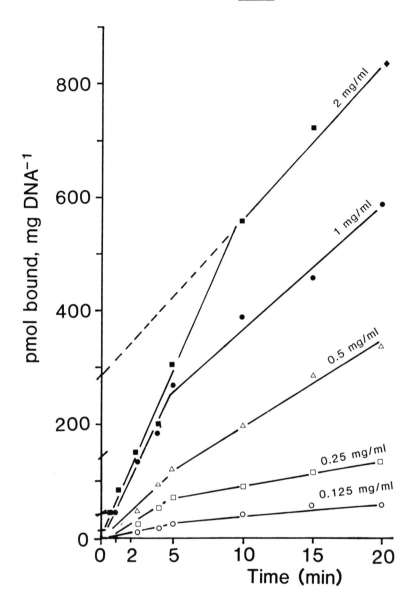

Figure 4. Biphasic reaction kinetics for the acetyl coenzyme A-dependent binding of [^{3}H]N-hydroxy-2-aminofluorene to calf thymus DNA. Standard 1 ml assays contained 100 μM N-hydroxy-arylamine, 1 mM acetyl coenzyme A, 2 mg DNA/ml, 1 mM DTT, 50 mM pyrophosphate buffer pH 7.4), and rat cytosol protein as indicated. Incubations at 37°, 15 min.

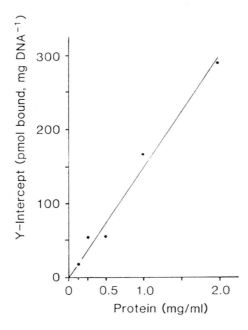

Figure 5. Replot of the y-intercepts from Figure 4 according to Hartley and Kilby (1954)

GENETIC REGULATION OF O-ACYLATION

The genetic control of the polymorphic response of acetyl coenzyme A-dependent arylamine N-acetyltransferase is the subject of a recent monograph (Weber, 1987) and review (Weber and Hein, 1985). This polymorphism was originally discovered through the differential response of humans to drugs which caused clinically observable toxicities. More recently, associations have been made between the N-acetylator phenotype and diseases including bladder cancer and colorectal cancers in humans; an excess of slow N-acetylators are present in populations of bladder cancer patients (Cartwright, 1984; Mommsen et al., 1985), while an excess of fast N-acetylators are found in groups of colorectal cancer patients (Ilet et al., 1987; Lang et al., 1986); no prevalence in acetylator status was detected in a population of lung cancer patients (Philip et al., 1988) or in a preliminary report on a population of German colorectal cancer patients (Drakoulis et al., 1988).

Acetylator genotype-dependent N-acetylation of arylamines has been shown in humans (Glowinski et al., 1978), rabbits (Hein et al., 1982a), mice (Hein et al., 1988; Mattano and

Weber, 1987) and hamsters (Hein et al., 1982b). The genetic polymorphism in N-acetyltransferase activity is determined by two codominant alleles (R, rapid; r, slow) at a single gene locus; the slow acetylator allele is inherited as an autosomal recessive trait (Weber, 1987). Thus the highest N-acetyltransferase activity is detected in the homozygous rapid, RR, genotype, the lowest activity is in the homozygous slow, rr, genotype and the heterozygous, Rr, genotype expresses an intermediate activity.

Genetic polymorphism in animal models

The relationship of arylamine N-acetyltransferase to O-acyltransferases has been studied in genetic models of rabbit (Glowinski et al., 1980), mouse (Hein et al., 1988) and hamster (Hein et al., 1987). In the rabbit model, the capacity of the hepatic cytosol obtained from the homozygous rapid, RR, genotype vs the homozygous slow, rr, genotype was at least 100-fold, 20-fold and 10-fold greater, respectively, for the N-acetylation of sulphamethazine and 2-aminofluorene, and for N,O-acyltransferase catalyzed binding of N-hydroxy-N-acetyl-2-aminofluorene (Glowinski et al., 1980). Furthermore, these same activities co-chromatographed in partial purification experiments. These results indicated that a single genetically determined polymorphism could be responsible for both the activation and detoxification pathways of arylamine carcinogens.

The relationship of arylamine N-acetyltransferase to O-acyltransferases in the hamster model appears to be more complicated (Hein et al., 1987). In these results (Figure 6), the N-acetylation of 2-aminofluorene and p-aminobenzoic acid demonstrate a gene dose effect in the cytosol of liver, kidney, intestine and lung obtained from the inbred rapid acetylator (Bio. 87.20), the inbred slow acetylator (Bio.73/H), and their F_1 progeny. The same result was also observed in the tissue cytosol for the acetyl coenzyme A-dependent binding of N-hydroxy-2-aminofluorene to DNA. In contrast, the genotype-dependent response of 4-aminobiphenyl N-acetyltransferase, N-hydroxy-4-aminobiphenyl and N-hydroxy-3,2'-4-aminobiphenyl acetyl coenzyme A-dependent O-acetyl-transferase and N-hydroxy-N-acetyl-2-aminofluorene N,O-acetyltransferase was not statistically different (Hein et al., 1987). However, partial purification of the hepatic cytosol resolved two peaks of acetyltransferase activities that differed in genetic regulation as previously described (Hein et al., 1987). The early eluting peak (Figure 7) demonstrates the genotype-dependent polymorphic response for

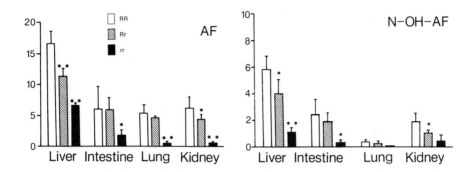

Figure 6. Acetyltransferase activities from tissue cytosol obtained from male inbred Syrian hamsters. Open bars, homozygous rapid, RR, acetylators, Bio. 87.20; solid bars, homozygous slow, rr, acetylators, Bio. 82.73/H; crossed bars, heterozygous, Rr, acetylators, (87.20 X 82.73/H progeny). Left panel: 2-aminofluorene N-acetyltransferase, nmoles N-acetyl-2-aminofluorene formed/min/mg cytosol protein. Right panel: acetyl coenzyme A-dependent binding of [^{3}H]N-hydroxy-2-aminofluorene to DNA, nmoles bound/15 min/mg DNA/mg cytosol protein. Determinations from 4 to 6 animals $\pm$ SD; * and **, (P<0.05) to bar on immediate left.

the acetyl coenzyme A-dependent metabolic activation of all three N-hydroxy-arylamines. It is also apparent that the late eluting peak, which demonstrates N-acetyltransferase activity that is independent (monomorphic response) of the N-acetylator genotype (Hein et al., 1987), also demonstrates a monomorphic response to the acetyl coenzyme A-dependent metabolic activation of the N-hydroxy-arylamines. In fact the larger contribution of this enzyme to the capacity of unfractionated liver to catalyze the binding reaction would obscure the genotype-dependent response for the two 4-aminobiphenyl derivatives in the unfractionated cytosol. Both the polymorphic and monomorphic enzymes catalyze the N,O-acyltransferase mediated binding of N-hydroxy-N-acetyl-2-aminofluorene to DNA but the majority of the binding capacity was in the late eluting, monomorphic peak (Hein et al., 1987). Two enzymes have also been fractionated from the hepatic cytosol of phenotypically-selected hamsters (Kato and Yamazoe, 1988). The purified enzyme, "AT-1", catalyzed all three acetyltransferase activities as expected and it is probably the monomorphic enzyme as judged by its

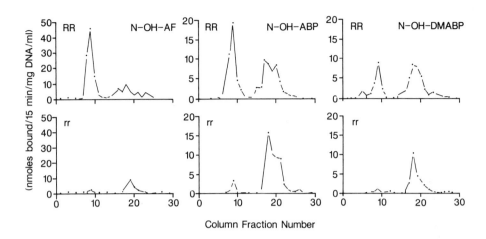

Figure 7. Metabolic activation of [^{3}H]N-hydroxyarylamines to bind DNA as catalyzed by hamster liver cytosol fractionated by FPLC ion exchange chromatography on a Mono Q* column (*Pharmacia Inc.). N-OH-AF = N-hydroxy-2-aminofluorene, N-OH-ABP = N-hydroxy-4-aminobiphenyl, N-OH-DMABP = N-hydroxy-3,2'-dimethyl-4-aminobiphenyl. The genotype designations, RR, Rr, and rr are described in Figure legend 6.

late elution on the ion exchange column; a second, early-eluting enzyme peak catalyzed only N-acetylation reactions (Kato and Yamazoe, 1988). Hanna and coworkers have also differentiated hamster N-acetyltransferases by affinity techniques (Smith and Hanna, 1986).

The relationship of acetyl coenzyme A-dependent N-acetyltransferases and O-acetyltransferases is just beginning to emerge in the mouse model (Hein et al., 1988). Hepatic cytosol obtained from the rapid acetylator, C57BL/6J (RR), which was fractionated by ion exchange chromatography, showed a dominant peak of activity which demonstrated 3- to 5-fold greater capacity than fractionated cytosol from the slow acetylator congenic mouse, B6.A.Nat^r (rr), to N-acetylate 4-aminobiphenyl and 2-aminofluorene, respectively. In contrast, the same peak appeared to demonstrate a monomorphic response for the acetyl coenzyme A-dependent binding of N-hydroxy-3,2'-dimethyl-4-aminobiphenyl to DNA. This result does not necessarily establish that the O-acetyltransferase is not controlled by the same genetic locus as the N-acetyltransferase. As the results in hamster discussed above indicate, this N-hydroxy-arylamine may be a poor substrate for polymorphic enzymes. However, we also found the same result with N-hydroxy-3,2'-dimethyl-4-aminobiphenyl using the hepatic cytosol from the C57BL/653Hf/Nctr mouse as compared to the A/J mouse (rr); in contrast, when N-hydroxy-2-aminofluorene was the substrate, the level of binding appeared higher in the Nctr mouse (515 ± 176 vs 462 ± 39, respectively; Flammang , 1985). A preliminary report has also indicated that the level of hepatic 2-aminofluorene-DNA adducts is significantly higher in the rapid N-acetylator C57BL/6J than in the slow N-acetylator A/J strain, 2 hours after a single i.p. dose of 2-aminofluorene (Levy and Weber, 1988).

Genetic polymorphism in humans

A genetic polymorphism in the enzymatic N-acetylation of drugs is well known and it has been associated with individual differences in susceptibility to drug toxicities and diseases, including cancer (see above). In comparison with animal models, little is known about the metabolism and disposition of carcinogens in humans. In vivo and in vitro studies have shown that sulphamethazine N-acetyltransferase in humans is also under the control of a single gene locus containing two codominant alleles: R, rapid and r, slow (Weber, 1987). Glowinski and coworkers (1978) have further demonstrated in vitro that the levels of 2-aminofluorene N-acetyltransferase activity in human liver are consistent with their sulphamethazine N-acetyltransferase phenotype.

We have further examined the role of acetyltransferases in carcinogen metabolism using frozen cytosol prepared from thirty-five human liver samples obtained from organ donors (Flammang et al., 1987). The capacity of these liver cytosols for N^4-sulphamethazine and 2-aminofluorene N-acetyltransferase activity varied over a 20-fold and a 60-fold range, respectively, in the same manner as previously reported (Glowinski et al., 1978). Furthermore, a comparison of these two acetyltransferase activities by linear regression analysis showed a correlation coefficient, r = 0.81, indicating that both reactions are catalyzed by the same or by co-regulated enzymes.

Each of the same human liver cytosols were also assayed for their capacity to activate [^{3}H]N-hydroxy-2-aminofluorene to a DNA binding metabolite (Flammang et al., 1987). Similar to the capacity of the livers for 2-aminofluorene N-acetyltransferase, the levels of the aminofluorene binding to DNA that were catalyzed by the cytosols were highly correlated to the capacity of the livers for N^4-sulphamethazine N-acetyltransferase (r = 0.80). In contrast, we did not detect any significant N,O-acyltransferase activity in the frozen cytosols using [^{3}H]N-hydroxy-N-acetyl-2-aminofluorene as the substrate (<20 pmol bound/mg protein/mg DNA/5 min). However, a recent report has indicated that N,O-acyltransferase activity is present in autopsy samples of human liver and that this activity is lost (50-100%) after freezing the cytosol (Land et al., 1988).

In studies of populations of colon cancer patients, we and others observed an excess of the fast N-acetylator phenotype as compared to the age-matched control population (Lang et al., 1986; Ilet et al., 1987). This has raised the question as to the connection, if any, between a clinically-determined N-acetylator phenotype and the levels of the acetyltransferase enzymes in target tissues. We have obtained data for the first part of these studies by assaying the acetyltransferase activities in fourteen surgical specimens of human colon. As observed in Figure 8, the colon cytosol levels of 2-aminofluorene N-acetyltransferase and acetyl coenzyme A-dependent binding of N-hydroxy-2-aminofluorene to DNA are also correlated (r = 0.80). This study is presently being extended to obtain the clinical N-acetylator phenotype and colon tissue samples from patients who are scheduled for resection.

These correlations do not prove that all of the acetyltransferase reactions in human tissue can be catalyzed by the same polymorphic enzyme but this is possible as has been demonstrated for hamster liver (Hein et al., 1987). The question still remains whether or not the various human acetyltransferase activities are controlled by the same polymorphic gene.

CONCLUSIONS

Interindividual differences in our response to the environment has made it difficult for both the epidemiologist to detect environmental toxins responsible for disease and for the clinician to treat the susceptible individual safely. In the case of cancer, our genetic variability in the metabolic activation of chemical carcinogens, e.g. oxidation and acetylation, and genetic variability in sensitivity factors, e.g. immune response, are thought to affect the tumorigenic process. From animal models it is predicted that a genetic polymorphism for arylamine N-acetyltransferases is involved both in the detoxification of chemical carcinogens and in the metabolic activation of carcinogens leading to the formation of covalently bound carcinogen-DNA adducts. These models further indicate that N-acylation and O-acylation metabolic

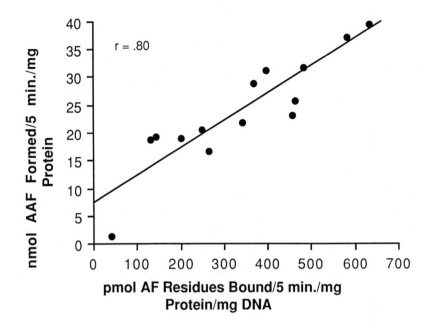

Figure 8. Linear regression analysis of the levels of 2-aminofluorene N-acetyltransferase vs acetyl coenzyme A-dependent binding of [^{3}H]N-hydroxy-2-aminofluorene to DNA in fourteen human colon cytosols

pathways are under a common genetic control. The hamster model further demonstrates that both polymorphic and monomorphic acetyltransferase enzymes can effect the overall metabolic response to the arylamine toxin.

Epidemiologists have shown that phenotypic slow N-acetylators are at increased risk for bladder cancer while the phenotypic rapid N-acetylator is at increased risk for colorectal cancers. In the studies with human tissue presented here it was observed that, in both liver and colon, the in vitro N-acetylator phenotype was also a predictor of the capacity of the tissue to metabolically activate the N-hydroxy-arylamine carcinogen through an acetyl coenzyme A-dependent pathway. Although the liver certainly mediates a large portion of xenobiotic metabolism, the colon tissue retains a high capacity to catalyze the binding reaction. In as much as colon cancer may be associated in part with the foods we eat, it is important to note that a large number of the the highly mutagenic compounds isolated from foods cooked at high temperature, are in the aromatic amine class. It therefore remains incumbent upon us to determine the relationship of the clinically-determined N-acetylator phenotype to the colon N-acetylator phenotype.

These studies support the hypothesis that the N-acetylator phenotype of an individual represents an indicator of genotoxic risk through the acetyl coenzyme A-dependent metabolic activation of N-hydroxy-arylamines. Studies in this laboratory, and others, are continuing to determine whether or not the N-acetylator genotype is related to the formation of carcinogen-DNA adducts in treated animals.

REFERENCES

Bartsch, H., Dworkin, M., Miller, J.A. and Miller, E.C., 1972, Electrophilic N-acetoxyaminoarenes derived from carcinogenic N-hydroxy-2-acetylaminoarenes by enzymatic deacetylation and transacetylation in liver. Biochimica et Biophysica Acta, 286, 272-298.

Bartsch, H., Dworkin, C., Miller, E.C. and Miller, J.A., 1973, Formation of electrophilic N-acetoxyarylamines in cytosols from rat mammary gland and other tissues by transacetylation from the carcinogen N-hydroxy-4-acetyl-aminobiphenyl. Biochimica et Biophysica Acta, 304, 42-55.

Beland, F.A. and Kadlubar, F.F., 1985, Formation and persistence of arylamine DNA adducts in vivo. Environmental Health Perspectives, 62, 19-30.

Beland, F.A., Allaben, W.T. and Evans, F.E., 1980, Acyl-transferase mediated binding of N-hydroxyarylamides to nucleic acids. Cancer Research, 40, 834-840.

Bryant, M.S., Skipper, P.L., Tannenbaum, S.R. and Maclure, M., 1987, Hemoglobin adducts of 4-aminobiphenyl in smokers and nonsmokers. Cancer Research, 47, 602-608.

Cartwright, R.A., 1984, Epidemiological studies on N-acetylation and C-center oxidation in neoplasia. In Genetic Variability in Responses to Chemical Exposure, Vol. 16, Banbury Reports, edited by G.S. Omenn and H.V. Gelboin (Cold Spring Harbor Laboratory) pp. 359-368.

Cerniglia, C.E., Freeman, J.P., Franklin, W. and Pack, D.L., 1982, Metabolism of azodyes derived from benzidine 3,3'-dimethylbenzidine and 3,3'-dimethoxybenzidine to potentially carcinogenic aromatic amines by intestinal bacteria. Carcinogenesis, 3, 1255-1260.

DeBaun, J.E., Rowley, J.Y., Miller, E.C. and Miller, J.A., 1968, Sulfotransferase activation of N-hydroxy-2-acetyl-aminofluorene in rodent livers susceptible and resistant to this carcinogen. Proceedings of the Society of Experimental Biology and Medicine, 129, 268-273.

Djuric, Z., Fifer, E.K. and Beland, F.A., 1985, Acetyl coenzyme A-dependent binding of carcinogenic and mutagenic dinitropyrenes to DNA. Carcinogenesis, 6, 941-944.

Drakoulis, N., Cuprunov, M., Ploch, A., Hirner, A. and Roots, I., 1988, Acetylator Phenotype as possible risk factor in colon, gastric, pharynx, bronchial, and bladder cancer. In Xenobiotic Metabolism and Disposition, IInd International ISSX Meeting-ISSX-88, May 16-20, 1988, Kobe, Japan, Abstract III-405-P2.

Flammang, T.J., 1985, Acetyl CoA-dependent metabolism of N-hydroxy-3,2'-dimethyl-4-aminobiphenyl and carcinogenic N-hydroxy arylamines to form carcinogen DNA-bound adducts. Ph.D. Dissertation. (University of Arkansas for Medical Sciences, Little Rock, AR, 1985).

Flammang, T.J. and Kadlubar, F.F., 1986, Acetyl coenzyme A-dependent metabolic activation of N-hydroxy-3,2'-dimethyl-4-aminobiphenyl and several carcinogenic N-hydroxyarylamines in relation to tissue and species differences, other acetyl donors, and aryl hydroxamic acid-dependent acyltransferases. Carcinogenesis, 7, 919-926.

Flammang, T.J., Westra, J.G., Kadlubar, F.F. and Beland, F.A., 1985, DNA adducts formed from the probable proximate carcinogen, N-hydroxy 3,2'-dimethyl-4-aminobiphenyl, by acid catalysis or S-acetyl coenzyme A-dependent enzymatic esterification. Carcinogenesis, 6, 251-258.

Flammang, T.J., Yamazoe, Y., Guengerich, F.P. and Kadlubar, F.F., 1987, The S-acetyl coenzyme A-dependent metabolic activation of the carcinogen N-hydroxy-2-aminofluorene by human liver cytosol and its relationship to the aromatic amine N-acetyltransferase phenotype. Carcinogenesis, 8, 1967-1970.

Glowinski, I.B., Radke, H.E. and Weber, W.W., 1978, Genetic variation in N-acetylation of carcinogenic arylamines by human and rabbit liver. Molecular Pharmacology, 14, 940-949.

Glowinski, I.B., Weber, W.W., Fysh, J.M., Vaught, J.B. and King, C.M., 1980, Evidence that arylhydroxamic acid N,O-acetyltransferase and the genetically polymorphic N-acetyltransferase are properties of the same enzyme in rabbit liver. Journal of Biological Chemistry, 225, 7883-7890.

Hartley, B.S. and Kilby, B.A., 1954, The reaction of p-nitrophenyl esters with chymotrypsin and insulin. Biochemical Journal, 56, 288-297.

Hein, D.W., Omichinski, J.G., Brewer, J.A. and Weber, W.W., 1982a, A unique pharmacogenetic expression of the N-acetylation polymorphism in the inbred hamster. Journal of Pharmacology and Experimental Therapeutics, 220, 8-15.

Hein, D.W., Smolen, T.N., Fox, R.R. and Weber, W.W., 1982b, Identification of genetically homozygous rapid and slow acetylators of drugs and environmental carcinogens among established inbred rabbit strains. Journal of Pharmacology and Experimental Therapeutics, 223, 40-44.

Hein, D.W., Flammang, T.J., Kirlin, W.G., Trinidad, A. and Ogolla, F., 1987, Acetylator genotype-dependent metabolic activation of carcinogenic N-hydroxyarylamines by S-acetyl coenzyme A-dependent enzymes of inbred hamster tissue cytosols: relationship to arylamine N-acetyltransferase. Carcinogenesis, 8, 1767-1774.

Hein, D.W., Trinidad, A., Yerokun, T., Ferguson, R.J., Kirlin, W.G. and Weber, W. W., 1988, Genetic control of acetyl coenzyme A-dependent arylamine N-acetyltransferase, hydrazine N-acetyltransferase, and N-hydroxy-arylamine O-acetyltransferase enzymes in C57BL/6, A/J, AC57F$_1$, and the rapid and slow acetylator A.B6 and B6.A congenic inbred mouse. Drug Metabolism and Disposition, 16, 341-347.

Ilet, K.F., David, B.M., Detchon, P. and Castlede, W.M., 1987, Acetylator phenotype in colorectal-carcinoma. Cancer Research, 47, 1466-1469.

Kadlubar, F.F. and Beland, F.A., 1985, Chemical properties of ultimate carcinogenic metabolites of arylamines and arylamides, in Polycyclic Hydrocarbons and Carcinogenesis, ACS Symposium Series 283, edited by R.G. Harvey (Washington D.C.: American Chemical Society) pp 341-370.

Kadlubar, F.F., Miller, J.A. and Miller, E.C., 1976, Hepatic metabolism of N-hydroxy-N-methyl-4-aminoazobenzene and other N-hydroxy arylamines to reactive sulphuric acid esters. Cancer Research, 36, 2350-2359.

Kadlubar, F.F., Miller, J.A. and Miller, E.C., 1977, Hepatic microsomal N-glucuronidation and nucleic acid binding of N-hydroxylamines in relation to urinary bladder carcinogenesis. Cancer Research, 37, 805-814.

Kadlubar, F.F., Dooley, K.L., Benson, R.W., Roberts, D.W, Butler, M.A., Teitel, C.H., Bailey, J.R. and Young, J.F., 1988, Pharmacokinetic model of aromatic amine-induced urinary bladder carcinogenesis in beagle dogs administered 4-aminobiphenyl. In Carcinogenic and Mutagenic Responses to Aromatic Amines and Nitroarenes, edited by C. M. King, L. J. Romano, and D. Schuetzle (New York: Elsevier) pp. 173-180.

Kato, R., and Yamazoe, Y., 1988, N-Hydroxyarylamine O-acetyltransferase in Mammalian livers and Salmonella. In Carcinogenic and Mutagenic Responses to Aromatic Amines and Nitroarenes, edited by C.M. King, L.J. Romano, and D. Schuetzle (New York: Elsevier) pp. 125-136.

King, C.M., 1974, Mechanisms of reaction, tissue distribution, and inhibition of arylhydroxamic acid acyltransferase. Cancer Research, 34, 1503-1515.

King C.M. and Phillips, B., 1968, Enzyme-catalyzed reactions of the carcinogen N-hydroxy-2-fluorenylacetamide with nucleic acid. Science, 159, 1351-1353.

King, C.M. and Phillips, B., 1969, N-Hydroxy-2-fluorenylacetamide: reaction of the carcinogen with guanosine, ribonucleic acid, deoxyribonucleic acid, and protein following enzymatic deacetylation or esterfication. Journal of Biological Chemistry, 224, 2609-2616.

King, C.M. and Glowinski, I.B. 1983, Acetylation, deacetylation and acyl transfer. Environmental Health Perspectives, 49, 43-50.

Kriek, E., Miller, J.A., Juhl, U. and Miller, E. C., 1967, 8-(N-2-Fluorenylacetamido)guanosine, an arylamidation reaction product of guanosine and the carcinogen N-acetoxy-N-2-fluorenylacetamide in neutral solution. Biochemistry, 1, 177-182.

Lai, C-C., Miller, E.C., Miller, J.A. and Liem, A., 1988, The essential role of microsomal deacetylase activity in the metabolic activation, DNA-(deoxyguanosin-8-yl)-2-aminofluorene adduct formation and initiation of liver tumors by N-hydroxy-2-acetylaminofluorene in the livers of infant male B6C3F$_1$ mice. Carcinogenesis, 9, 1295-1302.

Land, S., Kukowski, K., Lee, M.-S., Debiec-Rychter, M. and King C.M., 1988, Metabolism of aromatic amines: relationships of O-acetylation, N,O-acetyltransfer, N-acetylation and deacetylation in human liver. Proceedings of the AACR, 29, 122, Abstract 486.

Lang, N.P., Chu, D.Z.J., Hunter, C.F., Kendall, D.C., Flammang, T.J. and Kadlubar, F. F., 1986, Role of aromatic amine acetyltransferase in human colo-rectal cancer. Archives of Surgery, 121, 1259-1261.

Levy, G.A. and Weber, W.W., 1988, HPLC analysis of ^{32}P-postlabeled DNA-2-aminofluorene adducts. In Carcinogenic and Mutagenic Responses to Aromatic Amines and Nitroarenes, edited by C. M. King, L. J. Romano, and D. Schuetzle (New York: Elsevier), pp. 283-287.

Lotlikar, P.D. and Luha, L., 1971, Enzymatic N-acetylation of N-hydroxy-2-aminofluorene by liver cytosol from various species. Biochemical Journal, 123, 287-289.

Lower, G.M. and Bryan, G.T., 1973, Enzymatic N-acetylation of carcinogenic aromatic amines by liver cytosol of species displaying different organ susceptibilities. Biochemical Pharmacology, 22, 1581-1588.

Malejka-Giganti, D., Gutmann, H.R. and Rydell, R.E., 1973, Mammary carcinogenesis in the rat by topical application of fluorenyl hydroxamic acids. Cancer Research, 33, 2489-2497.

Mattano, S.S. and Weber, W.W., 1987, Kinetics of arylamine N-acetyltransferase in tissues from rapid and slow acetylator mice. Carcinogenesis 8, 133-137.

Mommsen, S., Borfod, M.M. and Aagaard, J., 1985, N-Acetyltransfer phenotypes in the urinary bladder carcinogenesis of a low risk population. Carcinogenesis, 6, 199-201.

Patriankos, C. and Hoffman, D., 1979, Chemical studies on tobacco smoke, LXIV. On the analysis of aromatic amines in cigarette smoke. Journal of Analytical Toxicology, 3, 150-154.

Peters, J.H. and Gutmann, H.R., (1955), The acetylation of 2-aminofluorene and the deacetylation and concurrent reacetylation of 2-acetylaminofluorene by rat liver slices. Journal of Biological Chemistry, 216, 713-726.

Philip, P.A., Fitzgerald, D.L., Cartwright, R.A., Peake, M.D. and Rogers, H.J., 1988, Polymorphic N-acetylation in lung cancer. Carcinogenesis, 9, 491-493.

Poirier, L.A., Miller, J.A. and Miller, E.C., 1963, The N- and ring hydroxylation of 2-acetylaminofluorene and the failure to detect N-acetylation of 2-aminofluorene in the dog. Cancer Research, 23, 790-800.

Radomski, J.L., Hearn, W.L., Radomski, T., Moreno, H. and Scott, W.E., 1977, Isolation of the glucuronic acid conjugate of N-hydroxy-4-aminobiphenyl from dog urine and its mutagenic activity. Cancer Research, 37, 1757-1762.

Saito, K., Shinohara, A., Kamataki, T. and Kato, R., 1985, Metabolic activation of mutagenic N-hydroxyarylamines by O-acetyltransferase in Salmonella typhimurium TA98. Archives of Biochemistry and Biophysics, 239, 286-295.

Schuetzle, D., 1983, Sampling of vehicle emissions for chemical analysis and biological testing. Environmental Health Perspectives, 47, 65-80.

Shinohara, A., Saito, K., Yamazoe, Y., Kamataki, T. and Kato, R., 1986, Acetyl coenzyme A-dependent activation of N-hydroxy derivative of carcinogenic arylamines:

Mechanism of activation, species difference, tissue distribution, and acetyl donor specificity. Cancer Research, 46, 4362-4367.

Smith, T.J. and Hanna, P.E., 1986, N-Acetyltransferase multiplicity and the bioactivation of N-arylhydroxamic acids by hamster hepatic and intestinal enzymes. Carcinogenesis, 7, 697-702.

Sugimura, T. and Sato, S., 1983, Mutagens-carcinogens in foods. Cancer Research, 43, 2415s-2421s.

Weber, W.W., 1987, The Acetylator Genes and Drug Response. (New York: Oxford University Press).

Weber, W.W. and Hein, D. W., 1985, N-Acetylation pharmacogenetics. Pharmacological Reviews, 37, 25-79.

Weeks, C.E., Allaben, W T., Tresp, N.M., Louie, S.C., Lazear, E.J. and King, C. M., 1980, Effects of structure of N-acyl-N-2-fluorenylhydroxylamines on hydroxamic acid acyltransferases and deacylase activities, and on mutations in Salmonella typhimurium TA1538. Cancer Research, 40, 1204-1211.

Wei, E.T. and Shu, H. P., 1983, Nitroaromatic carcinogens in diesel soot: A review of laboratory findings. American Journal of Public Health, 73, 1085-1088.

Westra, J.G., Flammang, T.J., Fullerton, N.F., Beland, F.A., Weis, C.C. and Kadlubar, F.F., 1985, Formation of DNA adducts in vivo in rat liver and intestinal epithelium after administration of the carcinogen 3,2'-dimethyl-4-aminobiphenyl and its hydroxamic acid. Carcinogenesis, 6, 37-44.

Young, J.F. and Kadlubar, F.F., 1982, A pharmacokinetic model to predict exposure of the bladder epithelium to urinary N-hydroxyarylamine carcinogens as a function of urine pH, voiding intervals, and resorption. Drug Metabolism and Disposition, 10, 641-644.

SULPHATE ESTER AND GLUCURONIC ACID CONJUGATES AS
INTERMEDIATES IN XENOBIOTIC METABOLISM

D.H. Hutson

Shell Research Ltd., Sittingbourne Research Centre,
Sittingbourne, Kent, ME9 8AG, UK

INTRODUCTION
One of the most common phase I biotransformations at aliphatic, alicyclic and aromatic carbon is hydroxylation to form an alcohol or phenol function. Subsequently, either or both of the common phase II reactions, glucuronidation and sulphation may occur and facilitate the removal of the primary products from the body. Following their discovery last century, glucuronic acid conjugation and sulphate conjugation have featured as major reactions in xenobiochemistry and most scientists working in the area have experienced these conjugates as terminal metabolites. Thus, both processes have the status of classical reactions in drug metabolism. Both also have important roles in normal biochemistry and have received much attention in this context. Consequently there exists a voluminous literature on their chemistry, biochemistry, enzymology and biological significance. Useful reference works are available for glucuronidation (Dutton, 1980), sulphation (Mulder, 1981) and their relative importance in the context of these and other conjugations (Paulson et al., 1986). Less literature is available on the toxicological consequences of the reactions because their major role in pharmacology has been viewed as detoxification, i.e. promoting the inactivation and elimination of drugs, etc. However, in common with the other conjugation processes, examples of bioactivation have been steadily accumulating.

Glucuronidation and sulphation are often discussed together because they compete for the same substrates in many cases. The substrate specificity is illustrated in Table 1. Both are reactions in which the endocon is bioactivated (to uridine-5'-diphospho-D-glucuronic acid, UDPGA and 3'-

phosphoadenosine-5'-phosphosulphate, PAPS, respectively).

TABLE 1 - Structural requirements for glucuronide and sulphate conjugation; R = alkyl; R* - aryl

SUBSTRATES	GLUCURONIC ACID	SULPHATE
ROH	+	+
R*OH	+	+
R*NHOH	+	+
R*(RCO)NOH	+	+
R=N-OH	+	+
RNH_2	+	+
$R*NH_2$	+	+
RR'NH	+	-
RR'R''N	+	-
RSH	+	?
R*SH	+	-
RCOOH	+	-
R*COOH	+	-

There are, however, some important differences between the two processes:
(i) glucuronyl transferase is a membrane-bound enzyme. It is located on the exterior side of the endoplasmic reticulum. This has consequences for both the formation and the disposition of the conjugates which are often directed to the intracisternal space and thence out of the cell via the Golgi complex. Sulphotransferase, on the other hand, is a soluble enzyme present in the cytosol. Consequently, sulphate esters will normally be available to interact with components within the cell in which they are formed.
(ii) sulphate ester formation tends to have a lower capacity than that of glucuronidation, consequently, if other factors are equal, the former predominates at low doses and the latter at high doses.
(iii) substrates for glucuronidation include carboxylic acids (to form 'ester' glucuronides). These conjugates possess interesting properties not shown by the glucuronides of alcohols, phenols, etc. (the so-called 'ether' glucuronides). There is no equivalent in sulphate conjugation because the product would be an unstable acid

anhydride, though it is interesting to note that the formation of such a product would initiate a futile cycle (the pseudo hydrolytic destruction of PAPS).
(iv) glucuronide and sulphate conjugates differ in their physical properties (e.g. molecular weights and pK_a values) and in their chemical properties. These differences are important in relation to their relative disposition and bioactivity.

The most well-known case of the intermediacy of glucuronide conjugates is the role that they play in the enterohepatic circulation of drugs and/or their metabolites. This is induced by biological instability (to the β-glucuronidases of gut microorganisms). A clue that sulphate esters could also be intermediary metabolites is to be found in the research of Young and coworkers in the 1960s who discovered that arylmethyl sulphate esters were substrates for glutathione transferases (Gilham, 1971). The early work of the Millers on the carcinogenic action of 2-acetylaminofluorene (see below) also pointed to an important intermediate role for sulphate conjugates.

A conjugate may have a dual role as an intermediary metabolite and as a terminal metabolite in the same system. Often the difference is purely a consequence of the kinetics of disposition. Thus a major proportion of a relatively stable conjugate may be eliminated in the urine. This chapter is concerned with that portion (which may range from 0 to 100%) which is not so eliminated.

SULPHATE ESTER CONJUGATES
Formation and reactivity

Sulphate conjugates are formed by the enzyme-catalyzed transfer of SO_3^- from PAPS to the acceptor alcohol, phenol or hydroxamic acid. Arylamines can also act as acceptors to form sulphamates. The sulphotransferases form a group of soluble enzymes with different substrate specificities. Relatively little research effort has been expended on these enzymes until recently in comparison with that on the glucuronyl and glutathione transferases. Consequently, few have been purified to homogeneity. However, several sulphotransferases are now recognized by the Enzyme Commission (Roy, 1981) and Jakoby and coworkers (Jakoby et al., 1984) have made significant advances of late in characterizing the enzymes and their substrate specificities. Two major groups have now been clearly defined (Jakoby et al., 1984): the arylsulphotransferases (EC 2.8.2.1) and the alcohol sulphotransferases (EC 2.8.2.2). The latter group contains the well-known 3-β-hydroxysteroid sulphotransferases.

Sulphate esters of cyclic and acyclic alcohols and of phenols are reasonably stable at pH 7.4 and are commonly found as urinary metabolites. Those of benzylic alcohols and hydroxamic acids, however, tend to be labile and degrade yielding carbonium ions and nitrenium ions, respectively (Figure 1). Reactivity is related to the degree of charge delocalization in these ions or in the transition state associated with their formation. The benzyl sulphates are generally the more stable of the two classes and can be prepared and used experimentally as substrates. The kinetics of the reactions of sulphate esters in neutral aqueous media are complex and are apparently intermediate between SN_1 and SN_2 mechanisms. Thus both C-O and O-S bond cleavage can occur (Benkovic and Benkovic, 1966). Some examples of the proposed or proven intermediacy of sulphate ester conjugates are given below.

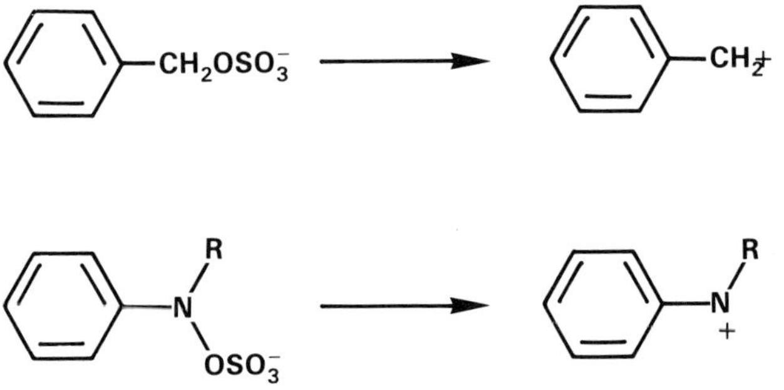

Figure 1. Carbonium ions and nitrenium ions formed from sulphate esters

Benzyl sulphates
Based on the known efficiency of 1-menaphthyl sulphate as a substrate for glutathione transferase (Gilham, 1971), a number of examples of sulphate esters as precursors of mercapturic acids have been identified. Benzyl acetate (used as a flavour and fragrance agent) is metabolized by hydrolysis, oxidation to benzoic acid and conjugation to N-benzoylglycine. A small proportion of the dose is excreted as benzyl mercapturic acid. The intermediacy of benzyl sulphate was proposed (Chidgey et al., 1986) and supported by the following evidence:

(i) co-administration of pyrazole (an inhibitor of alcohol dehydrogenase) increased the yield of benzyl mercapturic acid 11-fold,
(ii) pretreatment with pentachlorophenol (PCP) (an inhibitor of sulphotransferase activity) suppressed the formation of the mercapturate. These results indicate that benzyl sulphate was a precursor of the terminal metabolite. 4-Nitrotoluene is metabolised by rat hepatocytes to S-(4-nitrobenzyl)glutathione (de Bethizy and Rickert, 1984). The reaction is thought to occur via oxidation to 4-nitrobenzyl alcohol which is then sulphated to become the precursor for glutathione conjugation.

Polynuclear benzyl sulphates

In progressing from benzyl sulphate, through 1-menaphthyl sulphate to more complex polynuclear benzyl esters, greater reactivity would be expected. This would be gained either from greater stabilization of an SN_2 transition state or greater stabilization of (and tendency to form) a carbonium ion. Measurements of the half-lives in water at 37°C of a series of benzanthracene derivatives and a chrysene derivative (Table 2) would seem to confirm this (Watabe et al., 1987). All samples were potent mutagens.

TABLE 2 - Stabilities of polynuclear benzyl sulphates (sodium salts) in water at 37°C (Watabe et al., 1987)

Sulphate ester	Half-life (min)
Benz[a]anthracene-	
7-methyl-12-sulphooxymethyl-	<1
7-sulphooxymethyl-12-methyl-	<1
7-sulphooxymethyl-12-hydroxymethyl-	8
7-sulphooxymethyl-	3.5
Chrysene-	
5-sulphooxymethyl-	660

5-Methylchrysene (1) is as potent as benzo[a]pyrene in its carcinogenicity to mouse skin. It is metabolized by rat liver microsomes to 5-hydroxymethylchrysene (2), the major product, and to dihydrodiols and phenols. The carcinogenicity of 1 and 2 are approximately equal; chrysene has very little activity. Whilst the bay region dihydrodiol epoxide-type metabolites (cf. benzo[a]pyrene) have not been ruled out as ultimate carcinogens, there is strong evidence that 5-hydroxymethylchrysene sulphate (3) is a more potent genotoxicant. The microbial mutagenicity (S. typhimurium TA 98) of 2 was minimal with oxidative bioactivation, but was increased 5-6-fold when the 9000 g rat liver supernatant was replaced by dialyzed cytosol amended for PAPS generation. The sulphate ester (3), prepared biochemically, was a potent direct-acting mutagen. Furthermore, it covalently bound to calf thymus DNA through its 5-methyl carbon atom with loss of sulphate ion (Okuda et al., 1986). A study of this bioactivation in vivo is in progress. These reactions are summarized in Figure 2.

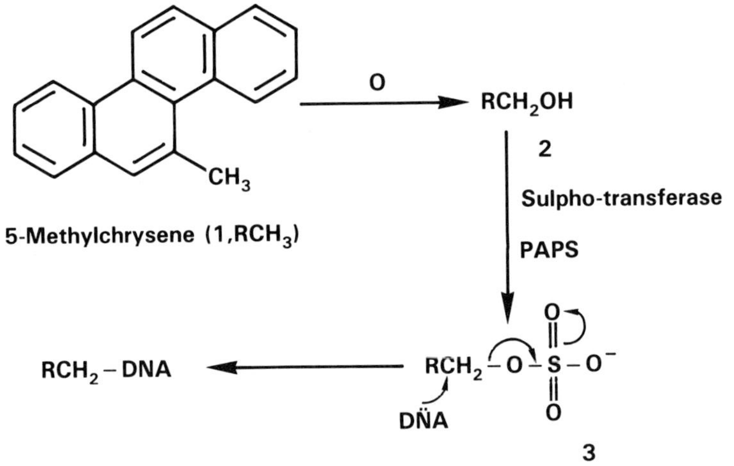

Figure 2. Metabolic activation of 5-methylchrysene to a DNA-reactive intermediate (Watabe et al, 1985)

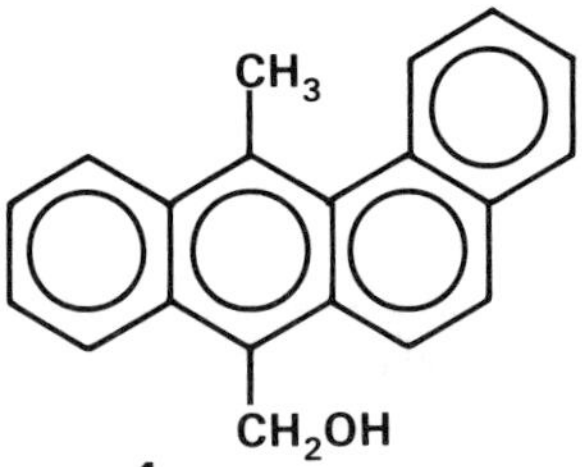

A similar study with 7-hydroxymethyl-12-methylbenz[a]-anthracene (4) by the same group (Watabe et al., 1985) revealed that this compound also bound to DNA via sulphation. $\underline{N}^2$-Guanine and $\underline{N}^6$-adenine adducts (via the methylene group) were identified. These adducts have recently been found in vivo in the livers of pre-weanling rats and mice given intraperitoneal injections of 7-hydroxymethyl- or 7-sulphooxymethyl-12-methylbenz[a]-anthracene (Surh et al., 1987). The sulphate ester afforded a several hundred-fold higher yield of adducts in comparison with the alcohol. Formation of adducts by the alcohol was reduced by pretreatment with the hydroxysteroid sulphotransferase inhibitor, dehydroepiandrosterone. Interestingly, pentachlorophenol was not a very effective inhibitor.

7,12-Dihydroxymethylbenz[a]anthracene presents a similar picture (Watabe et al., 1987, and earlier references therein) with the added interest that sulphation is regioselective for the 7-hydroxymethyl group.

The case for these esters as ultimate initiators of carcinogenesis has not yet been proven but they are undoubtedly good candidates for this role.

Alkenylbenzenes
Safrole (5), estragole (6) and methyleugenol (7) are carcinogenic prop-2-enylbenzenes which occur naturally with many other alkenylbenzenes in a variety of plant species that are used as sources of essential oils and spices. Safrole, for example, is a major component of oil of sassafras. These three compounds are weakly to moderately active hepatocarcinogens in rats and mice (Miller et al., 1983). Several related, but more highly substituted, analogues e.g. myristicin (8) and dill and parsley apiols (9 and 10) have little or no carcinogenic activity.

An obvious mechanism for the carcinogenic action of these compounds is oxygenation at the ethenyl bond to form 2,3-oxides. These metabolites are indeed mutagenic to S-typhimurium (Wislocki et al., 1977). However, metabolic and biochemical studies have revealed a mechanism involving sulphation. Safrole and estragole are hydroxylated at the 1'-position of the propenyl groups to form the 1'-hydroxy-derivatives. These metabolites are equally potent or more potent carcinogens than the parent chemicals (Miller et al., 1983). They are further metabolized to three types of electrophilic intermediate:1'-sulphooxy esters (11), 2',3'-epoxides (12) and 1'-oxo-derivatives (13). Several lines of research point to the sulphate esters as the ultimate

8, Myristicin, $R_1 = R_2 = H$
9, Dill apiol, $R_1 = OCH_3$, $R_2 = H$
10, Parsley apiol, $R_1 = H$, $R_2 = OCH_3$

carcinogens via their chemical reactivity (14). Firstly, only these esters form significant amounts of adducts with DNA in vivo (Phillips et al., 1981). The major of these (60%) was the 3'-isosafryl-deoxyguanosine adduct (15); the 1'-isomer comprised 20% of the adducts. 1'-Acetoxysafrole afforded the same adducts on incubation in vitro with DNA or with deoxyguanosine. 1'-Hydroxysafrole becomes covalently bound to RNA added to rat and mouse liver cytosols; this reaction was found to be PAPS-dependent (Wislocki et al., 1976). The sulphotransferase inhibitor pentachlorophenol (PCP) has again proved useful in implicating sulphate esters as intermediates. When 12-day old male B6C3F$_1$ mice were dosed with [2',3'-^{3}H]-1'-hydroxysafrole 45 minutes after a single dose of 0.04 µmoles of PCP per g body weight, DNA adducts were reduced to 15% of control levels. The average number of hepatomas per mouse at 10 months was reduced to less than 10% (see Table 3) (Boberg et al., 1983).

TABLE 3 - Carcinogenicity of 1'-hydroxysafrole (1'HS): effect of pretreatment with pentachlorophenol (PCP) on hepatic DNA adduct formation and on hepatoma initiation

Dose of 1'-HS (µmoles/g body weight)	Dose of PCP (0.04 µmoles/g body weight)	DNA adducts pmoles/mg	Hepatoma incidences (%)
0.2	-	190	97
0.2	+	24	10
0.1	-	6	86
0.1	+	9	12
0.2	- (normal mice)	110	43
0.2	- (brachymorphic mice)	16	6

When PAPS-deficient brachymorphic mice (B6C3F$_2$, 12 days old) were treated with 1'-hydroxysafrole, the level of DNA adducts in livers was only 15% of that found in their normal littermates (Table 3) and the hepatoma incidence was similarly reduced (Boberg et al., 1983). Taken together, these findings afford compelling evidence for the involvement of sulphation (after hydroxylation) in the carcinogenic action of safrole and its analogues. A recent study (Boberg et al., 1987), using the N,N-diethylnitrosamine-induced enzyme-altered foci test for promoting agents, suggests that sulphation is also important in the promoting activity of 1'-hydroxysafrole. Promotion was inhibited by treatment with PCP.

Miller et al. (1983) are of the opinion that, as the alkenylbenzenes occur naturally and as food additives at very low concentrations in total food intake, they probably make only a minor contribution to the exposure of man to exogenous carcinogens. In this context it is important to note that the metabolism of estragole is dose-dependent (Anthony et al., 1987). 1'-Hydroxy-estragole is excreted (as its glucuronide) at 1% of a dose of 0.05 mg/kg estragole, but at 11% of a dose of 500 mg/kg. However, further work using sensitive methods to detect reaction with DNA in vivo, e.g. the post-labelling technique of Randerath et al. (1985), is needed to assess the hazard posed by these chemicals.

2-Acetylaminofluorene (2AAF) and 2-aminofluorene (2AF)

The metabolism and mechanism of carcinogenic action of 2AAF and 2AF have been intensively studied by many workers for more than 20 years (Miller and Miller, 1981). They must vie with benzo[a]pyrene as the best studied model carcinogens. (One can only speculate whether or not progress would have been more rapid if two simpler models had been chosen 25 years ago).

N-Hydroxylation of 2AAF by cytochrome P450 to N-OH-2AAF is obligatory for carcinogenic action. The potential for the involvement of sulphation in its action was discovered around 1970. The sulphate ester of N-OH-2AAF is a highly reactive metabolite (Smith et al., 1987; van den Goorbergh et al., 1987) which reacts with water, protein, RNA and DNA and is mutagenic. These reactions occur because the sulphate group is a good leaving group in this situation and the resulting positive ion is stabilized by charge delocalization (forming carbonium-nitrenium canonical structures). The metabolism and action of 2AAF, however, appear still to be very complex involving N-hydroxylation, sulphation, de-sulphation, acetylation, de-acetylation and glutathione conjugation. This complexity is succinctly reviewed by Mulder et al. (1988) who, like the Millers, have carried out in-depth research on this compound. Perhaps the most convincing recent evidence for the involvement of sulphation in the mode of action of 2AF is that produced by the Millers (Lai et al., 1987) using PCP and brachymorphic mice as described above for safrole. Thus, the sole adduct in liver derived from the treatment of B6C3F$_1$ mice with 0.04 - 0.06 μmoles of N-OH-2AF per g body weight [N-(deoxyguanosin-8-yl)-2-aminofluorene] was found (9 hours after a single intraperitoneal dose) to be 1.0 - 1.7 pmoles/mg DNA. Pretreatment with a single i.p. dose of PCP (0.04 μmoles/g body weight) decreased this level by more than 80%. Furthermore, the livers of infant male brachymorphic B6C3F$_2$ mice contained only 0.3 pmoles of this adduct per mg DNA following an i.p. dose of 0.06 μmoles of N-OH-2AF; their normal littermates afforded 1.9 pmoles per mg DNA. In a tumorigenicity study, single i.p. doses of N-OH-2AAF of 0, 0.015, 0.03, 0.06 or 0.12 μmoles per g body weight induced, at 10 months, an average of 0.2, 2.5, 7, 11 or 14 hepatomas per B6C3F$_1$ mouse. Pretreatment with PCP reduced this incidence by more than 80%. This, together with other excellent work over 20 years, provides strong evidence for (i) the involvement of sulphation in the carcinogenicity of 2AF (and 2AAF) and (ii) the reaction of the nitrenium ion (16) at position 8 of a deoxyguanosine group in DNA, to form (17), as the initiation reaction.

4-Aminoazobenzene

Mono- and di-N-methyl derivatives of 4-aminoazobenzene (AAB) are hepatocarcinogens in rats and mice and AAB itself is active in young mice (Delclos et al., 1984). The role of sulphation in its action has been indicated for some time by experimentation and by analogy with AAF and AF. Once again, the use of PCP and brachymorphic mice has recently shown (Miller and Miller, 1986) that sulphation is critical to the formation of N-(deoxyguanosin-8-yl)-4-aminoazobenzene adducts in liver and for hepatoma formation.

Sulphation of other N-hydroxy groups

Sulphate esters of the N-hydroxy derivatives of singlering anilines in general do not possess the spectacular properties associated with the polynuclear analogues. However, sulphation may play a role in the toxic action (renal damage, etc.) caused by phenacetin (4-ethoxyacetanilide, 18) (Mulder et al., 1978). More recent studies, however, indicate that 4-nitrosophenetole (19) may be involved (Hinson and May, 1985). Certain other types of -N-O-SO$_3$ derivatives are stable metabolites and would not be expected to cause tissue damage. Examples of these include the sulphate esters of minoxidil (20) (Johnson et al., 1982) and 2-cyanoethylthioacetimidate (21) (Hutson et al., 1971).

16

17

18

19

20

21

Aliphatic hydroxylation and sulphation of acetanilides

Sulphation may be involved in the fate of N,N-dimethyl-4-cyanoaniline (CDA, 22, Figure 3) in rats and mice. An intermediate metabolite of CDA, 4-cyanoacetanilide (23), was bioactivated in the side-chain to a mercapturic acid precursor. When rats were given a single oral dose (18.5 mg/kg) of ^{14}C-CDA, 10% of the dose was excreted in the urine as the mercapturate, N-acetyl-S-[2-keto-2-(4-cyanoanilino)ethyl]cysteine (26, Figure 3). To explain the presence of this metabolite and of smaller amounts of 4-cyano-N-methylsulphinylacetylaniline (27) and its methylsulphonyl analogue (28), the intermediacy of a sulphate conjugate was postulated as shown in Figure 3 (Hutson et al., 1984). The C-2 hydroxylation of an N-acetyl group has precedents but the subsequent sulphation has not been unequivocally demonstrated. The toxicological significance of the formation of this conjugate is also unknown, but the exploitation of a species difference provides an indication that it may be important. The mercapturic acid (26) is not formed in mice (Logan et al., 1985). This is because the intermediate 4-cyanoaniline is not acetylated in this species; furthermore, the side-chain hydroxylation and, therefore, subsequent bioactivation also fails to occur when 4-cyanoacetanilide (23) is dosed to mice. It was of interest, therefore, to find that the binding of radioactivity derived from ^{14}C-CDA to hepatic DNA is 6-fold higher in rats than in mice (Waters et al., 1985). This suggests that the reaction sequence may be toxicologically significant with certain acetanilides.

We have recently attempted to confirm the involvement of sulphation in the route from CDA to the mercapturate (26) in the rat. Pretreatment of rats with PCP reduced the yields of 26 from 10% to 5% and of 27 from 1.4% to 0.7% (Hesk et al., 1988). The fate of 4-cyanoacetanilide with ^{14}C and ^{13}C labels in the acetyl group has also been investigated. The expected mercapturate (26) was formed. In addition, a polar metabolite fraction was slowly converted by incubation with arylsulphatase to 4-cyano-N-[1-^{14}C,1,2-^{13}C]glycolylaniline (24). This provides further evidence for the intermediacy of a sulphate conjugate (25) in the metabolism of the acetyl group of an acetanilide. Thus far, the reaction has not been reported for other acetanilides (e.g. phenacetin and acetaminophen). It would be surprising, however, if the reaction were restricted solely to 4-cyanoacetanilide.

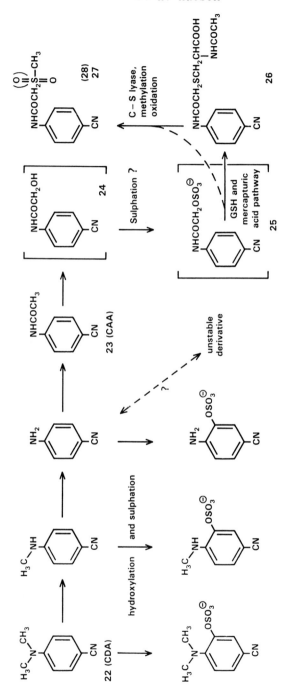

Figure 3. Metabolism of 4-cyano-N,N-dimethylaniline and 4-cyanoacetanilide to a mercapturic acid via a sulphate ester intermediate. (Hutson et al, 1984; Logan et al, 1985)

GLUCURONIC ACID CONJUGATES
Formation and reactivity

Glucuronic acid conjugates are formed by the enzyme-catalyzed transfer of the D-glucopyranosiduronyl group from UDPGA to the acceptor substrate (alcohol, phenol, hydroxylamine, amine, thiol, etc.). The glucuronyl transferases (EC 2.4.1.17) are located in the endoplasmic reticulum where they are ideally located to conjugate the hydroxylated metabolites of lipophilic chemicals. Indeed, the topographical relationship of the cytochromes P450 and the glucuronyltransferases is such that many lipophilic phase I metabolites are scarcely available within the cell to exert any biological action because they may leave the cell via the intracisternal space.

Glucuronic acid conjugates are more stable than sulphate conjugates. As the nature of the aglycone (exocon) changes from alcohol, to phenol, etc. to arylhydroxylamine and arylamine, some instability is observed. It is of consequence but it is generally of less toxicological significance than is the case with the analogous sulphates. An important difference between the two classes lies in the formation of acyl glucuronides (cf. acyl sulphates alluded to in the Introduction). Another difference is the somewhat higher molecular weight and greater tendency, at least in the rat, for biliary elimination (and thence enterohepatic circulation) of glucuronide conjugates.

Enterohepatic circulation

Much has been written about this interesting phenomenon (in which a conjugate is eliminated in the bile, deconjugated in the intestine and resorbed for further possible pharmacological action and metabolism) and it need not be repeated at length here. It is important to note, however, that this process affords one of the most common intermediary metabolites in xenobiochemistry: the $\underline{O}$-glucuronide conjugate. Enterohepatic circulation derives from two features: the tendency of glucuronides to be eliminated in the bile and the high activity of β-glucuronidase in gut flora. Often the glucuronide is not seen as a terminal metabolite but its formation is vital for the elimination of the ingested chemical or its metabolite. A variant was noted with 3-phenoxybenzoic acid. This pyrethroid metabolite is 4'-hydroxylated, eliminated in the bile as a mixture of glucuronides (4'-$\underline{O}$ and carboxylic), deconjugated, reabsorbed, sulphated and eliminated as the 4'-sulphoxy-derivative (Huckle et al., 1981). This process, enterohepatorenal disposition, may be quite common and is dependent on the relative availability of the re-absorbed substrate for sulphation and glucuronidation. Phenols are available to the hepatic cytosol for sulphation (or may even be sulphated during intestinal reabsorption). Perhaps the most important feature of enterohepatic circulation is that, when it occurs with a parent drug, therapeutic action (and potential toxicity) can be exerted for a much longer period than in the absence of such recycling.

Acyl glucuronides
These conjugates exhibit some interesting properties in addition to enterohepatic recycling.

Intramolecular acyl migration. During a study of the metabolism of the pyrethroid insecticide cypermethrin in the late 1970s, we encountered an apparent pH-dependent production of artefacts during the storage of urine containing a cyclopropanecarboxylic acid glucuronide. While we were puzzling over this problem, Heirwegh, Compernolle and coworkers (Compernolle et al., 1978; Blanckaert et al., 1978) presented the likely solution by describing the rearrangement of bilirubin IX monoglucuronide at pH values around 8-8.5 to a mixture of the 2-, 3- and 4-$\underline{O}$ isomers. Such intramolecular rearrangements are well-known in carbohydrate chemistry (Haines, 1976) where they are often utilized in synthetic reactions. Further reports followed. Probenecid affords a mixture of such isomers in the urine of treated humans (Eggers and Doust, 1981). Sinclair and Caldwell (1981, 1982) reported the pH-dependent rearrangement of the ester glucuronic acid conjugate of clofibric acid (29, 1-$\underline{O}$-clofibryl β-D-glucopyranosiduronate).

Also in 1982, Janssen et al. (1982) reported the rearrangement of the ester glucuronide of Wy-18251 [30, 3-(4-chlorophenyl)thiazolo[3,2-a]benzimidazole-2-acetic acid]. This carboxylic acid forms a quaternary $\underline{N}$-glucuronide in rhesus monkey but the ester glucuronide is the more general product (dog, rat and monkey). The glucuronide is unstable and above pH 6.5 is converted to a mixture of reducing sugars (i.e. C1-OH of the pyranose ring unsubstituted). These were considered to be the 2-, 3- and 4-positional isomers. Valproic acid (31, an antiepileptic drug) is also excreted as its acyl glucuronide in animals and man. It behaves similarly to those described above (Dickinson et al., 1984). Treatment for 3 hours at pH values between -0.8 and 12.9 revealed stability between 3 and 7 but rearrangements between 0 and 3 and between 7 and 11. GLC, trimethylsilylation and GC/MS revealed 7 products. Six were structural isomers and one was a dehydrated product. At the extreme pH values the situation becomes

Sulphation and glucuronidation 195

complicated by pyranose ring-opening, mutarotation and lactonization and more products than the simple 2-, 3- and 4-isomers would be expected. These other products, and the furanose ring structures reported from clofibryl glucuronide (Hignite et al., 1981) may have less biological significance.

Zomepirac (a non-steroidal anti-inflammatory drug) is eliminated in animals and man mainly as its acyl glucuronide conjugate. This metabolite has been shown by Benet and coworkers also to undergo acyl migration (Hasegawa et al., 1982). The products were revealed as isomers by fast-atom bombardment mass spectrometry but more recently (Smith and Benet, 1986) they have been unequivocally identified by high field proton NMR. The 2-, 3- and 4-$\underline{O}$-acyl isomers, each as α- and β-anomers (C_1-OH), were present after treatment of pure zomepirac glucuronide in 0.01M sodium phosphate buffer (pH 7.5) for 60 minutes at 37°C.

The significance of the acyl migration in glucuronic acid conjugates derives from the chemistry of the products: they are not glucuronides (i.e. they are not glycosides) and are, therefore, not labile to β-glucuronidase. This has consequences for conjugate disposition and at the purely technical level may lead to errors. β-Glucuronidase still plays a major part in the identification of glucuronides. These β-glucuronidase-resistant isomers can easily be characterized as "non-glucuronide polar conjugates of unknown structure", masking the possible 100% identity as one glucuronide conjugate when originally excreted. The consequence for drug disposition is equally (or maybe more) important. This is exemplified by a comparative study of two anti-inflammatory agents Wy-18251 (30) and Wy-41770 (32). The latter has a plasma half-life of 10 hours and an intestinal transport time of 3 days. The 32-glucuronide is much more resistant than 30-glucuronide to acyl migration. It is eliminated in the bile and undergoes classical enterohepatic recycling. 30-Glucuronide, on the other hand, undergoes acyl migration readily. The products are not cleaved by the β-glucuronidases of intestinal microorganisms and consequently the aglycone is not re-absorbed. Instead the rearranged conjugates accumulate in the caecum. There they are hydrolysed by microbial esterases (hydrolysis can be prevented by the administration of antibiotics; it cannot be inhibited by saccharonolactone, the β-glucuronidase inhibitor). The parent drug liberated at this point is excreted in the faeces. The result of this sequence of events is that the plasma half-life of Wy-18251 is only 2.9 hours and the intestinal transit time is about 24 hours (Ruelius et al., 1985). These results demonstrate that the chemistry of the acyl aglycone may be as important as the physiological and biochemical characteristics of the host.

Clearly acyl migration must be considered both in xenobiotic metabolism studies and in drug structure-activity relationships.

Electrophilic reactivity of acyl glucuronides

Intramolecular acyl migration reveals the electrophilic nature of the carbonyl group and its susceptibility to attack by the neighbouring hydroxyl group. Although the migration appears to be a stereochemically facile reaction this is not particularly so in the β-glucopyranoside series. All of the substituents are equatorially disposed and the cyclic intermediates are quite strained. (Acyl migration would be much more rapid in the β-mannopyranoside series where the C-2 hydroxyl is axial and in a 'cis' orientation with the C-1 and C-3 oxygen atoms). Given this situation, it perhaps should be no surprise to find other nucleophiles reacting with the carbonyl group of acyl glucuronides. Such reactions occur; for example, bilirubin glucuronide is transesterified with methanol (Salmon et al., 1975) (and acyl glucuronides decompose when methylated using diazomethane). Thus caution is required when alcohols are used in systems to extract acyl glucuronides from biological systems. Intermolecular reactions may also occur in vivo. The hypolipidemic drug clofibrate (see 29 for part structure) is eliminated in the urine partly as clofibryl mercapturic acid (33) as well as the ester glucuronide conjugate (29) (Stogniew and Fenselau, 1982). Experiments with synthetic 29 showed that it reacts with ethanethiol under physiological conditions in vitro and with the 4-nitrobenzylpyridine (NBP) reagent (used to detect alkylating agents). The source of the clofibryl mercapturate (33) is postulated to be the glucuronide (29) which reacts with glutathione and thence via the mercapturic acid pathway to the terminal product.

The acylation of albumin by 1-O-acyl glucuronides has also been reported (van Breeman and Fenselau, 1985). When the acyl glucuronides of the following acids were incubated with albumin (ratio 1.7:1.0) for 12 hours at 37°C the amounts of acylation (mmoles per mole albumin) were: flufenamic acid, 32; indomethacin, 43; clofibric acid, 14; benoxaprofen, 59. The glucuronides reacted with the NBP reagent at rates which correlated with their instabilities in aqueous solution but not with the pK_a values of the parent acids (van Breeman and Fenselau, 1986). Twenty-two other 1-O-acyl glucuronides were detected as electrophilic by reaction with the NBP reagent on TLC plates. These conjugates are not potent acylating agents but clearly they react with glutathione in vivo (the clofibrate experience) and with proteins and could potentially initiate cytotoxicity, immunogenicity and possibly mutagenicity. There is, as yet, no unequivocal evidence for the 1-O-acyl glucuronides as toxic intermediary metabolites. Clofibrate, however, is not well tolerated at high doses; such an intermediate could be involved. Furthermore, zomepirac has recently been withdrawn from use because of a high incidence of immunological reactions. Zomepirac glucuronide and its acyl-migrated isomers react with human serum albumin in vitro, causing acylation of the albumin. Zomepirac itself, when given to human volunteers as single 100 mg oral doses, irreversibly binds to albumin in vivo. The bond is base-labile and probably formed by acylation (Smith et al., 1986).

A large number of carboxylic acid drugs, herbicides and other xenobiotics have proved to be acceptable in lengthy toxicity trials and it would seem that the potential reactivity of the acyl glucuronides does not cause extensive problems. However, this is clearly a phenomenon which deserves further study and a higher profile.

Arylhydroxylamine O-glucuronides

In contrast to the arylhydroxylamine O-sulphates, the N-O-glucuronides are relatively stable because the glucuronyloxy group is a poor leaving group. The glucuronide of N-OH-2AAF, for example, is stable at pH7. However, when deacetylated to the N-O-glucuronide of 2AF, reactivity increases and binding to DNA has been demonstrated in vitro (Cardona and King, 1976). The involvement of this reaction in carcinogenicity has not been demonstrated.

Arylamine N-glucuronides

These compounds appear to be relatively stable transport forms of proximate carcinogens whose target toxicity is controlled by the hydrolysis of the conjugate at or in the target organ. The bladder carcinogen 2-naphthylamine (2-NA) is oxidized in the liver to N-OH-2NA and then converted to its N-glucuronide. This is transported to the kidney and actively secreted. In weakly acidic urine (e.g. of man and dogs) the N-glucuronide decomposes to 2-naphthylhydroxylamine which forms the reactive nitrenium ion directly or via further biochemical activation (e.g. sulphation) (Kadlubar et al., 1981) (see also Chapter 9).

1- and 2-Naphthylamines and 4-aminobiphenyl are directly N-glucuronylated (incidentally, specifically by the 3-α-hydroxysteroid UDP-glucuronyl transferase). This process may be a transport mechanism for the arylamines to the bladder and the intestine (Green and Tephly, 1987).

CONCLUSIONS

Clearly, these most 'traditional' of conjugates, the sulphate esters and the glucuronides, offer some interesting problems in detection, quantitative analysis and identification. The chemical reactivities of some members of both classes, particuarly the sulphates, confer toxic properties on the conjugates per se. The acyl glucuronides as a class may soon overhaul the sulphate esters in research interest if they are shown to have a role in the adverse reactions exhibited by some carboxylic acid analgesics and anti-inflammatory drugs. In addition, the disposition of the conjugates around the body as transport forms of proximate toxicants explains some forms of target organ toxicity. Even when found as terminal metabolites, the significance of that proportion which has escaped measurement must be considered.

REFERENCES

Anthony, A., Caldwell, J., Hutt, A.J. and Smith, R.L., 1987, Metabolism of estragole in rat and mouse and influence of dose size on excretion of the proximate carcinogen 1'-hydroxyestragole. Food and Chemical Toxicology, 25, 799-806.

Benkovic, S.J. and Benkovic, P.A., 1966, Studies on sulfate esters. I. Nucleophilic reactions of amines with p-nitrophenyl sulfate. Journal of the American Chemical Society. 88, 5504-5511.

de Bethizy, J.D. and Rickert, D.E., 1984, Metabolism of nitrotoluenes by freshly isolated Fischer 344 rat hepatocytes. Drug Metabolism and Disposition, 12, 45-50.

Blanckaert, N., Compernolle, F., Leroy, P., van Houtte, R., Fevery, J. and Heirwegh, K.P.M., 1978, The fate of bilirubin-IX glucuronide in cholestasis and during storage in vitro: intramolecular rearrangement to positional isomers of glucuronic acid. Biochemical Journal, 171, 203-214.

Boberg, E.W., Liem, A., Miller, E.C. and Miller, J.A., 1987, Inhibition by pentachlorophenol of the initiating and promoting activities of 1'-hydroxysafrole for the formation of enzyme-altered foci and tumours in rat liver, Carcinogenesis, 8, 531-539.

Boberg, E.W., Miller, E.C., Miller, J.A., Poland, A. and Liem, A., 1983, Strong evidence from studies with brachymorphic mice and pentachlorophenol that 1'-sulfooxysafrole is the major ultimate electrophilic and carcinogenic metabolite of 1'-hydroxysafrole in mouse liver. Cancer Research, 43, 5163-5173.

van Breeman, R.B. and Fenselau, C., 1985, Acylation of albumin by 1-O-acylglucuronides. Drug Metabolism and Disposition, 13, 318-320.

van Breeman, R.B. and Fenselau, C.C., 1986, Reaction of 1-O-acylglucuronides with 4-(p-nitrobenzyl)pyridine. Drug Metabolism and Disposition, 14, 197-201.

Cardona, R.A. and King, C.M., 1976, Activation of the O-glucuronide of the carcinogen N-hydroxy-N-2-fluorenylacetamide by enzymatic deacetylation in vitro: formation of fluorenylamine-tRNA adducts. Biochemical Pharmacology, 25, 1051-1056.

Chidgey, M.A.J., Kennedy, J.F. and Caldwell, J., 1986, Studies on benzyl acetate. II. Use of specific metabolic inhibitors to define the pathway leading to the formation of benzylmercapturic acid in the rat. Food and Chemical Toxicology, 24, 1267-1272.

Compernolle, F., van Hees, G.P., Blanckaert, N. and Heirwegh, K.P.M., 1978, Glucuronic acid conjugates of bilirubin-IX in normal bile compared with post-obstructive bile: transformation of the 1-O-acyl-glucuronide into 2-,3-, and 4-O-acylglucuronides. Biochemical Journal, 171, 185-201.

Delclos, K.B., Tarpley, W.G., Miller, E.C. and Miller, J.A., 1984, 4-Aminoazobenzene and N,N-dimethylaminoazobenzene as equipotent hepatic carcinogens in male C57BL/6x C3H/HeF$_1$ mice and characterization of N-(deoxyguanosin-8-yl)-4-aminoazobenzene as the major persistent hepatic DNA-bound dye in these mice. Cancer Research, 44, 2540-2550.

Dickinson, R.G., Hooper, W.D. and Eadie, M.J., 1984, pH-Dependent rearrangement of the biosynthetic ester glucuronide of Valproic acid to β-glucuronidase-resistant forms. Drug Metabolism and Disposition, 12, 247-252.

Dutton, G.J., 1980, Glucuronidation of Drugs and Related Compounds (Boca Raton, Fl.: CRC Press).

Eggers, N.J. and Doust, K., 1981, Isolation and identification of probenecid acyl glucuronide. Journal of Pharmacy and Pharmacology, 33, 123-124.

Gilham, B., 1971, The reaction of aralkyl sulphate esters with glutathione catalysed by rat liver preparations. Biochemical Journal, 121, 667-672.

van den Goorbergh, J.A.M., de Wit, H., Tijdens, R.B., Mulder, G.J. and Meerman, J.H.N., 1987, Prevention by thioethers of the hepatotoxicity and covalent binding to macromolecules of N-hydroxy-2-acetylaminofluorene and its sulfate ester in rat liver in vivo and in vitro. Carcinogenesis, 8, 275-279.

Green, M.D. and Tephly, T.R., 1987, N-Glucuronidation of carcinogenic aromatic amines catalyzed by rat hepatic microsomal preparations and purified rat liver uridine 5'-diphosphate-glucuronyltransferases. Cancer Research, 47, 2028-2031.

Haines, A.H., 1976, Relative reactivities of hydroxyl groups in carbohydrates. Advances in Carbohydrate Chemistry and Biochemistry, 33, 11-109.

Hasegawa, J., Smith, P.C. and Benet, L.Z., 1982, Apparent intramolecular acyl migration of zomepirac glucuronide. Drug Metabolism and Disposition, 10, 469-473.

Hesk, D., Hutson, D.H. and Logan, C.J., 1988, The fate of 4-cyanoacetanilide in rats and mice; mechanism of formation of a novel electrophilic metabolite. Xenobiotica, 18, 955-966.

Hignite, C.E., Txchanz, C., Lemons, S., Wiese, H., Azarnoff, D.L. and Huffman, D.H., 1981, Glucuronic acid conjugates of clofibrate: four isomeric structures. Life Sciences, 28, 2077-2081.

Hinson, J.A. and Mays, J.B., 1986, p-Nitrosophenetole: a reactive intermediate of phenacetin that binds to protein. In Biological Reactive Intermediates III, edited by J.J. Kocsis, D.J. Jollow, C.M. Witmer, J.O. Nelson and R. Snyder (New York: Plenum Press) pp. 691-696.

Huckle, K.R., Chipman, J.K., Hutson, D.H. and Millburn, P., 1981, Metabolism of 3-phenoxybenzoic acid and the enterohepatorenal disposition of its metabolites in the rat. Drug Metabolism and Disposition, 9, 360-368.

Hutson, D.H., Hoadley, E.C. and Pickering, B.A., 1971, The metabolism of S-2-cyanoethyl-N-[(methylcarbamoyl)oxy] thioacetimidate, an insecticidal carbamate, in the rat. Xenobiotica, 1, 179-191.

Hutson, D.H., Lakeman, S.K. and Logan, C.J., 1984, The fate of 4-cyano-N,N-dimethylaniline in rats; a novel involvement of glutathione in the metabolism of anilines. Xenobiotica, 14, 925-934.

Jakoby, W.B., Duffel, M.W., Lyon, E.S. and Ramaswamy, S., 1984, Sulfotransferases active with xenobiotics - comments on mechanism. In Progress in Drug Metabolism, edited by J.W. Bridges and L.F. Chasseaud (London: Taylor & Francis), Vol. 8, pp. 11-33.

Janssen, F.W., Kirkman, S.K., Fenselau, C., Stogniew, M. Hofmann, B.R. Young, E.M. and Ruelius, H.W., 1982, Metabolic formation of N- and O- glucuronides of 3-

(p-chlorophenyl)thiazolo[3,2-a]benzimidazole-2-acetic acid. Drug Metabolism and Disposition, 10, 599-604.

Johnson, G.A., Barsuhn, K.J. and McCall, J.M., 1982, Sulfation of minoxidil by liver sulfotransferase. Biochemical Pharmacology, 31, 2949-2954.

Kadlubar, F.F., Unruh, L.E., Flammang, T.J., Sparks, D., Mitchum, R.K. and Mulder, G.J., 1981, Alteration of urinary levels of the carcinogen, N-hydroxy-2-naphthylamine, and its N-glucuronide in the rat by control of urinary pH, inhibition of metabolic sulfation, and changes in biliary excretion. Chemico-Biological Interactions, 33, 129-147.

Lai, C-C., Miller, E.C., Miller, J.A. and Liem, A., 1987, Initiation of hepatocarcinogenesis in infant male B6C3F$_1$ mice by N-hydroxy-2-aminofluorene or N-hydroxy-2-acetylaminofluorene depends primarily on metabolism to N-sulfooxy-2-aminofluorene and formation of DNA-(deoxyguanosin-8-yl)-2-aminofluorene adducts. Carcinogenesis, 8, 471-478.

Logan, C.J., Hesk, D. and Hutson, D.H., 1985, The fate of 4-cyano-N,N-dimethylaniline in mice: the occurrence of a novel metabolite during N-demethylation of an aromatic amine. Xenobiotica, 15, 391-397.

Miller, E.C. and Miller, J.A., 1981, Searches for ultimate chemical carcinogens and their reactions with cellular macromolecules. Cancer, 47, 2327-2345.

Miller, J.A. and Miller, E.C., 1986, Electrophilic sulfuric acid ester metabolites as ultimate carcinogens. In *Biological Reactive Intermediates III*, edited by J.J. Kocsis, D.J. Jollow, C.M. Witmer, J.O. Nelson and R. Snyder (New York: Plenum Press) pp. 583-595.

Miller, E.C., Swanson, A.B., Phillips, D.H., Fletcher, T.L., Liem, A. and Miller, J.A., 1983, Structure-activity studies of the carcinogenicities in the mouse and rat of some naturally occurring and synthetic alkenylbenzene derivatives related to safrole and estragole. Cancer Research, 43, 1124-1134.

Mulder, G.J., 1981, *Sulfation of Drugs and Related Compounds* (Boca Raton, Fl.: CRC Press).

Mulder, G.J., Hinson, J.A. and Gillette, J.R., 1978, Conversion of the N-O-glucuronide and N-O-sulphate conjugates of N-hydroxyphenacetin to reactive intermediates. Biochemical Pharmacology, 27, 1641-1649.

Mulder, G.J., Kroese, E.D. and Meerman, J.H.N., 1988, The generation of reactive intermediates from xenobiotics by sulfate conjugation, and their role in drug toxicity.

In Metabolism of Xenobiotics edited by J.W. Gorrod,
H. Oelschlager and J. Caldwell (London: Taylor & Francis),
pp. 243-250.

Okuda, H., Hiratsuka, A., Nojima, H. and Watabe, T., 1986,
A hydroxymethyl sulphate ester as an active metabolite
of the carcinogen, 5-hydroxymethylchrysene. Biochemical
Pharmacology, 35, 535-538.

Paulson, G.D., Caldwell, J., Hutson, D.H. and J.J. Menn, 1986,
Xenobiotic Conjugation Chemistry, ACS Symposium Series
No. 299 (Washington,DC:American Chemical Society).

Phillips, D.H., Miller, J.A., Miller, E.C. and Adams, B.N.,
1981, N^2 atom of guanine and N^6 atom of adenine
residues as sites for covalent binding of metabolically
activated 1'-hydroxysafrole to mouse liver DNA in vivo.
Cancer Research, 41, 2664-2671.

Randerath, K., Randerath, E., Agrawal, H.P., Gupta, R.C.,
Schurdak, M.E. and Reddy, M.V., 1985, Postlabeling
methods for carcinogen-DNA adduct analysis. Environmental
Health Perspectives, 62, 57-65.

Roy, A.B., 1981, The chemistry of sulfate esters and related
compounds. In Sulfation of Drugs and Related Compounds,
edited by G.J. Mulder (Boca Raton, Fl.: CRC Press),
pp. 5-30.

Ruelius, H.W., Young, E.M., Kirkman, S.K., Schillings, R.T.,
Sisenwine, S.F. and Janssen, F.W., 1985, Biological fate
of acyl glucuronides in the rat. The role of rearrangement, intestinal enzymes and reabsorption. Biochemical
Pharmacology, 34, 451-452.

Salmon, M., Fenselau, C., Cukier, J.O. and Odell, G.B., 1975,
Rapid transesterification of bilirubin glucuronides in
methanol. Life Sciences, 15, 2069-2078.

Sinclair, K.A. and Caldwell, J., 1981, The pH-dependent
intramolecular rearrangement of glucuronic acid
conjugates of xenobiotics. Biochemical Society
Transactions, 9, 215.

Sinclair, K.A. and Caldwell, J., 1982, The formation of
β-glucuronidase-resistant glucuronides by the intramolecular rearrangement of glucuronic acid conjugates at
mild alkaline pH. Biochemical Pharmacology, 31, 953-957.

Smith, P.C. and Benet, L.Z., 1986, Characterisation of the
isomeric esters of zomepirac glucuronide by proton NMR.
Drug Metabolism and Disposition, 14, 503-505.

Smith, P.C., Mcdonagh, A.F. and Benet, L.Z., 1986,
Irreversible binding of zomepirac to plasma protein
in vitro and in vivo. Journal of Clinical Investigations,
77, 934-939.

Smith, B.A., Springfield, J.R. and Gutman, H.R., 1987, Solvolysis and metabolic degradation, by rat liver, of the ultimate carcinogen, $\underline{N}$-sulfonoxy-2-acetylaminofluorene. Molecular Pharmacology, 31, 438-445.

Stogniew, M. and Fenselau, C., 1982, Electrophilic reactions of acyl-linked glucuronides. Formation of clofibrate mercapturate in humans. Drug Metabolism and Disposition, 10, 609-613.

Surh, Y-J., Lai, C-C., Miller, J.A. and Miller, E.C., 1987, Hepatic DNA and RNA adduct formation from the carcinogen 7-hydroxymethyl-12-methylbenz[a]anthracene and its electrophilic sulfuric acid ester metabolite in pre-weanling rats and mice. Biochemical and Biophysical Research Communications, 144, 576-582.

Watabe, T., Fujieda, T., Hiratsuka, A., Ishizuka, T., Hakamata, Y. and Ogura, K., 1985, The carcinogen 7-hydroxymethyl-12-methylbenz[a]anthracene, is activated and covalently binds to DNA via a sulphate ester. Biochemical Pharmacology, 34, 3002-3005.

Watabe, T., Hiratsuka, A. and Ogura, K., 1987, Sulphotransferase-mediated covalent binding of the carcinogen 7,12-dihydroxymethylbenz[a]anthracene to calf thymus DNA and its inhibition by glutathione transferase. Carcinogenesis, 8, 445-453.

Waters, R., Edwards, S., Hesk, D. and Logan, C.J., 1985, The binding of 4-cyano-$\underline{N}$,$\underline{N}$-dimethyl-[^{14}C]aniline (CDA) metabolites in the rat. In Comparative Genetic Toxicology: The Second UKEMS Collaborative Study, edited by J.M. Parry and C.F. Arlett (London: MacMillan), pp. 533-536.

Wislocki, P.G., Borchert, P., Miller, J.A. and Miller, E.C., 1976, The metabolic activation of the carcinogen 1'-hydroxysafrole in vivo and in vitro and the electrophilic reactivities of possible ultimate carcinogens. Cancer Research, 36, 1686-1695.

Wislocki, P.G., Miller, E.C., Miller, J.A., McCoy, E.C. and Rosenkranz, H., 1977, Carcinogenic and mutagenic activities of safrole, 1'-hydroxysafrole and some known or possible metabolites. Cancer Research, 37, 1883-1891.

METABOLITES DERIVED FROM GLUTATHIONE CONJUGATION

J.E. Bakke

Metabolism and Radiation Research Laboratory,
Agricultural Research Service,
U.S. Department of Agriculture,
Fargo, ND 58105, USA

INTRODUCTION

Glutathione (GSH) conjugation is the most important process in animals for the detoxification of potentially damaging electrophilic compounds ingested as such or formed by the metabolism of a xenobiotic precursor. The initial conjugation reaction, catalyzed by the GSH transferases, has been the subject of several authoritative reviews (e.g. Meister, 1983; Reed and Meredith, 1984). GSH conjugates are intermediary xenobiotic metabolites in the sense that they are rarely, if ever, excreted unchanged from animals. Rather, they are considered to be eliminated as mercapturic acids (xenobiotic N-acetylcysteine conjugates) (Chasseaud, 1976). However, it has become apparent over the last 10 years that the mercapturic acid pathway (MAP) is only one of several routes for the catabolism of GSH conjugates (Bakke, 1986; Stevens and Jones, 1988). Thus the primary conjugate is the precursor of a number of other intermediary xenobiotic metabolites, some of which have toxicological significance. This chapter will emphasize recent and unpublished discoveries in order to highlight the number of pathways now known to exist in addition to mercapturic acid formation.

The narrative will be centered around the pathways outlined in Figures 1 and 2 which, in the author's opinion, summarize the major biotransformations occurring in GSH conjugation and catabolism. The biotransformations described in Figure 1 include conjugation processes, hydrolysis of GSH conjugates to cysteine connjugates (peptidase congeners), and acetylation of cysteine conjugates to mercapturic acids (mercapturic- and premercapturic acid pathways, MAP and preMAP). In addition to the MAP, Figure 2 includes the intermediary metabolism of

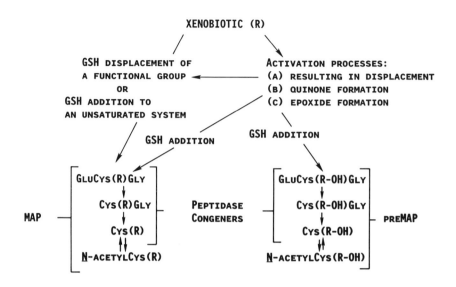

Figure 1. Routes of formation of glutathione (GSH) conjugates of xenobiotics (R) by displacement and addition reactions and their subsequent metabolism in the mercapturic acid (MAP) and premercapturic acid (preMAP) pathways by peptidase hydrolysis (peptidase congeners) and $\underline{N}$-acetyltransferase.

cysteine conjugates by transamination (transaminase congeners) and cysteine conjugate beta-lyase activity (lyase congeners). The lyase congeners include the methylthio oxidation congeners.

CONJUGATION WITH GLUTATHIONE

Conjugations between GSH and xenobiotics occur either by displacement or addition reactions to form either of the two types of conjugates shown in Figure 1. These addition and displacement reactions can occur with no prior metabolism of the xenobiotic or they can occur subsequent to metabolic transformations (activation processes A, B or C).

Specific examples of processes that "activate" xenobiotics prior to GSH displacement are given in Figure 3. Formation of the GSH conjugate of p-dimethylaminobenzonitrile (reaction 1, Figure 3) is an example where extensive metabolism took place prior to the proposed displacement of the sulphate ion by GSH (Hutson et al., 1984). Another activation process involves oxidation of sulphur.

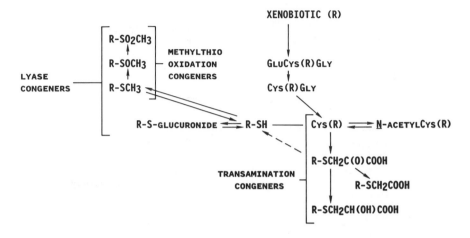

Figure 2. Catabolism of cysteine conjugates by oxidative transamination (transamination congeners) and cysteine conjugate-beta-lyase cleavage to thiols, S-glucuronides and methylthio congeners (lyase congeners).

Sulphoxidation of the methylthio-groups in the 2-methylthio-triazine (Crawford and Hutson, 1980) (reaction 2) and pentachlorothioanisole (PCTA, reaction 3) (Dulik et al., 1987) are required before displacement of the methylsulphinyl groups by GSH. Sulphoxidation is also required for GSH-displacement of chlorine in the metabolism of PCTA (reaction 3).

Activations that occur prior to addition reactions between GSH and xenobiotic metabolites are also given in Figure 3. These are exemplified by quinoneimine formation in phenacetin metabolism (reaction 4) (Calder et al., 1974) and epoxide formation in naphthalene metabolism (reaction 5) (Jerina et al., 1970).

Three conjugations that are probably examples of the epoxide addition pathway are the ortho-hydroxy GSH conjugates (reaction 6) (Bakke et al., 1988a), the isolation of an acetylated GSH conjugate of 1,2,4-trichlorobenzene (reaction 7) (Bakke et al., unpublished), and the formation of tetrahydrotrihydroxy GSH conjugates of naphthalene (reaction 8) (Bakke and Davison, 1988). The formation of the ortho-hydroxy GSH conjugate was prevalent in the metabolism of 2,6-dichlorobenzonitrile in rats and

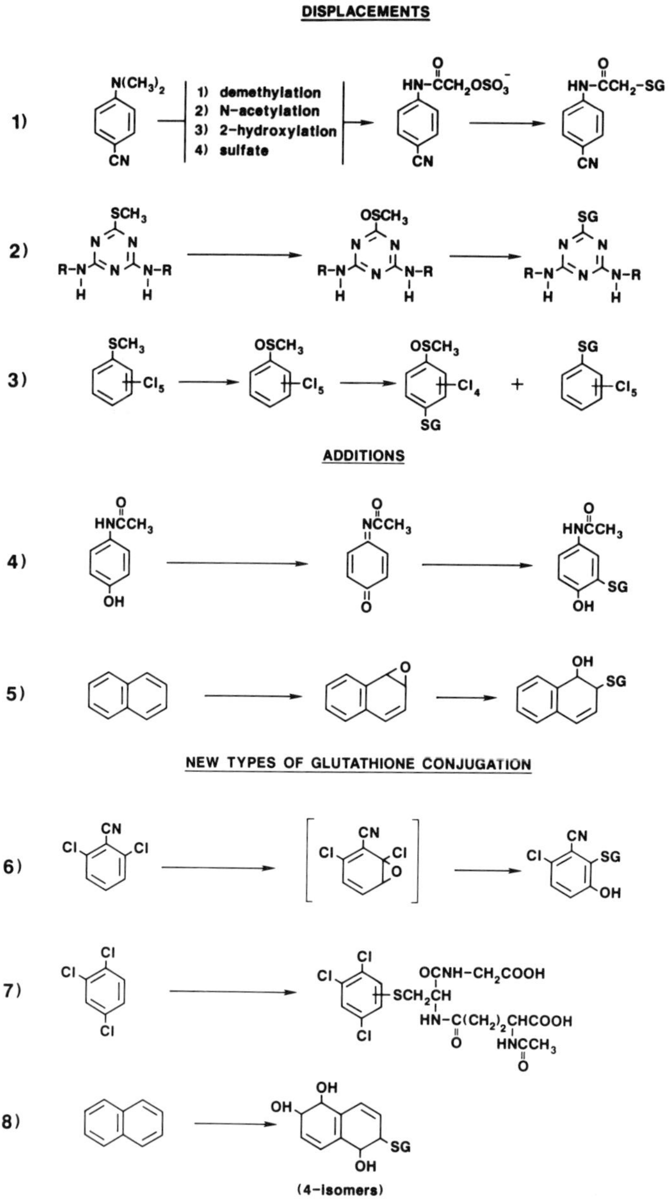

Figure 3. Examples where metabolic activation occurred prior to conjugation with glutathione by either displacement (reactions 1, 2 and 3) or addition (reactions 4 and 5) and examples of new conjugation pathways (reactions 6, 7 and 8).

goats (Bakke et al., 1988b) and was the major pathway in chickens (Davison and Bakke, 1988). It was also a major pathway of metabolism of 2,6-dichlorobenzamide (Bakke et al., 1988a) and 2,6-dichlorothiobenzamide (Bakke et al., 1988b) in the rat. In addition, other chlorinated benzenes give metabolites in which chlorine is removed and both a divalent sulphur and a hydroxyl group added; however, the positions of the ring substitutions were not determined. These include 1,3,5-trichlorobenzene (Mio and Sumino, 1985), trichlorobenzene, 3,5-dichlorothioanisole, 4-chlorothioanisole, and 2-chlorobenzonitrile (Bakke et al., unpublished).

We have proposed that <u>ortho</u>-hydroxy GSH conjugates form either by addition of GSH to a chlorine-substituted oxirane ring with subsequent aromatization by elimination of HCl, or by GSH displacement of the chlorine on the oxirane ring with rearrangement of the epoxide to a phenol. These proposed mechanisms are shown in Figure 4 for 2,6-dichlorobenzonitrile metabolism. In this case only one isomer was detected; such specificity would require a directed mechanism.

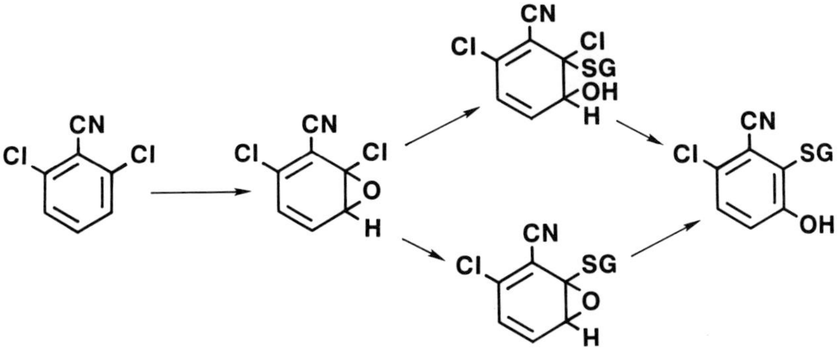

Figure 4. Pathways proposed for formation of an <u>ortho</u>-hydroxy glutathione conjugate in the metabolism of 2,6-dichlorobenzonitrile by chickens, rats and goats.

An acetylated GSH conjugate was isolated from the bile of rats dosed with 1,2,4-trichlorobenzene (reaction 7, Figure 3). The structure proposed in Figure 3 was assigned from positive ion fast atom bombardment mass spectrometry, and, after methylation with diazomethane, electron impact mass spectrometry which gave a fragmentation pattern compatible with the acetate being on the glutamic acid moiety. This was a major biliary metabolite and was present in concentrations similar to that of the GSH conjugate. Nothing is known of the site or mechanism of its formation. The acetylated GSH conjugate was isolated from urine unchanged when it was perfused through chicken kidney using methods reported by Davison et al. (1988). This indicates that the acetylated GSH conjugate was not a substrate for renal gamma-glutamyl-transpeptidase (GTPase). It is assumed, that in control rats, this biliary metabolite was metabolized by intestinal systems to conventional peptidase congeners that were translocated into the tissues.

Tetrahydrotrihydroxy GSH conjugates (isomers) were discovered in the metabolism of naphthalene in chickens (Bakke and Davison 1988). Urine from roosters contained tetrahydrotrihydroxy mercapturates along with the usual dihydrohydroxy mercapturate formed by rats. Bile contained both tetrahydrotrihydroxy-cysteinylglycine and -cysteine conjugates of naphthalene. The tetrahydrotrihydroxy GSH adducts indicate either diepoxide formation and/or sequential epoxide formation with subsequent addition of water and GSH. At least four isomers of these tetrahydrotrihydroxy conjugates were indicated because rigorous treatment with silylation reagents resulted in the elimination of the sulphur-containing moiety with the formation of four trimethylsilyl derivatives of dihydrodihydroxy-naphthols.

A discrimination in biliary excretion by chickens of GSH addition conjugates of naphthalene was also observed. Rooster bile contained tetrahydrotrihydroxy peptidase congeners but no dihydro-hydroxy conjugates were detected. They must have been formed because mercapturates from both the tetrahydrotrihydroxy- and dihydrohydroxy-GSH adducts were present in about equal amounts in urine from colostomized roosters dosed with naphthalene.

Multiple epoxide formations as described in the metabolism of naphthalene by chickens are of interest due to the role of dihydrodiol-epoxides in the carcinogenicity of polynuclear aromatic hydrocarbons. Chicken tissues may be a good medium for the study of their formation.

DISPOSITION OF GLUTATHIONE CONJUGATES

It is currently accepted that GSH conjugates are handled by organisms in the same way that they handle GSH, i.e., the conjugates are excreted from the cells in which they are formed before peptide hydrolysis is initiated. This means that only reactions involving GSH conjugates and not those involving subsequent metabolites can occur without intercellular transport. Examples of these pre-transport reactions are reductive displacements, episulphonium ion formation, aromatization and multiple conjugations. These transformations, which can also occur with other peptidase congeners, will be discussed later. The intracellular stability of the peptide bonds in GSH and GSH conjugates is attributed to the extracellular location of the first enzyme [gamma-glutmyl transpeptidase (GTPase)] to function in the stepwise hydrolysis of GSH conjugates to cysteine conjugates (peptidase congeners, Figure 1).

The hydrolysis of GSH conjugates to cysteine conjugates has significance for at least two related reasons. Cysteine conjugates can be transported into cells by amino acid transport systems (Christensen et al., 1986) where they are precursors for at least three metabolic pathways. These pathways are N-acetylation (mercapturic acid formation), oxidative transamination and thiol formation as shown in Figure 2. Factors that direct cysteine conjugates into these pathways are as yet poorly understood.

Mercapturic acid formation

N-Acetylation may have evolved as a conjugation reaction to prevent cysteine conjugates from entering toxic pathways. Acetylation can prevent or inhibit cellular uptake of cysteine conjugates by amino acid transport mechanisms and it can block tissue C-S lyase action, preventing the formation of toxic thiols in cells that have this activity. Evidence that N-acetylcysteine conjugates are probably not as readily transported into cells as cysteine conjugates comes from a comparison of the profiles of xenobiotic metabolites excreted in bile from rats with cannulated bile ducts with those excreted in faeces from germfree rats (Bakke et al., unpublished). The compounds studied were 2,6-dichlorobenzonitrile, pentachlorothioanisole, and 1,2,4-trichlorobenzene. The major biliary metabolites from these xenobiotics in the cannulated rats were peptidase congeners. The major faecal metabolites from germfree rats were mercapturates. Because acetylation has not been detected in the intestinal lumen, the mercapturates were probably not formed there. These results indicate a preferential absorption of peptidase congeners from the germfree lumen, leaving mercapturates to be excreted with the faeces. The absorbed peptidase congeners can then be biotransformed by tissue systems to mercapturates which can be excreted with either urine or bile. The levels of mercapturates derived from these compounds excreted with faeces by the germfree rats indicated that the transported cysteine conjugates were acetylated and in part re-excreted with the bile.

In rats there is evidence for selectivity in the absorption of peptidase congeners of different xenobiotics in the gut. 2,6-Dichlorobenzonitrile (DCBN) is metabolized to two GSH conjugates both of which are excreted in bile (Figure 5) (Bakke et al., 1988b). As described later under

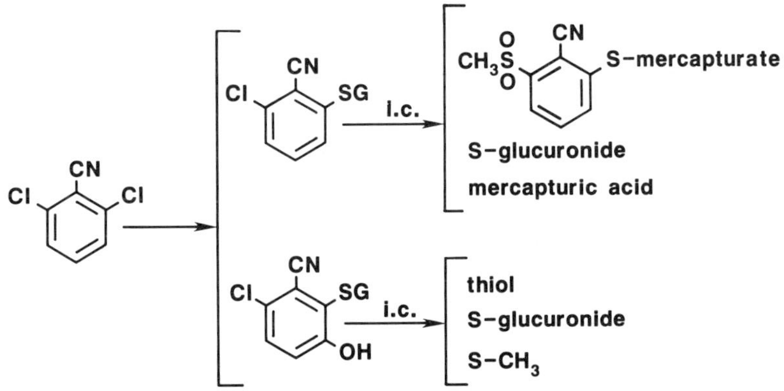

Figure 5. Structures of the two glutathione conjugates formed from dichlorobenzonitrile and the metabolism of the peptidase congeners when dosed intracaecally to rats.

"Thiol Formation", the cysteine conjugates derived from both GSH conjugates are substrates for microfloral C-S lyases; however, in control rats, only the ortho-hydroxy GSH conjugate is metabolized to C-S lyase congeners. It is assumed that in one case the peptidase congeners were absorbed from the gut before the cysteine conjugate reached the area of the gut that contained microfloral C-S lyase activity. A similar conclusion was reached for peptidase congeners of naphthalene (Bakke et al., 1985) and 1,2,4-trichlorobenzene (Bakke et al., unpublished). Mercapturic acids are not excreted with the faeces in the presence of conventional intestinal microflora because intestinal mercapturates can be deacetylated to cysteine conjugates (Larsen and Bakke, 1983) which can either be transported by amino acid transport systems or become substrates for microfloral C-S lyases.

Oxidative transamination

Oxidative transamination is a common pathway for the

metabolism of cysteine conjugates of aliphatic xenobiotics.
It is increasingly being detected as a minor pathway for
other xenobiotics, especially those formed in the preMAP
(for review see Bakke, 1986). No significance has been
attached to this pathway unless the enzyme reported by
Tomisawa et al. (1986) which cleaves S-substituted 3-
thiopyruvic acid conjugates (shown in Figure 2) is
functioning in vivo. In this case oxidative transamination
would also contribute to thiol formation in vivo.

Thiol formation
  Thiol formation from cysteine conjugates is catalyzed by
cysteine conjugate beta-lyase (C-S lyase) activities located
in liver (Tateishi et al., 1978, Stevens and Jakoby, 1983,
Stevens, 1985a), kidneys (Stevens, 1985b) and intestinal
microflora (Larsen and Bakke, 1983). The C-S lyase
activities in the microflora have a broader range of
substrate specificities than those in the tissues. The
tissue C-S lyase activities exhibit a specificity for
cysteine conjugates in which the sulphur is bonded to an
electron withdrawing group (e.g. aromatic). The microfloral
C-S lyases can apparently cleave most cysteine conjugates.
The thiols formed by C-S lyase activities are excreted as
free thiols, as conjugates with glucuronic acid and as
methylthio-oxidation products (Figure 3). Thiols of
suitable structure are now believed to be the precursors of
reactive metabolites, leading to toxicity and to bound
residues. Recent examples include the tissue damage caused
by S-(1,2-dichlorovinyl)-L-cysteine (Darnerud et al., 1988)
and hexachlorobutadiene (Jaffe et al., 1983). When
peptidase congeners are excreted with bile, there are often
non-extractable residues in the faeces that may arise from
thiols formed by microfloral C-S lyase activities. Thiols
can also be formed in the intestinal lumen by hydrolysis of
S-glucuronides that are excreted in bile.
  C-S lyase activity is either compartmentalized in the
intestinal tract or competes with mucosal transport systems
for the peptidase congeners as mentioned above in DCBN
metabolism. This is concluded from observations that some
biliary peptidase congeners do not become substrates for C-S
lyase activity in vivo, but when injected into the caecum
they are cleaved to thiols as determined by excretion of
methylthio-containing metabolites in urine (Figure 5). This
was observed in the metabolism of 2,6-dichlorobenzonitrile
(Bakke et al., 1988b), 1,2,4-trichlorobenzene (Bakke et al.,
unpublished) and naphthalene (Bakke et al., 1985) in rats.
  There is much to be learned about the conjugation of
thiols and the isolation of some of these conjugates intact

requires attention. Some S-glucuronides have been isolated quite readily, whereas a number of metabolites that are known by their chromatographic characteristics to be conjugates could only be isolated as thiols. We also believe that thiols can result as artefacts from derivatization of peptidase congeners. Often thiols and thioanisoles are isolated as minor components when chromatographic fractions that contain peptidase congeners are methylated with diazomethane.

Methylation of thiols

The methylation of xenobiotic thiols with a methyl group from S-adenosylmethionine (SAM) is catalyzed by thiol methyltransferase (S-MT) (Weisiger et al., 1980). The highest activities of S-MT in rat tissues are found in the mucosae of the caecum and colon. The activities in the liver, kidneys and mucosa of the small intestine were 70, 50, and 25%, respectively, of that in the colon. It is of interest that the highest activities of S-MT are located in organs that contain C-S lyase activities which produce the thiol-containing substrates.

It is, as yet, not established with certainty which organs are involved in the methylation of a particular thiol in vivo. For example, when the o-hydroxy GSH conjugate of 2,6-dichlorobenzonitrile was infused into the renal artery of pigs, the corresponding thiol was excreted with the urine but no thioanisole was formed. When the same conjugate was infused into the portal veins of pigs with cannulated bile ducts again the thiol was the major urinary metabolite but the thioanisole was also present in appreciable quantities. This indicated that renal S-MT was not functioning on the thiol formed in the kidney and that either the thioanisole was formed in the liver, or metabolism (thiol formation?) was required in the liver before renal methylation could occur (Bakke et al., unpublished).

The disposition of thiols in the intestine of rats appears to involve a compartmentalization or continuum of enzyme activities which are influenced by competitions and specificities of substrates for lumenal enzymes (peptidases, glucuronidases and C-S lyases) and mucosal activities (amino acid transport, thiol absorption, S-MT activity, and UDP-glucuronyl transferase). Several scenarios for thiol, glucuronide, and peptidase congener dispositions in the intestine are possible but as yet only a few have been observed.

In 1,2,4-trichlorobenzene metabolism in rats, thiols and peptidase congeners are excreted in bile (60% of the dose) (Bakke et al., unpublished) but the most prevalent urinary metabolites from control rats are the thiols and mercapturic acids shown in Figure 6 (Lingg et al., 1982). In this case it appears that the biliary thiols are absorbed and subsequently excreted by the kidney and that the peptidase congeners are absorbed (after conversion to cysteine conjugates?) acetylated and excreted with urine as mercapturates. This is another example where peptidase congeners are absorbed from the intestinal lumen before the cysteine conjugates reached the C-S lyase activity in the flora, however, it is known that the cysteine conjugates were substrates for C-S lyase because when the mercapturic acids were injected into the caecum much of the dose was excreted as mercapturic acids of the methylsulphonyl-containing dichlorobenzene shown in Figure 6. The same mercapturic acids were formed when rats were dosed orally with 2,4,5-trichlorothioanisole, which supports the probability that methylthio-containing intermediates were formed after intracaecal dosing of the original mercapturates (Bakke et al., unpublished).

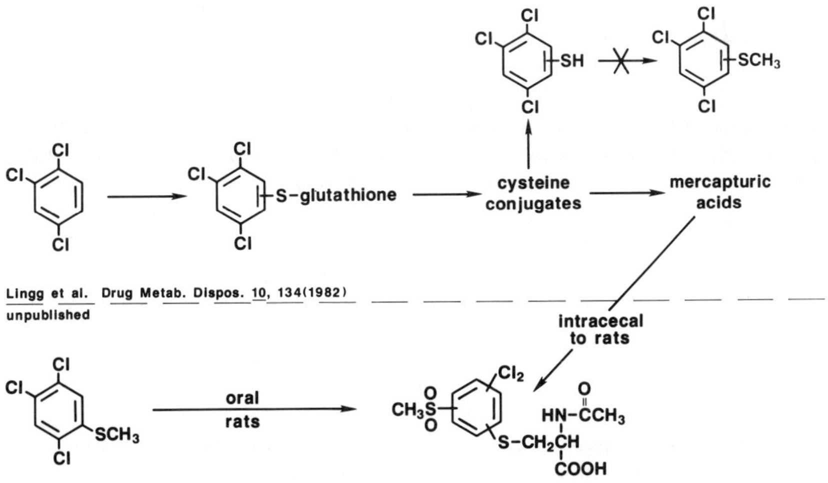

Figure 6. Metabolism of 1,2,4-trichlorobenzene and 2,4,5-trichlorothioanisole in rats, and the metabolism of the mercapturic acids of 1,2,4-trichlorobenzene dosed intra-caecally to rats.

It is also of interest that the biliary thiols formed in 1,2,4-trichlorobenzene metabolism do not become substrates

for S-MT when they obviously pass through at least three
tissues containing that enzyme (intestinal mucosa, liver and
kidney) during their translocation to the urine. The
results from the caecal injection of the mercapturic acids
indicate that the thiols are substrates for S-MT. It is
possible that the thiols are translocated as some other
conjugate that we have been unable to isolate. We have been
unable to detect the S-glucuronides in either the urine or
bile, however, the hplc fraction from which Lingg et al.
(1982) isolated these thiols had elution characteristics
very similar to those of the O-glucuronides they isolated as
monkey metabolites of 1,2,4-trichlorobenzene.

It appears that most biliary peptidase congeners that do
not become substrates for amino acid transport become
substrates for C-S lyases. The thiols that are formed are
then absorbed and usually methylated. The prevalence of
methylation is probably due to the juxtaposition of C-S
lyase and S-MT activities in the lower intestinal tract.
Why S-glucuronidation is not observed is not known.

Disposition of methylthio-containing metabolites (CH$_3$S-metabolites)

The methylation of thiols can further complicate the
metabolism of GSH conjugates. CH$_3$S-metabolites are often
oxidized to their sulphinyl and sulphonyl derivatives (CH$_3$S-oxidation congeners). These derivatives usually have
polarities similar to those of the original xenobiotic and
therefore undergo further metabolism to enable excretion.
This metabolism can involve any functional group on the
metabolite including methylthio-turnover as outlined in
Figure 7. Examples of metabolic processes subsequent to

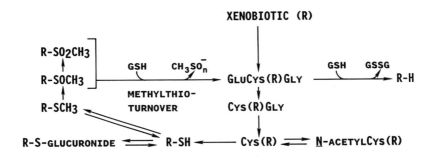

Figure 7. Methylthio-turnover and glutathione-mediated reductive displacement

methylthio-formation are given in Figure 8 for propachlor (Bakke and Price, 1979) and have been described for pentachlorothioanisole metabolism (Bakke et al., 1987).

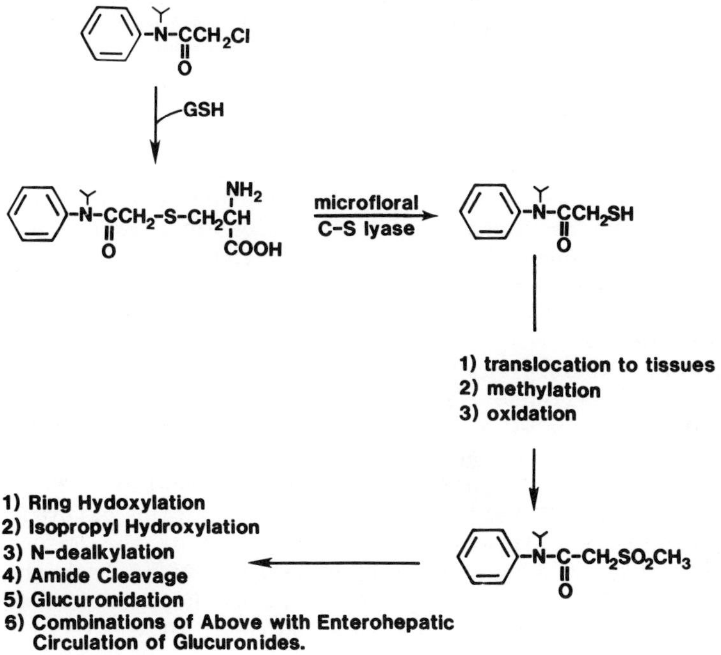

Figure 8. Complexities in the metabolism of 2-chloro-N-isopropylacetanilide subsequent to the mercapturic acid pathway.

Methylsulphonyl-containing metabolites from polychlorinated biphenyls and chlorinated benzenes (Brandt and Bergman, 1987; Kato et al., 1986) are known to occur as persistent but extractable tissue residues. Some of these residues have been shown to be associated with specific proteins in lung (Lund et al., 1985) and kidney (Larson, G.L., unpublished).

Demethylation of methylthio-groups *in vitro* was reported by Mazel et al (1964). A recent study has indicated that the original assay for S-demethylation (formaldehyde production) probably also measured GSH-mediated methylthio-turnover because the displaced methylthio group was metabolized to formaldehyde and sulphate (Larsen et al.,

1988). In the latter in vitro study, methylthio-
displacement was three to five times more prevalent than S-
demethylation. S-Demethylation has not been reported in
vivo whereas methylthio-turnover was shown to be a major
pathway (circa 50% of the dose) in the metabolism of
pentachlorothioanisole to mainly bis-methylthiotetra-
chlorobenzene by rats (Bakke et al., 1987).

S-Glucuronidation
S-Glucuronidation occurs following thiol formation from
catabolism of GSH conjugates (Colucci and Buyske, 1965;
Larsen and Bakke, 1978) but the sites of conjugation in vivo
are not established. Similarly, the factors which influence
glucuronidation as opposed to methylation of thiols are not
known. As mentioned above, it appears that some S-
glucuronides can be quite unstable.

Aromatizations
Aromatization of preMAP metabolites and the associated
lyase and transaminase congeners are not usually discussed in
connection with in vivo metabolism or transformation of GSH
conjugates. No dehydrogenases or dehydrases for preMAP
metabolites have been reported. If aromatization occurs in
vivo, it is most probably a spontaneous reaction determined
by the structure of the xenobiotic moieties of the adducts.
The occurrence of dehydration is usually deduced from the
presence of isomers as metabolites and/or the presence of
ortho-diols or dihydrodiols in the excreta. The
dehydrohalogenation shown in Figure 3 for the formation of
o-hydroxy GSH conjugates is another possible nonenzymatic
aromatization process.
Aromatizations other than those resulting from dehydration
of GSH-epoxide adducts have been observed, proposed or
predicted in the metabolism of benzo(a)pyrene [b(a)p].
Renwick and Drasar (1976) have shown that b(a)p was formed
when rat biliary metabolites of b(a)p were incubated with
intestinal contents. This bile most assuredly contained
products of GSH conjugation which could serve as precursors
for the b(a)p produced by the microflora. Studies on the
facile conversion of 9,10-dihydro-9-hydroxy-10-(S-
cysteinyl)-phenanthrene to phenanthrene lend credence to
such aromatizations (Feil et al., 1986). The reaction was
proposed to proceed through an episulphonium ion resulting
in the apparent elimination of cysteine sulphenic acid as
shown in Figure 9 (reaction A). Evidence that this
aromatization occurred in vivo came from the isolation of a
hydroxy-phenanthrene other than the 9-isomer as a rat

metabolite of the above cysteine conjugate (reactions A and C) (Struble et al., 1986).

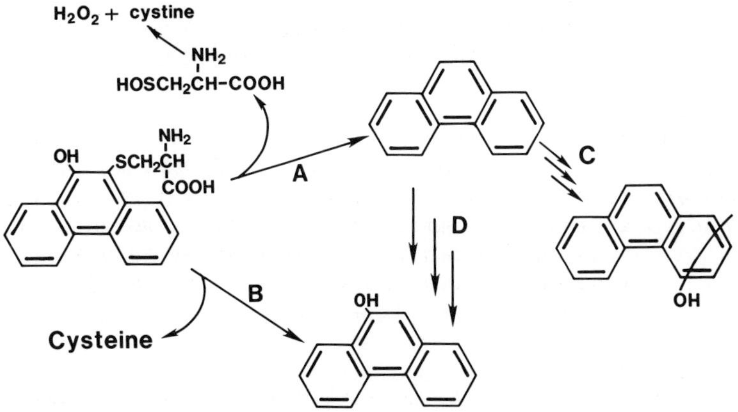

Figure 9. Proposed mechanisms for the formation of phenanthrene from the dihydrohydroxy cysteine conjugate *in vitro* and the formation of phenanthrols *in vivo*.

These reactions, which represent a reversal of the original GSH-epoxide adduct formation, need further study to determine if the conversion of GSH adducts to the original xenobiotics functions as a mechanism for transport of carcinogens to susceptible tissues such as the intestinal mucosa.

OTHER EXAMPLES OF THE INTERMEDIACY OF GLUTATHIONE CONJUGATES
Episulphonium ions

An episulphonium ion has been proposed as the immediate precursor for bound residues from dibromoethane (van Bladeren et al., 1980). The episulphonium ion was proposed to be formed from the GSH conjugate, however, the results from the studies with phenanthrene mentioned above suggest that almost any of the dihydro-hydroxy adducts involved in GSH conjugate catabolism could form episulphonium ions.

Reductive Displacements

Reductive dehalogenation of a phenacyl chloride has been shown to be a GSH-mediated process (Hutson et al., 1976). In this case the initial conjugate, an S-phenacyl glutathione, undergoes an enzyme-catalyzed attack at its now relatively electrophilic sulphur atom by GS⁻. GSSG and the dehalogenated product are formed. A number of other functional groups have also been shown to undergo reductive displacement in vivo (for a review see Bakke, 1986). These have usually been functional groups involved in the methylthio-turnover cycle shown in Figure 7. Further in vitro studies are needed to determine the immediate precursors for reductive displacement.

CONCLUSIONS

Although mercapturic acid formation is the best known indicator that a xenobiotic has been conjugated with GSH, it is not necessarily the major pathway for disposition of GSH conjugates. There is a wide variation in the amount of GSH conjugate converted to mercapturic acid. As examples, only 0.5% of a dose of pentachlorothioanisole was eliminated as the mercapturic acid whereas 90% was originally conjugated with GSH (Bakke et al., unpublished); the corresponding values for naphthalene were 35% and 45%, respectively (Bakke et al., unpublished).

In addition to mercapturic acids, other products formed from GSH conjugates may be summarized as follows: 1. $\underline{N}$-acetylcysteine- and cysteine conjugates; 2. $\underline{S}$-substituted thiopyruvic-, 3-thiolactic-, and 2-thioacetic acid conjugates resulting from oxidative transamination; 3. thiol-, methylthio-, methylsulphinyl-, and methyl-sulphonyl-containing metabolites; 4. extractable methyl-sulphonyl-containing tissue residues; 5. glucuronides of thiol-containing metabolites; 6. phenols and phenol glucuronides resulting from aromatization of dihydro-hydroxy-adducts; 7. methane-sulphinic and -sulphonic acids in urine resulting from methylthio-turnover; 8. bound tissue residues resulting from formation of reactive thiols; 9. nonextractable faecal residues possibly from reactive thiol formation; 10. metabolites in which a functional group has been displaced by hydrogen (reductive displacement).

It is difficult to assign any beneficial function to the biotransformations leading to metabolites of GSH conjugates other than the conservation of the two non-essential amino acids, glycine and glutamic acid. None of the processes conserve the essential amino acid cysteine. With all the metabolic energy spent on GSH conjugate catabolism it would appear more efficient for the organism to excrete the GSH conjugate intact. However, to effect the excretion of GSH conjugates would require the inhibition of gamma-glutamyl-transpeptidase. This would be even more wasteful because it would result in the excretion and loss of GSH.

REFERENCES

Bakke, J.E., 1986, Catabolism of Glutathione Conjugates. In Xenobiotic Conjugation Chemistry, edited by G.D. Paulson, J. Caldwell, D.H. Hutson and J.J. Menn, ACS Symposium Series 299, (Washington, D.C.: ACS) pp. 301-321.

Bakke, J.E. and Davison, K.L., 1988, Metabolism of naphthalene in roosters, Xenobiotica, 18, 1057-1062.

Bakke, J.E. and Gustafsson, J.-Å., 1986, Role of intestinal flora in metabolism of agrochemicals conjugated with glutathione, Xenobiotica, 16, 1047-1056.

Bakke, J.E., Larsen, G.L., Feil, V.J., Brittebo, E.B. and Brandt, I., 1988a, Metabolism of 2,6-dichlorobenzamide in rats and mice, Xenobiotica, 18, 817-829.

Bakke, J.E., Larsen, G.L., Struble, C., Feil, V.J., Brandt, I. and Brittebo, E.V., 1988b, Metabolism of 2,6-dichlorobenzonitrile, 2,6-dichlorothiobenzamide in rodents and goats, Xenobiotica, 18, 1063-1075.

Bakke, J.E., Mulford, D.J., Feil, V.J. and Larsen, G.L., 1987, Intermediary metabolism of mercapturic acid pathway metabolites. In Pesticide Science and Biotechnology, edited by R. Greenhalgh and T.R. Roberts (Oxford: Blackwell Scientific Publications) pp. 513-516.

Bakke, J.E. and Price, C.E., 1979, Metabolism of 2-chloro-$\underline{N}$-isopropylacetanilide (propachlor) in the rat. Journal of Environmental Science and Health, B14, 427-441.

Bakke, J., Struble, C., Gustafsson, J-Å. and Gustafsson, B., 1985, Catabolism of premercapturic acid pathway metabolites of naphthalene to naphthols and methylthio-containing metabolites in rats. Proceedings of the National Academy of Science of the United States of America, 82, 668-671.

Brandt, I. and Bergman, Å., 1987, PCB methyl sulphones and related compounds: identification of target cells and tissues in difference species. Chemosphere, 16, 1671-1676.

Calder, I.C., Creek, M.J. and Williams, P.J., 1974, N-Hydroxyphenacetin as a precursor of 3-substituted 4-hydroxyacetanilide metabolites of phenacetin. Chemico-Biological Interactions, 8, 87-90.

Chasseaud, L.F., 1976, Conjugation with glutathione and mercapturic acid excretion. In Glutathione Metabolism and Function, edited by I.M. Arias and W.B. Jakoby, (New York: Raven Press) pp. 77-114.

Christensen, H.N., Curthoys, N.P., Mortimore, G.E., Smith, R.J., Goldstein, L. and Harper, A.E., 1986, Interorgan transport. Federation Proceedings, 45, 2165-2183.

Colucci, D.F. and Buyske, D.A., 1965, The biotransformation of a sulfonamide to a mercaptan and to mercapturic acid and glucuronide conjugates. Biochemical Pharmacology, 14, 457-466.

Crawford, M.J. and Hutson, D.H., 1980, S-Oxygenation of an alkylmercaptotriazine herbicide by rat, rabbit and human liver enzymes. Xenobiotica, 10, 187-192.

Darnerud, P.O., Brandt, I., Feil, V.J. and Bakke, J.E., 1988, Factors affecting accumulation and tissue damage of S-(1,2-dichlorovinyl)-L-cysteine (DCVC) in the mouse kidney. Toxicology and Applied Pharmacology, in press.

Davison, K.L. and Bakke, J.E., 1988, Intermediary metabolism of 2,6-dichlorobenzonitrile (dichlobenil) in chickens and growth of chickens fed dichlobenil. Xenobiotica, 18, 941-948.

Davison, K.L., Bakke, J.E. and Larsen, G.L., 1988, A kidney perfusion method for metabolism studies with chickens using propachlor as a model. Xenobiotica, 18, 323-329.

Dulik, D.M., Huwe, J.K., Bakke, J.E. and Fenselau, C., 1987, Three types of transformation of pentachlorophenyl methyl sulfoxide and sulfone are catabolized by cytosolic and microsomal glutathione-S-transferases. Federation Proceedings, 46, 862.

Feil, V.J., Huwe, J.K. and Bakke, J.E., 1986, Degradation products of 9-alkylthio-9,10-dihydro-10-hydroxyphenanthrenes. In Polynuclear Aromatic Hydrocarbons: Chemistry, Characterization and Carcinogenesis, Ninth International Symposium, edited by M. Cook and A.J. Dennis, (Columbus: Batelle Press) pp. 893-900.

Hutson, D.H., Holmes, D.S. and Crawford, M.J., 1976, The involvement of glutathione in the reductive dechlorination of a phenacyl halide. Chemosphere, 5, 79-84.

Hutson, D.H., Lakeman, S.K. and Logan, C.J., 1984, The fate of 4-cyano-N,N-dimethylaniline in rats; a novel involvement of glutathione in the metabolism of anilines. Xenobiotica, 14, 925-934.

Jaffe, D.R., Hassall, C.D., Brendel, K., Gandolfi, J.A., 1983, In vivo and in vitro nephrotoxicity of the cysteine conjugate of hexachlorobutadiene. Journal of Toxicology and Environmental Health, 11, 857-867.

Jerina, D.M., Daly, J.W., Witkop, B., Zaltzman-Nirenberg, P. and Udenfriend, S., 1970, 1,2-Naphthalene oxide as an intermediate in the microsomal hydroxylation of naphthalene. Biochemistry, 9, 147-155.

Kato, U., Kogure, T., Sato, M., Murat, T. and Kimura, R., 1986, Evidence that methylsulfonyl metabolites of m-dichlorobenzene are causative substances of induction of hepatic microsomal drug-metabolizing enzymes by the parent

compound in rats. Toxicology and Applied Pharmacology, 82, 505-511.
Larsen, G.L. and Bakke, J.E., 1978, Mass spectral characterization of the glucuronide conjugates of terbutryn, 2-(t-butylamino-4-(ethylamino)-6-(methylthio)-s-triazine metabolites from rats and goat. Biomedical Mass Spectrometry, 5, 391-394.
Larsen, G.L. and Bakke, J.E., 1983, Metabolism of mercapturic acid pathway metabolites of 2-chloro-N-isopropylacetanilide (propachlor) by gastrointestinal bacteria. Xenobiotica, 13, 115-126.
Larsen, G.L., Bakke, J.E., Feil, V.J. and Huwe, J.K., 1988, In vitro metabolism of the methylthio group of 2-methylthiobenzothiazole by rat liver. Xenobiotica, 18, 313-322.
Lingg, R.D., Kaylor, W.H., Pyle, S.M., Kopfler, F.C., Smith, C.C., Wolfe, G.F. and Cragg, S., 1982, Comparative metabolism of 1,2,4-trichlorobenzene in the rat and rhesus monkey. Drug Metabolism and Disposition, 10, 134-141.
Lund, J., Brandt, I., Peollinger, L., Bergman, Å., Klasson-Wehler, E. and Gustafsson, J-Å., 1985, Target cells for the polychlorinated biphenyl metabolite 4,4'-bis (methylsulfonyl)-2,2',5,5'-tetrachlorobiphenyl; characterization of high affinity binding in rat and mouse lung cytosol. Molecular Pharmacology, 27, 314-323.
Mazel, P., Henderson, J.F. and Axelrod, J., 1964, S-De-methylation by microsomal enzymes. Journal of Pharmacology and Experimental Therapeutics, 143, 1-6.
Meister, A., 1983, Selective modification of glutathione metabolism. Science, 220, 472-477.
Mio, T. and Sumino, K., 1985, Mechanisms of biosynthesis of methylsulfones from PCBs and related compounds. Environmental Health Perspective, 59, 129-135.
Reed, D.J. and Meredith, M.J., 1984, Glutathione conjugation systems and drug disposition. In *Drugs and Nutrients*, edited by D.A. Roe and T.C. Campbell, (New York: Marcel Dekker, Inc.), pp. 179-224.
Renwick, A.G. and Drasar, B.W., 1976, Environmental carcinogens and large bowel cancer. Nature, 263, 234-235.
Stevens, J.L., 1985a, Isolation and characterization of a rat liver enzyme with both cysteine conjugate beta-lyase and kynureninase activity. The Journal of Biological Chemistry, 260, 7945-7950.
Stevens, J.L., 1985b, Cysteine conjugate beta-lyase activities in rat kidney cortex: subcellular localization

and relationship to the hepatic enzyme. Biochemical and Biophysical Research Communications, 129, 499-504.
Stevens, J.L. and Jakoby, W.B., 1983, Cysteine conjugate beta-lyase. Molecular Pharmacology, 23, 761-765.
Stevens, J.L. and Jones, D.P., 1988, The mercapturic acid pathway: biosymthesis, intermediary metabolism, and physiological disposition. In Coenzymes and Cofactors, edited by D. Dolphin, R. Paulson and A. Avramovic (New York: John Wiley & Sons), in press.
Struble, C.V., Larsen, G.L., Feil, V.J. and Bakke, J.E., 1986, Metabolism of 9-hydroxy-9,10-dihydro-10-cysteinyl-phenanthrene in rats. In Polynuclear Aromatic Hydrocarbons: Chemistry, Characterization and Carcinogenesis, edited by M. Cooke and A.J. Dennis, (Columbus: Battelle Press) pp. 893-900.
Tateishi, M., Suzuki, S. and Shimizu, H., 1978, Cysteine conjugate beta-lyase in rat liver: a novel enzyme catalyzing formation of thiol-containing metabolites of drugs. Journal of Biological Chemistry, 253, 8854-8859.
Van Bladeren, P.J., Breimer, D.D., Rotteveel-Smijs, G.M.T., Dejong, R.A.W., Buijs, W., Van Der Gen, A. and Mohn, G.R., 1980, The role of glutathione conjugation in the mutagenicity of 1,2-dibromoethane. Biochemical Pharmacology, 29, 2975-2982.
Weisiger, R.A., Pinkus, L.M. and Jakoby, W.B., 1980, Thiol S-methyltransferase: suggested role in detoxication of intestinal hydrogen sulfide. Biochemical Pharmacology, 29 2885-2887.

FOOTNOTE
No warranties are herein implied by the U.S. Department of Agriculture.

REACTIVE INTERMEDIATES AND THEIR REACTION
WITH MACROMOLECULES

D.R. Hawkins

Huntingdon Research Centre,
Huntingdon,
Cambridgeshire, PE18 6ES, UK

INTRODUCTION

Intermediary metabolites are those which have short half-lives due to chemical instability, rearrangement, chemical reactivity or further rapid metabolism or those which lack the physicochemical properties for excretion from the body such as the lipophilic conjugates (Quistad and Hutson, 1986). Reactive metabolites are of particular interest since by their nature they are capable of reacting with tissue macromolecules to form covalently-bound adducts. These adducts can also be considered as a class of intermediary metabolites. Formation of such adducts with nucleic acids and proteins can be important since it is recognised to be an initial event in several kinds of chemically-induced toxicity such as carcinogenesis, mutagenesis, cellular necrosis and hypersensitivity reactions. Identification of the structure of these adducts presents a demanding challenge and since there are relatively few examples they merit classification as novel or unusual metabolites. The occurrence of non-extractable or bound tissue residues is a relatively frequent event for those researchers involved in animal metabolism studies but investigation of the chemical structure of the adducts is seldom undertaken.

Identification of metabolites derived by a glutathione conjugation pathway implies the existence of electrophilic metabolites which generally are also capable of reacting with nucleophilic centres in proteins and nucleic acids. The extent of such covalent binding will depend on many factors such as the site and rate of formation of the intermediate, its intrinsic chemical reactivity and electronic and steric characteristics of the electrophilic centre. Glutathione generally performs a protective role by removing the reactive intermediate but high concentrations of the intermediate can deplete glutathione and hence a dose threshold exists for covalent binding. A classic example of the formation of a reactive intermediate and its detoxification and mechanism of toxicity is illustrated in the metabolism of bromobenzene. The electrophilic intermediate is an epoxide which undergoes the classical reactions of conversion to a dihydrodiol by epoxide hydratase, glutathione conjugation and rearrangement to p-bromophenol. At low doses to rodents, a small amount also reacts with liver proteins but as the dose level is increased there is a disproportional increase in the amount bound which coincides with the onset of centrilobular necrosis.

This paper will give illustrations of some unusual or novel reactive intermediates and in particular those which have been shown to react with glutathione. The types of reactive functional groups which have been shown to form chemically-bound adducts with nucleophilic centres in proteins and nucleic acids and the structures of these adducts will be described.

REACTIVE INTERMEDIATES

N-Methylformamide (1) is an experimental antitumour agent and a potent hepatotoxin and metabolic activation is thought to be a prerequisite for its observed toxicity. A major rat urinary metabolite has been identified as N-acetyl-S-(N-methylcarbamoyl)cysteine (2) (Tulip et al., 1986). Evidently, this metabolite is formed by reaction of glutathione with a reactive intermediate. It has been postulated that this intermediate is methyl isocyanate formed by N-hydroxylation and dehydration of the resulting hydroxamic acid. The metabolite was readily synthesised by reaction of methyl isocyanate with N-acetylcysteine at room temperature.

$$CH_3NHCH=O \longrightarrow \left[\begin{array}{c} CH_3N-CH=O \\ | \\ OH \end{array} \right]$$

(1)

$$\longrightarrow [CH_3N=C=O] \underset{RS^-}{\longrightarrow} \longrightarrow CH_3NHCSCH_2 \underset{CO_2H}{CHNHCCH_3}$$

(where the final product shows two C=O groups on the carbamoyl and acetyl carbons)

(2)

Isothiocyanates are also very reactive compounds for which a major biotransformation pathway is glutathione conjugation (Brusewitz et al., 1977). The thiocarbamate fungicides represent a class of compound believed to act via formation of isothiocyanates which combine readily with thiol groups of essential enzymes. Similar formation of intermediate isothiocyanates may well occur in animals in which case it would be expected that cysteine conjugates would be found as metabolites. Ethylene thiourea (4) is a known metabolite of dithiocarbamates (3) in animals which is responsible for the observed effects on thyroid function. This metabolite could be formed by an internal cyclisation of aminoethylisothiocyanate.

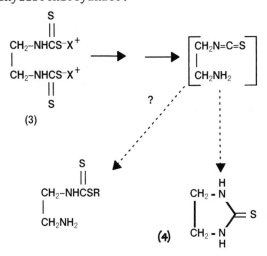

The biotransformation of ethylene dibromide and other 1,2-dihaloalkanes involves glutathione conjugation since N-acetyl-S-(2-hydroxyethyl)-cysteine is a major urinary metabolite. There are two possible routes to the formation of this metabolite (Figure 1), one involving direct conjugation with glutathione and the other, microsomal oxidation followed by loss of hydrogen bromide to give bromoacetaldehyde. The glutathione conjugate is thought to form a novel reactive intermediate by an intramolecular nucleophilic displacement of the remaining halogen to give an episulphonium ion followed by hydrolysis to an alkyl alcohol. Evidence for this pathway was demonstrated by showing that a proportion of the terminal metabolite retained all the deuterium when the tetradeuterated dibromide was the substrate (van Bladeren et al., 1981). The metabolite was also formed with retention of only one deuterium atom indicating that the oxidative pathway also occurred.

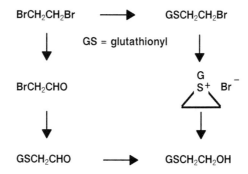

Figure 1. Alternative biotransformation pathways for ethylene dibromide

Busulfan (5) is a bifunctional alkyl compound which can also react with nucleophiles such as glutathione. The major excreted metabolites in rats are the sulphoxide and sulphone of tetrahydrothiophene (7). The formation of this compound has been shown to occur via a monoglutathione conjugate which undergoes internal cyclisation to give a novel sulphonium ion intermediate (6) (Hassan and Ehrsson, 1987). This intermediate was the major metabolite formed in an experiment with isolated perfused rat liver and was shown to decompose to tetrahydrothiophene at physiological pH.

$$CH_3SO_2OCH_2(CH_2)_2CH_2OSO_2CH_3 \longrightarrow CH_3SO_2OCH_2(CH_2)_2CH_2SG$$

(5)

(6)   (7)   GS = glutathionyl

Thiols are metabolic intermediates which are often formed by degradation of glutathione conjugates. Thiols, being nucleophiles are potentially reactive intermediates capable of alkylating tissue components leading to necrosis. Two examples where glutathione conjugation and thiol formation are implicated in nephrotoxicity are o-bromophenol and hexachloro-1,3-butadiene. However, there is now the possibility that another stage is involved, namely formation of S-sulphates. PAPS-dependent sulphate conjugation of a xenobiotic thiol has been demonstrated using the model compound 4-nitrobenzylmercaptan (Miwa et al., 1987). The sulphate reacted very readily with the thiol group of the substrate to form bis-(4-nitrobenzyl) disulphide. The sulphate also reacted readily with the sulphydryl group of cytosolic proteins. This is the first example of an S-sulphate and may represent a new type of reactive intermediate.

Formation of electrophilic intermediates can be established by the use of nucleophilic trapping agents, a technique which has been used with the model compound 1-benzylpyrrolidine (8) (Ho and Castagnoli, 1980). Incubation of the compound with rabbit liver microsomes in the presence of cyanide ion led to formation of 1-benzyl-2-cyanopyrrolidine (10). The cyano adduct was thought to arise from nucleophilic attack of cyanide on the metabolically generated iminium ion. The iminium ion could be formed via the carbinolamine (9). The enamine (11) was only detected in the absence of cyanide indicating that it was formed via the iminium ion. It was also demonstrated that some material was covalently-bound to microsomal protein; this could involve the electrophilic intermediate.

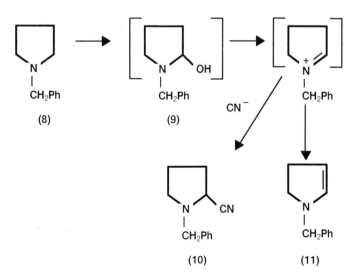

Formation of dihydrodiols is an established pathway for the biotransformation of aromatic hydrocarbons via aryl epoxides. There is now evidence for the conversion of these metabolites to o-quinones by the action of dihydrodiol dehydrogenase. Incubation of naphthalene trans-1,2-dihydrodiol with the dehydrogenase produced a mixture of components but when the trapping agent 2-mercaptoethanol was included in the incubation a single component (13) was formed which was identical to the adduct produced by reacting naphthoquinone (12) with the thiol (Smithgall et al., 1988). The mechanism of adduct formation involved nucleophilic attack at the 4-position, keto-enol rearrangement and autoxidation. The o-quinone was also found to react readily with glutathione and cysteine. This type of metabolite has not been identified for polycyclic aromatic hydrocarbons but may occur in some of the unidentified water-soluble fractions. Thus, o-quinones represent a new type of reactive intermediate electrophilic metabolite.

Metabolites derived by deamination have been reported for arylamines such as the sulphonamides, sulphadiazine and sulphamethazine and an insight into the mechanism of formation was provided by the observation that interaction with nitrite was a contributing factor (Paulson et al., 1987; Paulson and Feil, 1987). Evidence was obtained that the desamino compound was formed via a diazonium ion intermediate produced by the action of nitrite in the acidic medium of the stomach. When sulphamethazine (14) plus nitrite or the pre-formed diazonium salt (15) were administered orally to rats or swine, the desamino compound was a major metabolite. Replacement of a diazonium group by hydrogen is a known chemical process which can occur by use of reducing agents in aqueous solution. A diazonium compound is a potentially reactive intermediate and it was shown that for sulphamethazine there was extensive covalent binding to serum albumin when it was incubated with either a combination with nitrite or as the diazonium salt. Benzene diazonium compounds have been reported to react with both purines and amino acids.

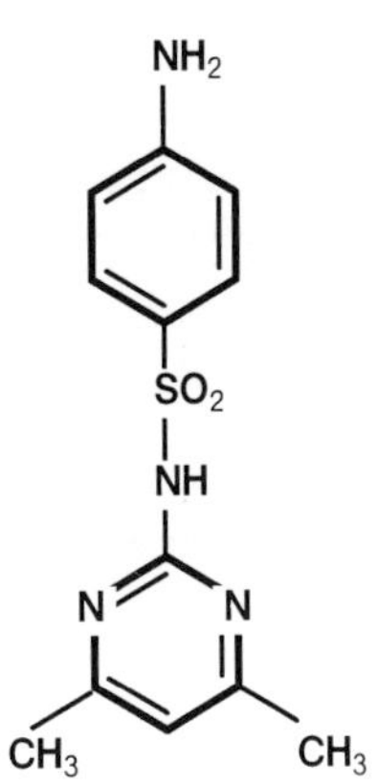

(14)

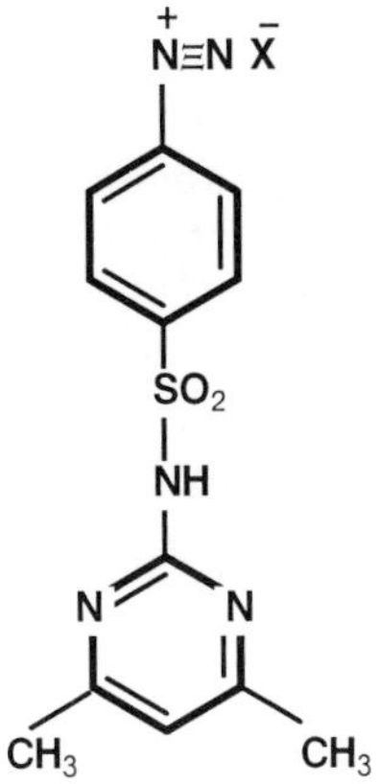

(15)

Nitrosamines are the presumed intermediates in the diazotization of amines. Alkyl nitrosamines are known to form alkylated nucleic acid in vivo; this has been postulated as the mechanism for the carcinogenicity of this class of compound. It has been proposed that the reactive intermediates involved are diazohydroxides or diazonium ions. For a dialkylnitrosamine, α-hydroxylation is believed to be the process leading to formation of the reactive intermediate (Figure 2). The diazonium hydroxide or diazonium ion is able to react with nucleophilic functional groups such as hydroxyls or amines to form intermediates which decompose to give the O- or N-alkylated product (Figure 3). Administration of N,N-di-n-propylnitrosamine

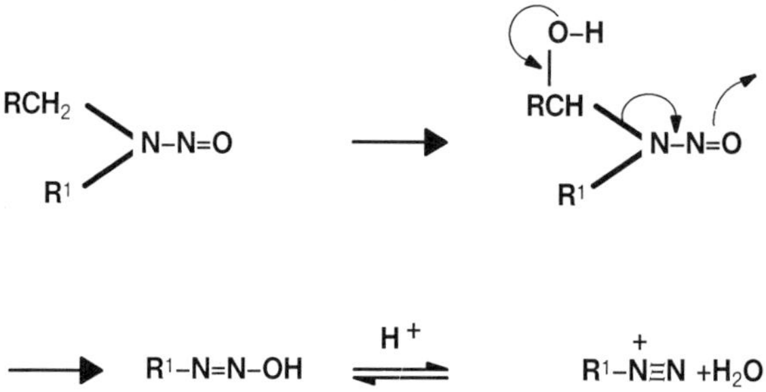

$$RCH_2{\overset{R^1}{>}}N-N=O \longrightarrow RCH{\overset{R^1}{>}}N-N=O \text{ (with O-H)}$$

$$\longrightarrow R^1-N=N-OH \underset{}{\overset{H^+}{\rightleftharpoons}} R^1-N\equiv N^+ + H_2O$$

Figure 2. Activation of alkylnitrosamines

to rats has been shown to result in formation of 7-n-propylguanine in hepatic DNA and other nitrosamines have been shown to alkylate protein.

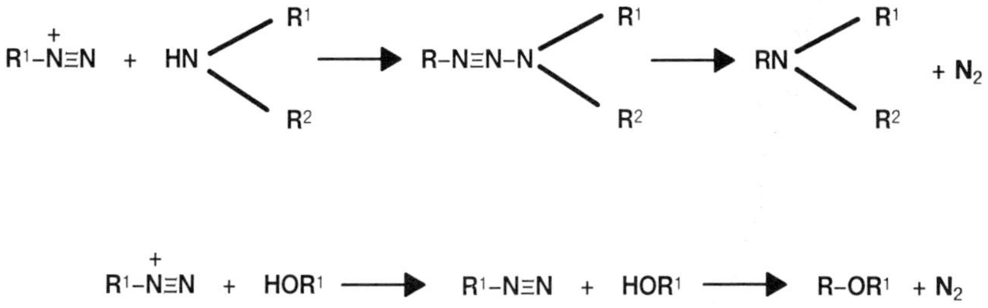

$$R^1-N\equiv N^+ + HN{\overset{R^1}{\underset{R^2}{<}}} \longrightarrow R-N\equiv N-N{\overset{R^1}{\underset{R^2}{<}}} \longrightarrow RN{\overset{R^1}{\underset{R^2}{<}}} + N_2$$

$$R^1-N\equiv N^+ + HOR^1 \longrightarrow R^1-N\equiv N + HOR^1 \longrightarrow R-OR^1 + N_2$$

Figure 3. Reaction of diazonium compounds with alcohols and amines

COVALENTLY-BOUND ADDUCTS

The potential toxicity of aromatic amines is well established as is their ability to form bound adducts. After a single 5 mg/kg intraperitoneal dose of 4-aminobiphenyl to rats, 8% of the dose is bound to red blood cells (Green et al., 1984). The amount of the adduct formed was directly proportional to dose over the range 0.5 µg/kg to 5 mg/kg and chronic administration showed an accumulation of the bound material of about thirty-fold. Mild acid treatment of the bound material released about 60% as

4-aminobiphenyl. In vitro experiments with the authentic N-hydroxy compound showed that it bound to haemoglobin with the loss of free sulphydryl groups. Based on these observations, a mechanism was proposed involving formation of the nitroso compound, reaction with thiol groups and rearrangement of the intermediate to a more stable sulphinyl adduct (Figure 4).

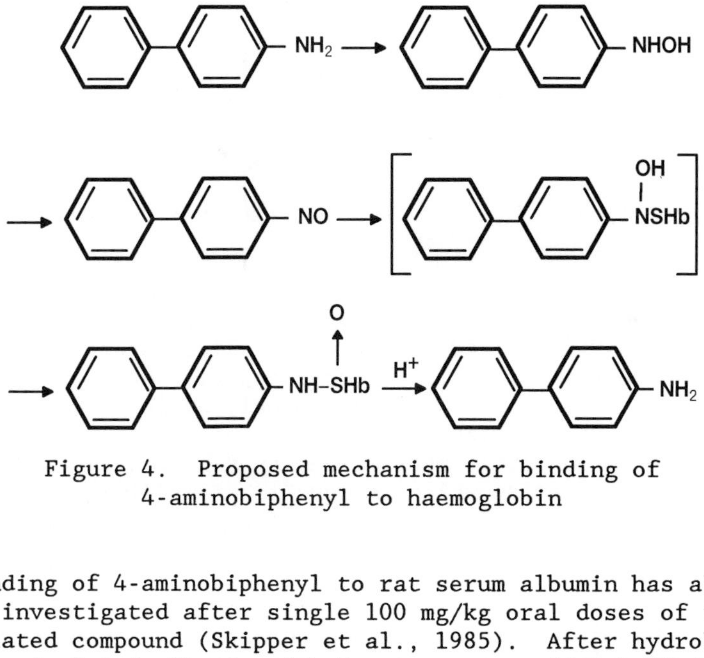

Figure 4. Proposed mechanism for binding of 4-aminobiphenyl to haemoglobin

Binding of 4-aminobiphenyl to rat serum albumin has also been investigated after single 100 mg/kg oral doses of the tritiated compound (Skipper et al., 1985). After hydrolysis of the isolated albumin with Pronase, one radiolabelled peptide adduct was isolated. Amino acid analysis indicated a tetrapeptide containing one valine and two alanine residues with an additional amino acid, identified by NMR as tryptophan, to which an aminobiphenyl group was bound. Since the complete amino acid sequence for rat serum albumin is known it was possible to devise a structure (16) for the peptide which was confirmed by FAB-MS. NMR indicated that the adduct was formed by 3-substitution in the biphenyl ring. The position of substitution in the indole nucleus of tryptophan was less clear but evidence was obtained that it occurred either at N-1 or C-3.

(16)

Imidazoquinolines and imidazoquinoxalines are structurally-related members of a class of carcinogenic heterocyclic aromatic amines which have been identified in cooked meat and fish. The quinoline (17) is an important member of this class and it is partly metabolized by N-hydroxylation to give a hydroxylamine which is capable of binding to macromolecules. Analysis of macromolecular carcinogen adducts provides a method of monitoring exposure and, to investigate this possibility, the binding of the quinoline to rat blood proteins has been investigated (Turesky et al., 1987). Serum albumin and haemoglobin were selected as the most appropriate proteins due to their abundance and biological half-life. After oral administration of radiolabelled compound, radioactivity bound to both proteins increased proportionally with dose. Albumin bound three to five times more radioactivity than haemoglobin per mole of protein. The identity of the major bound adducts was achieved by generation of chromatographically identical components from in vitro incubation of the compound with a microsomal activation system in the presence of rat serum albumin. The radiolabelled adducts were isolated after proteinase treatment of albumin. Synthetic compounds were also prepared by reacting N-hydroxy-(17) with albumin. Spectroscopic (NMR and MS) analysis and amino acid analysis provided good evidence that two major adducts consisted of a sulphinamide involving the cysteine moiety of the tripeptide cys-pro-try (18). The two compounds were identified as

(17) [structure]

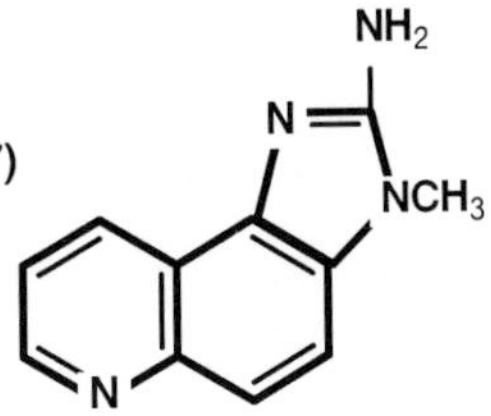

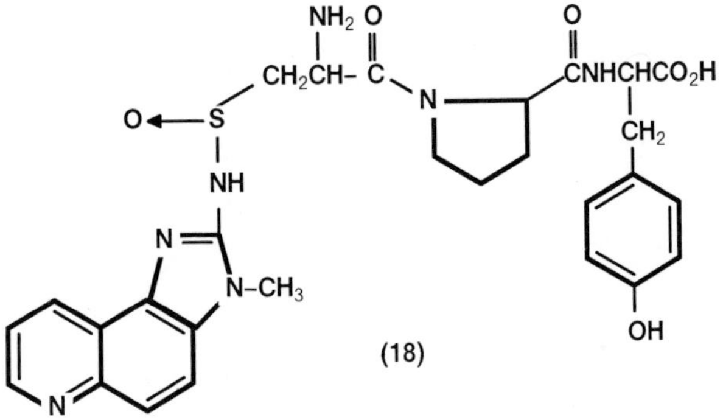

(18) [structure]

diastereoisomers differing only in the absolute configuration about the asymmetric sulphur. It was concluded that the nitroso derivative, which could be formed non-enzymatically from the hydroxylamine, was the most likely intermediate that reacted with the cysteine residue. It has been shown that various arylnitroso compounds react non-enzymatically with glutathione to form sulphinamides. A sulphinamide conjugate with haemoglobin was not observed which indicated that the hydroxylamine was probably not transported from the liver to blood.

N-Hydroxy-6-aminochrysene, a potential proximate carcinogen of the 6-amino and 6-nitro compounds has been shown to bind to DNA in rat hepatocytes. Two of the main adducts were isolated and identified as nucleoside adducts with deoxyguanosine (Delclos et al., 1987). The adducts were formed between the C-2 amino group and C-8 linked to the ortho position (19) and the amino group respectively in chrysene (20).

Acetylaminofluorene or its N-hydroxy derivative has been shown to bind to liver protein in rats (De Baun et al., 1970). The isolated protein adduct was hydrolysed with ethanolic alkali which released two components identified as the 1- and 3-methylmercapto derivatives. Formation of the

3-substituted isomer was favoured in vivo. The same adducts
were formed from an in vitro incubation of the N-hydroxy-
acetylamino sulphate with methionine. The proposed
mechanism for adduct formation involves N-hydroxylation and
sulphation to give an electrophilic carbonium ion intermediate which reacts with the nucleophilic sulphur in methionine.

Acetaminophen (21) is a widely used analgesic and is one
of the most studied examples of a compound which, at high
doses, becomes bound covalently to liver protein causing
extensive hepatic necrosis in both laboratory animals and
man. Biotransformation of the drug includes oxidation to a
reactive intermediate, N-acetyl-p-benzoquinoneimine (22),
which is inactivated by formation of a glutathione conjugate
(23). The nature of the bound protein adduct has now been
investigated incorporating bovine serum albumin into a mouse
liver microsomal preparation incubated with the ^{14}C-
substrate (Hoffmann et al., 1985). The protein adduct was
isolated, hydrolysed to the constituent amino acids and the
^{14}C-adduct isolated and identified after derivatization, by
mass spectrometry, as a 3-cysteinyl derivative. Hence it was
concluded that the structure of the adduct could be defined

as (24) involving the only free sulphydryl group in the albumin associated with cysteine-34.

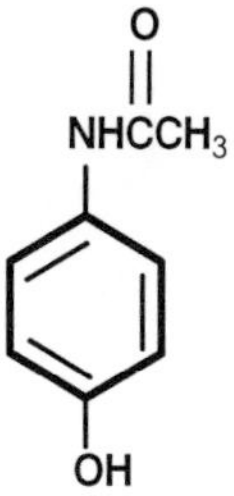

(21) 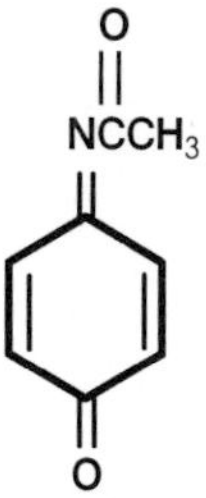 (22) 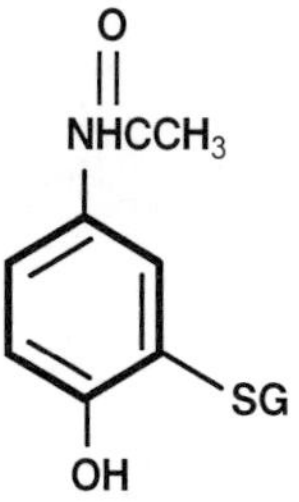 (23)

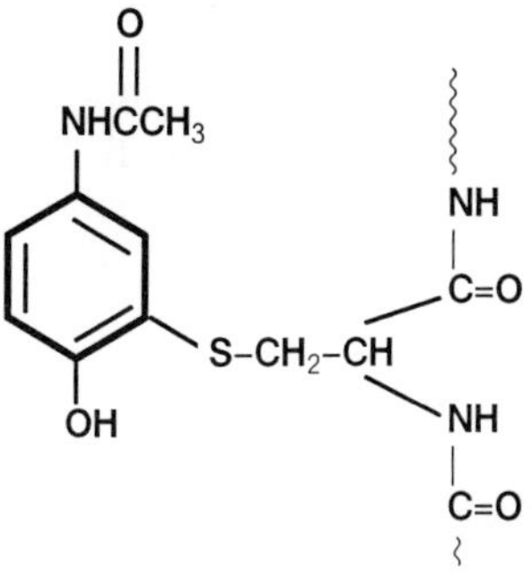

(24)

Ethylene oxide is an alkylating agent and the simplest example of the epoxide class. Epoxides are commonly encountered reactive intermediates in biotransformation processes and ethylene oxide is an intermediate metabolite of ethylene. Besides inactivation by hydrolysis and glutathione conjugation, they are also capable of reacting with tissue macromolecules and ethylene oxide has been shown to react with DNA and proteins. After exposure of mice to ethylene or ethylene oxide, the adduct $\underline{N}^7$-(2-hydroxyethyl) guanine (25) was the main product (Segerback, 1983). Similar hydroxyethylated adducts with cysteine and histidine in haemoglobin were also detected. The amount of the $\underline{N}$-(2-hydroxyethyl)histidine adduct (26) was shown to be virtually linearly related to the dose of ethylene oxide and it could also be correlated with the amount of DNA adduct formed in tissues.

Demonstration that the monomer vinyl chloride is an animal carcinogen prompted extensive investigations on its metabolism. Biotransformation of the compound proceeds via epoxidation to give chloroethylene oxide which spontaneously rearranges to chloroacetaldehyde. Both these intermediates

(25)   (26)

are highly reactive and bind to nucleic acid. Reaction of
chloroacetaldehyde with DNA in vitro showed the formation of
two major adducts, ethenodeoxyguanosine (27) and
ethenodeoxycytidine (28) (Green and Hathway, 1978). The
same two adducts were isolated from the liver DNA of rats
that had been exposed to vinyl chloride in their drinking
water. These adducts are formed by attack of the
bifunctional intermediates at the N-1 and C-6 amino group of
adenine and at the N-3 and C-4 amino group of cystosine.

(27)   (28)

Captopril (29) is an unusual drug which contains a
sulphydryl group, but studies on its biotransformation and
covalent binding may also be relevant for compounds in which
a sulphydryl group is introduced as a result of metabolism.
Incubation of the compound with plasma and blood of various
species showed that material bound mainly to albumin and was
greatest in the rat (Wong et al., 1981). Addition of
glutathione reduced the amount of bound material and
resulted in formation of captopril disulphide and disulphide
conjugates with glutathione and cysteine (30). Similar
results have been obtained from in vivo studies which
indicate that captopril forms disulphide links with cysteine
groups in plasma protein possibly involving the breaking of
existing disulphide links in albumin. It is of interest
that although haemoglobin contains more sulphydryl groups
than albumin, no extensive binding to this component occurs.

$$\underset{(29)}{\text{HSCH}_2\text{CH}-\overset{\overset{\text{CH}_3}{|}}{\underset{}{\text{C}}}-\text{N}\diagup\text{CO}_2\text{H}} \qquad \underset{(30)}{\text{HO}_2\text{CCHCH}_2-\text{S}-\text{SCH}_2\text{CH}-\overset{\overset{\text{CH}_3}{|}}{\underset{}{\text{C}}}-\text{N}\diagup\text{CO}_2\text{H}}$$

The major biotransformation pathway for cyanatryn (31) in rats involves glutathione conjugation by replacement of the methylthio group (Crawford et al., 1980). Conjugation proceeded via the S-oxide which was shown to react spontaneously with glutathione. A specific uptake into blood cells of material, which had a long biological half-life (about ten days) was also observed. Analysis showed that it was not extractable and was associated with haemoglobin. Only the S-oxide was found to react with the haemoglobin thiol groups. It is of interest that the extent of binding (in vitro) was species-dependent being very low in the human where formation of a glutathione conjugate predominated. Acid hydrolysis of the bound material yielded a component which corresponded to the hydroxyl derivative (32), the carboxylic acid function presumably being formed by hydrolysis of the nitrile. Extensive investigations have also been carried out on the binding of related triazines including simetryn (33) (Hamboeck et al., 1981) where evidence was obtained for the binding of an S-oxide to cysteine residues in haemoglobin. A reactive cysteine

(31) — triazine with SCH_3, C_2H_5HN, $NH-C(CH_3)_2CN$

(32) — triazine with OH, C_2H_5HN, $NH-C(CH_3)_2CO_2H$

(33) — triazine with SCH_3, C_2H_5HN, NHC_2H_5

(β-93) is present in many haemoglobins including man but it appeared that the only other cysteine in rat haemoglobin (β-125) was responsible for the binding. The rationale and explanation for the species differences was that only the β-125 located on the surface of the haemoglobin provided a conformationally acceptable binding site for the triazines.

Oxidation of aryl methyl groups to benzyl alcohols is a common biotransformation pathway; however, conjugation of the alcohol with sulphate affords compounds which are susceptible to nucleophilic attack. The sulphate esters of the 1'-hydroxy derivatives of estragole and safrole have similarly been implicated as intermediary metabolites in the binding of these carcinogenic natural products to DNA (see Alkenylbenzenes and structures 5-15 on pages 185-188).

CONCLUSIONS

When bound adducts are formed in metabolism studies they represent one of the most difficult types of metabolite to characterise. However, by increasing our knowledge and understanding of the types of interaction of reactive intermediates with components in macromolecules it should be possible to devise rational approaches and analytical methodology for the characterisation of these residues. Formation of bound adducts can be considered to be either a method of detoxification for a reactive intermediate or an initiating event for toxicity. More investigations on the nature of adducts and correlation with toxicity should help to make predictions on the implications of formation of particular types of adduct. Since these adducts have long biological half-lives, development of methods for their measurement provides a means of monitoring systemic exposure and also the extent of formation of potentially toxic metabolites. Application of these procedures to animal toxicity studies could provide valuable information to assist with interpretation of the results. Similar investigations in humans, potentially exposed to the same compounds, would provide a way of assessing systemic exposure and also allow a better extrapolation from animals to man with respect to toxicological risk (Farmer et al., 1987).

REFERENCES

Brusewitz, G., Cameron, B.D., Chasseaud, L.F., Gorler, K., Hawkins, D.R., Koch, H. and Mennicke, W.H., 1977, The metabolism of benzyl isothiocyanate and its cysteine conjugate. Biochemical Journal, <u>162</u>, 99-107.

Crawford, M.J., Hutson, D.H. and Stoydin, G., 1980, The metabolic fate of a herbicidal methylmercapto-s-triazine (cyanatryn) in the rat. Xenobiotica, 10, 169-185.

DeBaun, J.R., Miller, E.C. and Miller, J.A., 1970, N-Hydroxy-2-acetylaminofluorene sulfotransferase: Its probable role in carcinogenesis and in protein-(methion-S-yl) binding in rat liver. Cancer Research, 30, 577-595.

Delclos, K.B., Miller, D.W., Lay, J.O., Cascians, D.A., Walker, R.P., Fu, P.P. and Kadlubar, F.F., 1987, Identification of C8-modified deoxyinosine and N^2- and C8-modified deoxyguanosine, as major products of the in vitro reaction of N-hydroxy-6-aminochrysene with DNA and the formation of these adducts in isolated rat hepatocytes treated with 6-nitrochrysene and 6-aminochrysene. Carcinogenesis, 8, 1703-1709.

Farmer, P.B., Neumann, H.-G and Henschler, D., 1987, Estimation of exposure of man to substances reacting covalently with macromolecules. Archives of Toxicology, 60, 251-260.

Green, T. and Hathway, D.E., 1978, Interactions of vinyl chloride with rat-liver DNA in vivo. Chemico-Biological Interactions, 22, 211-224.

Green, L.C., Skipper, P.L., Turesky, R.J., Bryant, M.S. and Tannenbaum, S.R., 1984, In vivo dosimetry of 4-aminobiphenyl in rats via a cysteine adduct in haemoglobin. Cancer Research, 44, 4254-4259.

Hamboeck, H., Fischer, R.W., Di Iorio, E.E. and Winterhalter, K.H., 1981, The binding of s-triazine metabolites in rodent haemoglobins appears irrelevant to other species. Molecular Pharmacology, 20, 579-584.

Hassan, M. and Ehrsson, H., 1987, Metabolism of ^{14}C-busulfan in isolated perfused rat liver. European Journal of Drug Metabolism and Pharmacokinetics, 12, 71-76.

Ho, B. and Castagnoli, N., 1980, Trapping of metabolically generated electrophilic species with cyanide ion: Metabolism of 1-benzylpyrrolidine. Journal of Medicinal Chemistry, 23, 133-139.

Hoffmann, K.-J., Streeter, A.J., Axworthy, D.B. and Baillie, T.A., 1985, Structural characterisation of the major covalent adduct formed in vitro between acetaminophen and bovine serum albumin. Chemico-Biological Interactions, 53, 155-172.

Miwa, K., Okuda, H. and Watabe, T., 1987, The S-sulphate formation from 4-nitrobenzyl-mercaptan by rat liver cytosolic sulfotransferase and its covalent binding to the cytosolic proteins. Journal of Pharmaceutical Sciences, 76, S44.

Paulson, G.D., Feil, V.J. and Macgregor, J.T., 1987, Formation of a diazonium cation intermediate in the metabolism of sulphamethazine to desaminosulphamethazine in the rat. Xenobiotica, 17, 697-707.

Paulson, G.D. and Feil, V.J., 1987, Evidence for diazotisation of ^{14}C-sulfamethazine (4-amino-N-(4,6-dimethyl-2-pyrimidinyl)benzene[U-^{14}C]sulfonamide) in swine. Drug Metabolism and Disposition, 15, 841-845.

Quistad, G.B. and Hutson, D.H., 1986, Lipophilic xenobiotic conjugates. In Xenobiotic conjugation chemistry, edited by G.D. Paulson, J. Caldwell, D.H. Hutson and J.J. Menn (Washington: American Chemical Society), pp. 204-213.

Segerback, D., 1983, Alkylation of DNA and haemoglobin in the mouse following exposure to ethene and ethene oxide. Chemico-Biological Interactions, 45, 139-151.

Skipper, P.L., Obiedzinski, M.W., Tannenbaum, S.R., Miller, D.W. and Mitchum, R.K., 1985, Identification of the major serum albumin adduct formed by 4-aminobiphenyl in vivo in rats. Cancer Research, 45, 5122-5127.

Smithgall, T.E., Harvey, R.G. and Penning, T.M., 1988, Spectroscopic identification of ortho-quinones as the products of polycyclic aromatic trans-dihydrodiol oxidation catalysed by dihydrodiol dehydrogenase. Journal of Biological Chemistry, 263, 1814-1820.

Tulip, K., Timbrell, J.A., Nicholson, J.K. Wilson, I. and Troke, J., 1986, A proton magnetic resonance study of the metabolism of N-methylformamide in the rat. Drug Metabolism and Disposition, 14, 746-749.

Turesky, R.J., Skipper, P.L. and Tannenbaum, S.R., 1987, Binding of 2-amino-3-methylimidazol[4,5-f]quinoline to haemoglobin and albumin in vivo in the rat. Identification of an adduct suitable for dosimetry. Carcinogenesis, 8, 1537-1542.

Van Bladeren, P.J., Breimer, D.D., van Huijgevoort, J.A.T.C.M., Vermeulen, N.P.E. and van der Gen, A., 1981, The metabolic formation of N-acetyl-S-hydroxymethyl-L-cysteine from tetradeutero-1,2-dibromoethane. Relative importance of oxidation and glutathione conjugation in vivo. Biochemical Pharmacology, 30, 2499-2502.

Wong, K.K., Shih-Jung, L. and Migdalof, B.H., 1981, In vitro biotransformations of [^{14}C]captopril in the blood of rat, dogs and humans. Biochemical Pharmacology, 30, 2643-2650.

3. Methods for Studying Intermediary Xenobiotic Metabolism

CHEMICAL MODULATION OF XENOBIOTIC METABOLISM IN VIVO

David J. Jollow, Veronica F. Price, Stanley A. Roberts[1]
and Stephen L. Longacre[2]

Department of Pharmacology, Medical University
of South Carolina, Charleston, South Carolina 29425;
[1]Pre-Clinical Safety Assessment Department, Bldg. 404,
Sandoz Research Institute, East Honover, New Jersey 07936;
[2]Toxicology Department, Rohm and Haas Co., 727
Norristown Road, Spring House, Pennsylvania 19477, USA

INTRODUCTION
    It is well known that plant cells contain a wide variety of structurally diverse, lipophilic organic compounds which have no counterpart in animal cells, and appear to play no role in the intermediary metabolism of the plant cell. These compounds, termed xenobiotics, are not only non-nutrients but are potentially toxic for insects and higher organisms, and are considered to represent part of the defence of the plant kingdom against animal predators. To survive, herbivores and omnivores have had to evolve elaborate mechanisms for the detoxification and elimination of these potential toxins. With the advent of the industrial revolution, man-made xenobiotics, such as drugs and other industrial products, have been added to the biosphere. Since the xenobiotics present in the plant material are structurally diverse, it is not surprising that the enzyme systems of the mammalian liver, which evolved to convert these lipophilic and "non-excretable" compounds to water soluble and hence readily "excretable" derivatives, are equally diverse. These enzymes are known collectively as the xenobiotic metabolizing system (XMS) or drug metabolizing system (since drugs are an important subset of xenobiotics).
    The XMS is composed of a number of different types of enzymes which differ in regard to their reaction mechanism and the groups for which they are specific. The individual types of enzyme; viz, cytochrome P-450-dependent mixed function oxidase, glutathione transferase (GSH-T), UDP-glucuronosyltransferase (G-T), sulphotransferase (S-T), etc., consist in turn of a number of isozymes which differ in regard to their substrate specificity. Considerable overlap in substrate specificity exists between isozymes

such that a given xenobiotic may be metabolized to a single product by more than one isozyme.

Further, a xenobiotic molecule entering the liver is likely to be attacked by more than one component of the XMS. For example, depending on the number and variety of functional groups, the xenobiotic will be alternatively or sequentially acted on by various P-450 isozymes and the product of these reactions subjected to conjugation with glutathione, water, glucuronic acid, etc. Typically, a single xenobiotic molecular species will yield multiple metabolic products. Thus, the antihypertensive drug, propranolol, is known to yield well over 80 urinary metabolites. Even as simple a molecule as bromobenzene, can yield more than 12 urinary metabolites.

The identification of active and reactive metabolites of xenobiotics and the assessment of their contribution to the pharmacological or toxic effects of a xenobiotic thus presents a formidable challenge. An important experimental approach in these types of studies has been the use of chemical modulators to enhance or depress the activity of various components of the XMS, and thus allow a comparison of alteration in metabolite profile with alteration in biological response. The bulk of our present information on the effects of modulators on the activities of the various components of the XMS has been obtained from *in vitro* enzyme activity assays. Much less is known as to the selectivity and utility of chemical modulators in the intact animal where ultimately the comparisons of metabolite formation and biological response must be made.

THEORETICAL ASPECTS

Certain general features of xenobiotic metabolism are readily apparent. As illustrated in Figure 1, a xenobiotic may be cleared by linear sequences (I→II→III), branched sequences (I→X→XI and XIII), reversible reactions (I⇌XX) which may set up quasi-equilibria, and via reactive metabolites (RM). Such reactive species are usually entirely detoxified, but may, on occasion, bind covalently to tissue macromolecules. The significance of the covalent binding reaction for tissue toxicity now has an extensive literature. Of importance, a single reaction product (e.g. XXI) may arise from more than one pathway.

Most commonly, human exposure to individual xenobiotics is to low or extremely low levels and tissue levels rarely reach micromolar concentrations. Even in the situation of drug dosage, where a relatively large amount of a single chemical entity is administered, equilibrium tissue levels

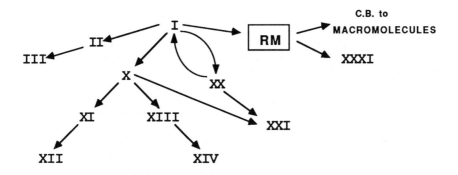

- linear sequences
- branched sequences
- quasi-equilibria
- reversibility
- reactive metabolites
- multiple pathways to single product

Figure 1. General features of pathways of xenobiotic metabolism. C.B. = covalent binding.

are seldom above low micromolar concentrations. Since the K_m values of the enzymes of the XMS are usually in the 100 to 400 micromolar range, most xenobiotic metabolism occurs with xenobiotic concentrations of less than 10% of the K_m value. In this situation, the first order rate constant of the reaction (k) approximates to V_{max}/K_m. Provided that the elimination is not delivery-limited, the first order rate constant for overall elimination (β) is then the sum of the rate constants for the individual pathways of metabolism. That is,

$$\beta = k_1 + k_2 + \ldots\ldots = \frac{V_{max_1}}{K_{m_1}} + \frac{V_{max_2}}{K_{m_2}} + \ldots\ldots$$

Under conditions of overall first order elimination, and total or near total recovery of the dose as urinary metabolites, apparent values for the first order rate constants for the individual primary pathways of metabolism may be estimated from relationships such as,

$$k'_{12} = \frac{II_{AM}}{I_{AM}} \cdot \beta$$

where k'_{12} is the apparent rate constant for conversion of I to II; II_{AM} is the amount of metabolite II found in the urine, and I_{AM} is the dose of compound I.

Estimation of the apparent rate constants for metabolite formation is of particular value for studies on the effects of chemical modulators on the XMS since it allows assessment of the effects of the modulator on the individual primary pathways of metabolism. Induction is reflected in an increase in the k' value since enhancement of the amount of enzyme in the liver acts to increase the V_{max} value. Conversely, inhibition of a pathway results in a decrease in the k' value secondary to an increase in K_m value (competitive inhibition) or decrease in V_{max} value (non-competitive inhibition).

CLASSIFICATION OF CHEMICAL MODULATORS

Chemical modulators of xenobiotic metabolic pathways may be conveniently classified on the basis of mechanisms of action (Table 1). Agents in category I act to alter the amount of enzyme in the tissue. Subgroup IA comprises the inducing agents. Examples include phenobarbitone, polycyclic aromatic hydrocarbons such as 3-methyl-cholanthrene (3MC), clofibrate and analogues, ethanol and isoniazid, and cyanopregnenolone and other steroids (Siest et al., 1988).

An illustration of the effect of the classical inducers, phenobarbitone and 3MC on the overall elimination of acetaminophen (paracetamol) in the hamster and on the apparent rate constants for its major primary pathways of metabolism, is shown in Figure 2. As indicated, acetaminophen is cleared by first order processes. Except for minor mechanical losses, the dose of acetaminophen is recoverable in the urine as unchanged drug and metabolites. Determination of the overall elimination rate constant (β) and the urinary metabolite composition for individual animals thus permits estimation of the apparent rate constants for each pathway (Jollow et al., 1982). In this model system, the apparent rate constant for acetaminophen mercapturate formation is a measure of the activity of the P-450 mixed function oxidase which activates acetaminophen to its reactive metabolite, since the glutathione $\underline{S}$-transferase activity is not rate limiting.

As shown in Figure 2, both phenobarbitone and 3MC significantly enhanced the elimination of acetaminophen in the hamsters. The increase in the overall rate constant (β)

TABLE 1 - Classification of chemical modulators of xenobiotic metabolism

CATEGORY	EXAMPLE
I. Alter Amount of Enzyme in Tissue	
A. Inducing Agents	PB, 3MC, ethanol
B. Suicide inhibitors	AIA, CCl_4, CS_2
C. Suppressors of Prosthetic Group Synthesis	$CoCl_2$, Cobalt haem
II. Alter Catalytic Activity of Enzyme	
A. Competitive Inhibitors	Alt. Substrates, Pip butoxide DCNP, SKF525A, BNPP, TCPO,
B. Non-Competitive Inhibitors	Ethanol, α-NF
C. Allosteric Modulators	Unknown
III. Alter Availability of Co-Substrates	
A. Alternate Substrates	PABA
B. Depleters	DEM, GALN
C. Suppressors of Synthesis	BSO

Abbreviations: PB, Phenobarbitone; 3MC, 3-methylcholanthrene; AIA, allylisopropylacetamide; Cobalt haem, cobalt protoporphyrin IX; Pip butoxide, piperonyl butoxide; DCNP, Dichloronitrophenol; SKF525-A, Prodafen; TCPO, trichloropropylene oxide; α-NF, α-naphthoflavone; PABA, p-aminobenzoic acid; DEM, diethylmaleate; BSO, buthionine sulphoximine.

was, in each case, due to increases in both the P-450 pathway (K'_{MA}) and the UDP-glucuronosyltransferase pathway (K'_G). These data clearly illustrate both the responsiveness of the XMS to inducing agents and the lack of specificity of the inducing agents for P-450 activity alone. In analogous experiments with rats, we have observed that phenobarbitone and 3MC pretreatments do not enhance the value of K'_G,

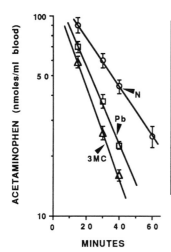

	β	K'$_G$	K'$_S$	K'$_{MA}$	K$_E$
N	1.75 ±0.10	0.70 ±0.03	0.68 ±0.06	0.25 ±0.02	0.07 ±0.04
Pb	*3.03 ±0.15	*1.53 ±0.12	0.90 ±0.12	*0.37 ±0.04	0.09 ±0.06
3MC	*3.20 ±0.21	*1.24 ±0.13	0.60 ±0.12	*0.84 ±0.13	0.11 ±0.03

* Significantly increased from controls.

Figure 2. Effects of phenobarbitone (PB) and 3-methylcholanthrene (3MC) pretreatments on the elimination of acetaminophen (20 mg/kg i.p.) in hamsters. The hamsters were pretreated with phenobarbitone (80 mg/kg i.p. x 3) or 3MC (20 mg/kg i.p. x 3) in standard regimens and then given ^{3}H-acetaminophen (20 mg/kg i.p.). Blood half-life and urinary metabolite composition were determined and used to calculate the kinetic parameters (Jollow et al., 1982) (β = overall elimination rate constant, K' values are apparent rate constants for formation of glucuronide, K'$_G$, sulphate, K'$_S$, and mercapturate, K'$_{MA}$, and renal elimination rate constant, K$_E$). The values are means ± SD, N=4.

indicating that in this species acetaminophen glucuronosyltransferase is not inducible (Price and Jollow, 1987). These in vivo data are in agreement with the microsomal assay data of Ulrich and Bock (1984).

However, additional studies in rats on the effect of diet on the activity of the enzymes responsible for acetaminophen clearance (Figure 3), indicated that after 20 days of feeding a completely synthetic diet (AIN-76), the rats had significantly lower apparent rate constants for glucuronidation and the P-450 dependent pathway (Price and Jollow, 1988). These data clearly suggest that the standard laboratory chow (Lab Blox) contains an inducing agent which maximally induces acetaminophen glucuronosyltransferase activity and partially induces the P-450 dependent pathway in the rat.

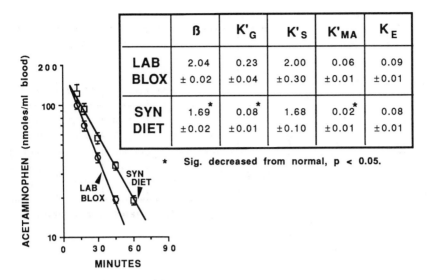

Figure 3. Effect of a synthetic diet on the elimination of acetaminophen (20 mg/kg i.p.) in rats. The rats were fed lab blox or completely synthetic diet (American Institute of Nutrition, AIN-76) for 20 days. Growth rates were optimal and identical. The rats were then given ^{14}C-acetaminophen and blood half-life and urinary metabolite composition determined. The kinetic parameters were calculated as previously described (Jollow et al., 1982) and as defined under Figure 2. The values are means $\pm$ SD, N=4.

When fed the synthetic diet, the rats responded to chronic ethanol exposure in a complex fashion (Table 2). Chronic ethanol caused an induction of the P-450-dependent metabolism of acetaminophen. Acetaminophen-sulphotransferase and -glucuronosyltransferase activities were also enhanced. However, at the dose of acetaminophen used in this experiment (300 mg/kg), these conjugative pathways are partially capacity-limited due to decreased cosubstrate availability. The observed increase in activity could thus be due to either an increase in amount of enzyme (that is, induction) or an increase in cosubstrate availability, or both. Since acetaminophen-sulphotransferase is thought not to be an inducible enzyme, it seems likely that these observed enhancements in <u>in vivo</u> activity reflect in part a non-specific change in the metabolic state of the cell. Collectively, the data indicate that in the intact animal, modulating agents may influence the activity of the enzymes of the XMS by mechanisms in addition to the

TABLE 2 - Effect of chronic and acute ethanol on the kinetic parameters of elimination of a high dose of acetaminophen (300 mg/kg) in rats fed a synthetic diet.

Diet	β (hr^{-1})	Apparent rate constant (hr^{-1})			
		K'_G	K'_S	K'_{MA}	K_E
AIN-76	0.35^a ±0.06	0.14^a ±0.05	0.14^a ±0.03	0.02^a ±0.003	0.04^a ±0.01
AIN-76 + Chronic EtOH	0.59^b ±0.03	0.23^b ±0.03	0.27^b ±0.02	0.04^b ±0.01	0.05^a ±0.01
AIN-76 + Acute EtOH	0.18^c ±0.04	0.07^c ±0.03	0.08^c ±0.02	0.005^c ±0.002	0.01^b ±0.01

Rats were fed AIN-76 diet for 20 days, with or without ethanol (12g/kg/day). On day 21, all animals received AIN-76 diet alone. On day 22, the rats received saline or acute ethanol (2.7 g/kg), 45 min prior to acetaminophen (300 mg/kg). Blood half-life and urinary metabolite composition were determined and used to calculate the kinetic parameters (Jollow et al., 1982). Values with different superscripts within a column are significantly different, $p < 0.05$, n=4.

classical "direct" effects of induction and inhibition.

Agents in subgroup IB (Table 1) depress the activity of cytochrome P-450 isozymes by selective destruction (for a review, see Ortiz de Montellano and Correia, 1983). This type of inhibition has been termed "suicidal" because the P-450 isozymes which accept the inhibitor as a substrate act on it to convert it to a chemically reactive metabolite. This reactive species immediately attacks the haem prosthetic group of the isozyme which formed it and causes inactivation. This suicidal nature of the inhibition was first recognized in early studies on the mechanism of P-450 destruction by allylisopropylacetamide (AIA) and related compounds. Analogous destructive behaviour toward P-450 species is now known to occur with a wide range of chemical structures including haloakanes (e.g. CCl_4), internal acetylenes, alkenes, dihydropyridines and dihydroquinolines (Ortiz de Montellano and Reich, 1986).

Little information is currently available about the selectivity of suicidal type inhibitors for specific P-450 isozyme-dependent metabolism in the whole animal. Since the specificity for P-450 inactivation is conferred by the substrate specificity of the isozymes, this type of inhibition is potentially highly specific. The design of suicidal-type inhibitors to have minimal substrate-specificity overlap with other P-450 isoforms, would permit the deletion of individual P-450 isoform-dependent activity from the livers of the intact animals, and hence allow the development of animal models to mimic specific P-450 isoform genetic deficiencies of man.

The utility of suicidal-type inhibitors for chronic studies on xenobiotic metabolism is more problematic. For example, AIA which is known to promote the destruction of phenobarbitone-inducible isozymes of P-450, causes a drastic reduction in cellular haem and hence promotes a marked P-450 induction phenomenon (De Matteis, 1978). However, it seems plausible that the greater the specificity of a suicidal-type inhibitor for an individual isoform of P-450, the less will be the rebound P-450 synthesis phenomenon and the greater will be the usefulness of the chemical modulator for chronic studies.

Agents in group IC (Table 1) act by depressing the availability of enzyme prosthetic groups, such as haem. The prototype compound in this group is cobalt chloride, recommended by Tephly and colleagues (Tephly et al., 1973) for selective inhibition of P-450 activities as compared with other XMS-catalyzed reactions. Subsequent studies however have indicated that cobalt chloride has multiple effects on the liver cell and especially on XMS activities. Cobalt ion effects include marked enhancement of hepatic glutathione content (Maines and Kappas, 1977, Sasame and Boyd, 1978), suppression of acetaminophen sulphotransferase and enhancement of acetaminophen glucuronosyltransferase activities (Roberts et al., 1986). The enhancement of acetaminophen glucuronosyltransferase activity was seen only at high doses of the drug and appeared to be secondary to a generalized effect on carbohydrate homeostasis which resulted in enhanced production of UDPGA, the co-substrate of the enzyme.

More recently, Drummond and Kappas (1982) reported that cobalt protoporphyrin IX (cobalt haem) decreased hepatic P-450 levels in the rat by over 80% and that the decrease lasted over 10 days. They reported that the decrease in cytochrome P-450 levels was secondary to enhancement of haem destruction and proposed that cobalt haem was a particularly useful tool for the study of the effects of long-term P-450 depletion.

Spaethe and Jollow (1988) assessed the specificity of cobalt haem for P-450 activity as compared with phase II and other metabolic pathways in the hamster and rat. Antipyrine and lidocaine clearances were markedly suppressed in both species. Sulphotransferase, glutathione-$\underline{S}$-transferase and glucuronosyltransferase were unaffected. The flavoprotein microsomal mixed function oxidase activity was modestly increased in the hamster. Intermediary carbohydrate metabolism was perturbed in both species, though to a lesser extent than that seen with cobalt chloride. Collectively, the data supported the proposal of Drummond and Kappas (1982) that cobalt haem is a valuable tool for the study of the roles of hepatic P-450 in the intact animal.

Category II compounds alter the activity of the XMS enzymes in tissues, by combining with the enzymes in a reversible fashion with a resultant decrease in catalytic activity. This category of compounds is by far the largest, since all substrates for XMS enzymes are by virtue of their affinity for the active site of the enzymes, potential inhibitors of other substrates for the same enzyme. In practice, useful inhibitors for in vivo studies are much fewer in number. To be most useful, a compound needs to be absolutely selective for one pathway (or one type of pathway, e.g. P-450 oxidation) or to show at least an order of magnitude difference in K_i value towards one (or one type of) pathway as compared with other pathways of xenobiotic metabolism.

Prototype inhibitors in this category include the insecticide synergists such as piperonyl butoxide and prolonging agents such as SKF 525-A. While these compounds are effective inhibitors of P-450 activity in vivo and have been used extensively for this purpose, it should be noted that they are not highly selective. Thus SKF 525-A and its analogues not only inhibit various P-450 oxidative reactions but also suppress microsomal esterase, monoamine oxidase, and glucuronyltransferase (Mannering, 1971; Anders, 1971).

It should also be noted that while the classification of competitive (group IIA) and non-competitive (group IIB) is highly useful when dealing with enzyme preparations in vitro, the utility of the distinction is lost when the compounds are given to the intact animals. For example, the competitive inhibitor, SKF 525-A, is dealkylated in vivo to its secondary and primary amine analogues, which may act as non-competitive inhibitors. The inhibitory activity of SKF

525-A in vivo results from the combined effects of all three analogues and at best the inhibitory activity should be regarded as mixed.

The lack of selectivity of inhibitory activity in vivo may be illustrated by the effect of isosafrole on acetaminophen elimination in hamsters (Figure 4). While isosafrole administered immediately prior to acetaminophen was highly effective in suppression of the cytochrome P-450-dependent metabolism (K'_{MA}), significant blockage was also seen with the sulphotransferase and glucuronosyltransferase pathways.

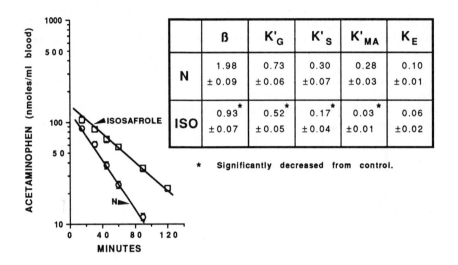

Figure 4. Effect of isosafrole (150 mg/kg i.p.) 30 min prior to acetaminophen (20 mg/kg i.p.) on the metabolic elimination of acetaminophen in hamsters. Blood half-life and urinary metabolite composition were determined and used to calculate the kinetic parameters (Jollow et al., 1982) as defined under Figure 2. The values are means $\pm$ SD, N=4.

A more successful example of selective inhibition was seen in studies on dichloronitrophenol (DCNP) inhibition of acetaminophen sulphation (Miller and Jollow, 1987). DCNP, proposed for this purpose by Mulder and colleagues (Koster et al., 1979), was also found to inhibit the glucuronidation and P-450 activation of acetaminophen in freshly isolated hepatocytes. However, the K_i value for inhibition of acetaminophen-sulphotransferase activity was about ten-fold lower than the K_i values for inhibition of the other pathways. A dose of 10 mg DCNP/kg to acetaminophen-treated hamsters was observed to decrease K'_s without discernible effect on the apparent rate constants for other major pathways of metabolism.

The difficulty of predicting the usefulness of XMS substrates as selective chemical modulators was illustrated to us by studies on the inhibitory activity of structural analogues of salicylamide towards the three major pathways of acetaminophen metabolism in isolated hepatocytes. As shown in Figure 5, salicylamide simultaneously inhibited all three pathways, (glucuronidation, sulphation, and P-450 activation [as measured by glutathione conjugate formation]), with selectivity towards the non-cytochrome P-450 pathways. Addition of a second phenolic hydroxyl to the molecule in either the 4 or 6 position did not markedly improve the selectivity. In contrast, methylation of the phenolic hydroxyl groups decreased inhibitory action towards the P-450 pathway, yet retained significant activity towards the conjugative pathways, for which it was not a direct substrate. 2,4-Dimethoxyacetophenone was found to be a useful selective inhibitor of acetaminophen metabolism in the intact hamster in that it blocked the conjugative pathways without inhibiting the P-450 activity.

Allosteric activation and inhibition of enzyme activity (group IIC) is a well known phenomenon. In regard to XMS activity, in vitro studies using purified enzyme have established that the flavoprotein microsomal mixed function oxidase activity is subject to allosteric activation by aliphatic primary amines and related compounds (Ziegler, 1980). However, the efficacy of these compounds in modulating flavoprotein mixed function oxidase activity in vivo is yet to be assessed.

The third major category of chemical modulators contains those compounds which alter XMS activity by depressing the availability of the co-substrates for the enzyme-catalyzed reactions. Group IIIA compounds act as alternate substrates. For example, it is well known that the N-acetyltransferase activity of the liver is due to the action of a family of enzymes with differing substrate specificities (Weber and Glowinski, 1980). However, acetyl CoA is the common co-substrate. Thus a substrate for one N-acetyltransferase could inhibit the activity of a different N-acetyltransferase activity by competing for available acetyl CoA. p-Aminobenzoic acid (PABA) is a relatively non-toxic compound which, when given in large doses, appears to drain acetyl CoA pools effectively and suppress other N-acetylations.

Group IIIB compounds serve to deplete co-substrate pools by mechanisms other than, or in addition to, competing for

INHIBITOR		GLUTATHIONE CONJUGATE	GLUCURONIDE CONJUGATE K_i (mM)	SULFATE CONJUGATE
SALICYLAMIDE	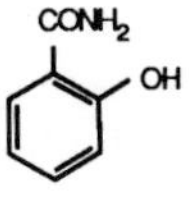 (structure)	0.30 (C)	0.035 (C)	0.015 (C)
2,4-DIHYDROXYBENZAMIDE	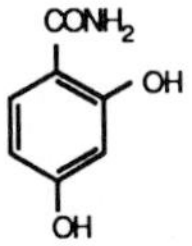 (structure)	0.26 (C)	0.082 (C)	0.25 (C)
2,6-DIHYDROXYBENZAMIDE	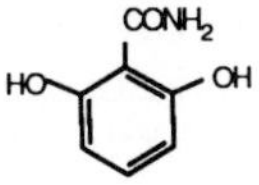 (structure)	0.66 (C)	0.037 (C)	0.037 (C)
2,6-DIMETHOXYBENZAMIDE	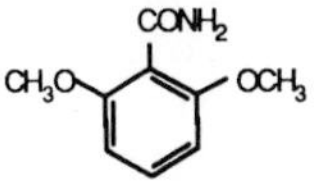 (structure)	NI	1.40 (NC)	0.092 (C)
o-HYDROXYBENZYL ALCOHOL	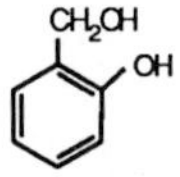 (structure)	0.74 (C)	0.40 (C)	0.87 (C)
2,6-DIHYDROXYACETOPHENONE	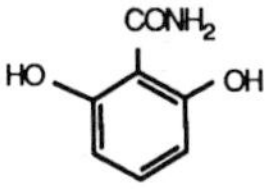 (structure)	0.37 (C)	0.17 (C)	0.059 (C)
2,6-DIMETHOXYACETOPHENONE	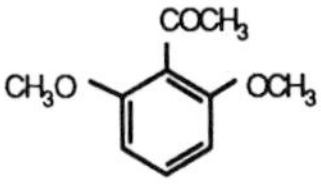 (structure)	NI	0.26 (NC)	0.11 (C)

Figure 5. Inhibition of acetaminophen metabolism by salicylamide and related compounds in hepatocytes freshly isolated from hamsters. Various concentrations of ^{3}H-acetaminophen and inhibitors were incubated with hamster hepatocytes for 30 min at 37°. The amounts of acetaminophen-glutathione conjugate, acetaminophen-glucuronide and acetaminophen-sulphate formed were determined and used to construct Lineweaver-Burk and Dixon plots. The K_i values were derived from the plots. NI = Not inhibited, C = competitive inhibition and NC = non-competitive inhibition.

co-substrates. For example, galactosamine is known to trap uridine nucleotides as UDP-galactosamine (Keppler and Decker, 1969). UDP-Glucuronic acid levels in the liver thus decrease with resultant inhibition of glucuronidation (Abou-El-Makarem et al., 1975, Singh and Schwartz, 1981). In the intact animal, co-administration of galactosamine with acetaminophen selectively blocked the glucuronidation of the drug (Jollow et al., 1982).

Diethyl maleate (DEM) is a substrate for some glutathione-S-transferase enzymes and hence could be regarded as a competing substrate (group IIIA). However, DEM can act as a Michael acceptor and deplete glutathione even in the absence of glutathione-S-transferase. Since its ability to block glutathione-S-transferase activity is not dependent on the presence in the tissue of the particular isoform(s) of the transferase which use it as a substrate, it is more useful to classify it as a general depleter (group IIIB).

Group IIIC compounds act by blocking the synthesis of essential co-substrates. The prototype of this group is buthionine sulphoximine (BSO) which inhibits γ-glutamyl-cysteine synthetase and hence depresses cellular glutathione levels (Griffith and Meister, 1979).

It is important to note that since the K_m values of the XMS enzymes towards the endogenous co-substrates are usually low, depletion of the co-substrates needs to be extensive before inhibition of xenobiotic metabolism occurs. In practice, utility of cosubstrate depletion as a chemical modulation is greatest when the dose of xenobiotic administered to an animal is high.

CONCLUSIONS

A number of general conclusions may be drawn from this brief review. Firstly, a large number of chemical modulators suitable for administration to laboratory animals is available and these compounds can significantly alter the in vivo activity of the XMS enzymes. The modulators exert their effects by a variety of mechanisms. It is experimentally feasible to combine modulators in ways such that effects on individual pathways may be more than additive. For example, the combined regimens in hamsters, of 3-MC induction (IA) plus DEM depletion of hepatic glutathione (IIIB) plus borneol competitive inhibition of glucuronidation and sulphation (IIA) act to decrease the minimum hepatotoxic dose of acetaminophen in the animals from about 250 mg/kg to therapeutic dose levels (ca. 30 mg/kg).

Secondly, the chemical modulators currently available as a group are only moderately selective. This is particularly a

TABLE 3 - Suggested modulators of pathways of xenobiotic metabolism *in vivo*

Pathway	Modulator	Specificity	Comment
Cytochrome P-450	Inducers	Low	Also induce GSH-T, G-T, EH.
	Inhibitors (alternative substrates)	Variable	Frequently also inhibit G-T, S-T.
	Cobalt protoporphyrin IX	High	Long-lasting; suitable for chronic studies.
Individual isozymes	Selective inducers and inhibitors	Variable	Area of need.
Flavoprotein MFO	Methimazole	?	Weak P-450 inhibition?
Glutathione-transferase (GSH-T)	Diethyl maleate Buthionine sulphoximine	Moderate	Lowers GSH concentration
Sulpho-transferase (S-T)	Dichloro-nitrophenol	High	Weak inhibitor of P-450, G-T
Glucuronyl-transferase (G-T)	Alternative substrates (acetaminophen, borneol)	Low	Also inhibits S-T
	Galactosamine	High	Depletes UDPGA
Epoxide hydrolase (EH)	Trichloro-propylene oxide Cyclohexene oxide	?	
N-Acetyl-transferase	p-Aminobenzoic acid	Moderate	Weak inhibitor of P-450
Carboxyl-esterase	Bis-p-nitro-phenyl phosphate	High	

problem in category II where cross inhibition between different types of pathways tends to be the rule rather than the exception. There is a need for the identification of inhibitors which are selective in the in vivo setting for individual pathways, and ultimately, for individual isoforms of the enzymes.

Table 3 presents a list of suggested modulators for the various pathways of xenobiotic metabolism, together with some comment on the specificity of the modulator. In the present state-of-the-art, we are able to alter markedly the activity of various pathways of xenobiotic metabolism, though the selectivity of the effects is suboptimal. We have very little ability to modulate individual isoforms within a given pathway.

Thirdly, it is apparent that since xenobiotics are simultaneously metabolized by multiple members of the XMS families, suppression of one primary activity will result in increased amounts of other metabolites being formed. Thus in experiments aimed at correlating metabolite formation with biological effect, it may be necessary to consider the possible significance of enhanced amounts of the other metabolites.

Lastly, and perhaps most important, our current available information indicates that effects can occur in vivo which are not predictable from in vitro studies on inhibition/enhancement of individual pathways of xenobiotic metabolism. As clearly documented by the cobalt chloride studies, effects on general intermediary homeostasis can have a profound effect on the activity of individual pathways of xenobiotic metabolism.

Acknowledgements. The authors thank Ms. Jennifer Schulte for excellent technical assistance and for preparation of the Figures. This work was supported by grants from the U.S. Public Health Service (GM 30546 and GM 36687).

REFERENCES

Abou-El-Makarem, M.M., Otani, G. and Bock, K.W., 1975, Glucuronidation of 1-naphthol and bilirubin in intact liver and microsomal fractions: Influence of the uridine diphosphate glucuronic acid content. Biochemical Society Transactions, 3, 881-883.

Anders, M.W., 1971, Enhancement and inhibition of drug metabolism. Annual Reviews of Pharmacology, 11, 37-56.

De Matteis, F., 1978, Hepatic porphyrias caused by 2-allyl-2-isopropylacetamide, 3,5-diethoxy-carbonyl-1,4-dihydrocollidine, griseofulvin, and related compounds. In Heme and Hemoproteins, Handbook of Experimental Pharmacology,

edited by F. De Matteis and W.N. Aldridge (New York: Springer-Verlag), 44, 129-137.
Drummond, G.S. and Kappas, A., 1982, The cytochrome P-450-depleted animal: An experimental model for in vivo studies in chemical biology. Proceedings of the National Academy of Sciences, 79, 2384-2390.
Griffith, O.W. and Meister, A., 1979, Potent and specific inhibition of glutathione synthesis by buthionine sulfoximine. Journal of Biological Chemistry, 254, 7558-7560.
Jollow, D.J., Roberts, S., Price, V.F., Longacre, S. and Smith, C., 1982, Pharmacokinetic consideration in toxicity testing. Drug Metabolism Reviews, 13, 983-1007.
Keppler, D. and Decker, K., 1959, Studies on the mechanism of galactosamine hepatitis: accumulation of galactosamine-1-phosphate and its inhibition of UDP-glucose pyrophosphorylase. European Journal Biochemistry, 10, 219-225.
Koster, H., Scholtens, E. and Mulder, G.J., 1979, Inhibition of sulfation of phenols in vivo by 2,6-dichloro-4-nitrophenol. Selectivity of its action in relation to other conjugations in the rat in vivo. Medical Biology (Helsinki), 57, 340-344.
Maines, M.D. and Kappas, A., 1977, Regulation of heme pathway enzymes and cellular glutathione content by metals that do not chelate with tetrapyrroles: blockade of metal effects with thiols. Proceedings of the National Academy of Sciences USA, 74, 1875-1878.
Mannering, G.J., 1971, Inhibition of drug metabolism. In Handbook of Experimental Pharmacol, edited by B.B. Brodie and J.R. Gillette (New York: Springer-Verlag), 28, 452-476.
Miller, M.G. and Jollow, D.J., 1987, Relationship between sulfotransferase activity and susceptibility to acetaminophen hepatotoxicity. Drug Metabolism and Disposition, 15, 143-150.
Ortiz de Montellano, P.R. and Correia, M.A., 1983, Suicidal destruction of cytochrome P-450 during oxidative drug metabolism. Annual Reviews of Pharmacology and Toxicology, 23, 481-503.
Ortiz de Montellano, P.R. and Reich, N.O., 1986, Inhibition of cytochrome P-450 enzymes. In Cytochrome P-450 edited by P.R. Ortiz de Montellano (New York: Plenum Press) pp. 273-314.
Price, V.F. and Jollow, D.J., 1987, Regulation of acetaminophen glucuronidation. 7th International Symposium of Microsomes and Drug Oxidation, Adelaide, S.A., Australia.

Price, V.F. and Jollow, D.J., 1988, Effect of sulfur deficient diet on acetaminophen metabolism and toxicity in rats. Toxicologist, 8, 32.

Roberts, S.A., Price, V.F. and Jollow, D.J., 1986, The mechanisms of cobalt chloride-induced protection against acetaminophen hepatotoxicity. Drug Metabolism and Disposition, 14, 25-33.

Sasame, H.A. and Boyd, M.R., 1978, Paradoxical effects of cobalt chloride and salts of other divalent metals on tissue levels of reduced glutathione and microsomal mixed-function oxidase components. Journal of Pharmacology Experimental Therapeutics, 205, 718-724.

Siest, G., Batt, A.M., Fournel-Gigleux, S., Galteau, M.M., Wellman-Bednawska, M., Minn, A. and Amar-Costesec, A., 1988, Induction of plasma and tissue enzymes by drugs: significance in toxicological studies. Xenobiotica, 18, Supp. 1, 21-34.

Singh, J. and Schwartz, L.R., 1981, Dependence of glucuronidation rate on UDP-glucuronic acid levels in isolated hepatocytes. Biochemical Pharmacology, 30, 3252-3254.

Spaethe, S.M. and Jollow, D.J., 1988, Effect of cobalt heme on hepatic drug metabolizing enzymes: specificity for cytochrome P450. Biochemical Pharmacology, submitted.

Tephly, T.R., Webb, C., Trussler, P., Kniffen, F., Hasegawa, E. and Piper, W., 1973, The regulation of heme synthesis related to drug metabolism. Drug Metabolism and Disposition, 1, 259-266.

Ulrich, D. and Bock, K.W., 1984, Glucuronide formation of various drugs in liver microsomes and in isolated hepatocytes from phenobarbital- and 3-methylcholanthrene-treated rats. Biochemical Pharmacology, 33, 97-101.

Weber, W.W. and Glowinski, I.B., 1980, Acetylation. In *Enzymatic Basis of Detoxication* Vol. II edited by W.B. Jakoby, (New York: Academic Press) pp. 169-182.

Ziegler, D.M., 1980, Microsomal flavin-containing monooxygenase: Oxygenation of nucleophilic nitrogen and sulfur compounds. In *Enzymatic Basis of Detoxication* Vol. I edited by W.B. Jakoby, (New York: Academic Press) pp. 201-225.

GERMFREE RATS IN THE STUDY OF THE METABOLISM OF
XENOBIOTIC COMPOUNDS

Peter Goldman

The Department of Nutrition, Harvard School of Public Health
and the Department of Biological Chemistry and Molecular
Pharmacology, Harvard Medical School, 665, Huntingdon Ave.,
Boston, Massachusetts 02115, USA

INTRODUCTION
    The recognition that toxic effects may be mediated by the
metabolites of a xenobiotic compound, as well as by the
xenobiotic itself, provides a practical incentive for
studying the metabolism of xenobiotic compounds. Although
metabolic reactions occurring in mammalian organs,
particularly the liver, receive most attention, it is now
recognized that many reactions of xenobiotic metabolism may
be carried out by the intestinal microflora and that some of
these reactions may have toxicological significance. A
first step in understanding such reactions is to ascertain
whether or not they are indeed carried out by the intestinal
bacteria.
    Experiments with germfree[1] animals are particularly useful
for distinguishing metabolic reactions occurring in the
flora from those occurring in the mammalian tissues. Thus
if a metabolite of a xenobiotic compound appears in the
excreta or tissues of a conventional animal, but not in its
germfree counterpart, it is reasonably certain that the
flora have an obligatory role in the compound's metabolism.
The methods of gnotobiotic research have now been adapted to
suit the needs and facilities of laboratories whose major
focus of research is nutrition, pharmacology or toxicology
(McLafferty and Goldman, 1981). Unfortunately, however,
germfree rats, the animals for which these methods have been

[1]The terms "germfree" and "gnotobiotic" have limitations
because they refer only to <u>demonstrable</u> forms of life and
therefore depend on the sophistication of the diagnostic
techniques used. In this review the terms are used in a
somewhat restricted sense to refer mainly to bacteria and
their host.

devised, are no longer readily obtained from commercial suppliers in the United States and so this key technology in establishing the role of the flora is not as readily accessible as it was a few years ago. Nevertheless, gnotobiotic rats will continue to be helpful in distinguishing between metabolism in mammalian tissues and that occurring in the flora.

When a xenobiotic compound yields a metabolite in conventional rats that does not appear in germfree rats, it is presumptive evidence that reactions of the flora are obligatory for the compound's *in vivo* metabolism. Other evidence, however, can make this conclusion more secure. First of all, past experience has indicated that the biochemical transformations carried out by the flora have certain chemical characteristics (Scheline, 1973; Goldman, 1978). Thus, the intestinal bacteria tend to carry out reduction of xenobiotic compounds, as might be expected in the normally anaerobic environment of these bacteria where compounds other than oxygen must serve as a final electron acceptor. For reactions observed *in vivo* a useful generalization is that reductive reactions are likely to occur in the flora, whereas such oxidative reactions as those carried out by the cytochrome P450 system are carried out in mammalian tissues. Another contrast between the reactions carried out by the flora and those that occur in mammalian tissues concerns conjugations reactions. The liver and other mammalian organs conjugate compounds to glucuronic acid, sulphate and glutathione, whereas the flora contain enzymes that hydrolyse these conjugates (Scheline, 1973; Goldman, 1978; Bakke and Gustafsson, 1986).

Of course, these two generalizations provide only limited guidance in predicting the location of a metabolic reaction. The flora's role in a reaction sequence is more firmly established if it can be demonstrated that the reaction missing in the germfree rat is carried out in an *in vitro* system consisting of bacteria derived from the flora. In building this logical chain, however, consideration must be given to the anaerobic environment in which the intestinal flora normally exist. Hence reactions carried out by bacteria isolated from the intestinal tract must not be attributed to the normal flora, unless they can be demonstrated to take place under anaerobic conditions.

The following sections provide some illustrations of experiments conducted with drugs and other xenobiotic compounds in gnotobiotic animals which served to distinguish the metabolic reactions of the flora from those carried out in mammalian tissues.

REDUCTIVE REACTIONS
Nitro group reduction

Cancer, methaemoglobinemia and other toxic manifestations of compounds containing the nitro group appear in many cases to be mediated by intermediates formed during metabolic reduction of the nitro group. The site at which nitro group reduction takes place, however, may not be clear, particularly when the reaction can be demonstrated both by enzyme preparations derived from liver and by bacteria isolated from the flora. In such instances the gnotobiotic rat has served as a means of deciding whether the flora or mammalian enzymes are responsible for the in vivo reaction (Goldman, 1978).

Gnotobiotic animals in proving that a reaction is due to the flora. A prototype for this use of gnotobiotic rats is a study done to establish the site of reduction of 4-nitrobenzoic acid. 4-Nitrobenzoic acid is a good compound for this type of study because a single metabolite, 4-aminobenzoic acid, forms when the nitro group is reduced; 4-aminobenzoic acid is then readily acetylated in vivo. 4-Nitrobenzoic acid itself and its two metabolites, 4-aminobenzoic acid and 4-acetamidobenzoic acid, are eliminated in the urine and can easily be measured. In the conventional rat approximately 25% of a test dose of 4-nitrobenzoic acid is eliminated in the urine in the form of either 4-aminobenzoic acid or its acetylated derivative.

Misconceptions existed in the past about the site at which 4-nitrobenzoic acid is reduced. Initially, the liver seemed to be the site of this reaction when it was found that the conversion of 4-nitrobenzoic acid to 4-aminobenzoic acid was catalyzed by a preparation of rat liver microsomes. This interpretation is unwarranted, however, as might have been anticipated from the observation that liver microsomes carry out nitro group reduction only when incubated under rather unphysiological anaerobic conditions. That the flora participated in the reductive metabolism of 4-nitrobenzoic acid was indicated by the finding that the reaction diminished in rats treated concomitantly with antibiotics and that the reduction of 4-nitrobenzoic acid was mediated by intestinal contents (Zachariah and Juchau, 1974).

Gnotobiotic rats provided the definitive evidence on the relative role of the flora and mammalian enzymes in the metabolism of 4-nitrobenzoic acid. Thus, in germfree rats only approximately 2 per cent of a test dose of 4-nitrobenzoic acid was converted to reduced metabolites, in contrast to the 20-25% conversion that was found in conventional rats. Furthermore the fraction of 4-nitrobenzoic acid converted to reduced metabolites increased progressively as germfree rats were deliberately associated with an increasing variety of intestinal bacteria with the capacity to reduce 4-nitrobenzoic acid in culture (Wheeler et al. 1975). Thus, there was a proportionality between the amount of a test dose of 4-nitrobenzoic acid reduced and the metabolic activity of the anaerobic flora selectively associated with the gnotobiotic rats. A final aspect of these studies was the demonstration that the reduction of 4-nitrobenzoic acid was negligible in tissues isolated from the germfree rat (Gardner and Renwick, 1978). Studies with 4-nitrobenzoic acid therefore serve to illustrate how experiments with gnotobiotic rats can define the role of the flora in a metabolic reaction.

Gnotobiotic animals for determining whether or not a reaction occurs in mammalian tissues. The gnotobiotic rat is also useful for determining the extent to which nitro group reduction takes place in mammalian tissues. As indicated above, liver microsomes can carry out reduction of 4-nitrobenzoic acid, but only under anaerobic conditions. How applicable are such results to the metabolism of the compound that occurs *in vivo*? The answer appears to depend, at least in part, on the oxygen tension of the tissue. Thus when the reduction of nitrofurantoin was examined in the perfused rat liver, it appeared that the extent of nitro group reduction was determined simply by the oxygen tension maintained in the perfusate (Jonen, 1980). Clearly, then, oxygen tensions must be maintained at physiological levels in an experimental preparation in order to be assured that the results obtained are applicable to the *in vivo* situation. Although most of the reduction of 4-nitrobenzoic acid appears to be the result of the metabolic activity of the flora, reduction to the extent of 1-2% does occur in the germfree rat. For other compounds, nitro group reduction may occur to an even greater extent. Thus 5-6% of a test dose of 4-nitrobenzenesulphonamide is reduced by the germfree rat (Wheeler et al., 1975). It seems reasonable to assume that the tissues of the germfree rat are maintained at normal oxygen tensions and that reduction of a xenobiotic compound by a germfree rat is therefore a reflection of how the compound is normally metabolized in the tissues.

With this assumption the germfree rat has been used to estimate the role of mammalian tissues in the reduction of several nitroheterocyclic drugs. The compounds studied were the 5-nitroimidazole, metronidazole, and the 2-nitroimidazole, misonidazole, as well as the nitrofuran, nitrofurazone

(Yeung et al., 1983). To estimate nitro group reduction of these compounds, it was necessary to identify stable metabolites that gave an indication that nitro group reduction had occurred. All three of the nitroheterocyclic compounds mentioned above are reduced by the flora. Thus, when metronidazole was incubated with either pure or mixed cultures of the flora, N-(2-hydroxyethyl)oxamic acid and acetamide were isolated (Koch et al., 1979). Under similar conditions misonidazole yielded its 2-aminoimidazole functionality as well as other metabolites indicating cleavage of the imidazole ring (Koch et al., 1982) and nitrofurazone was reduced to yield 4-cyano-2-oxybutyraldehyde semicarbazone (Yeung et al., 1983). When each of these nitroheterocyclics was administered to conventional rats it was possible to recover its respective metabolite(s) characteristic of the reduction of its nitro group (Koch et al., 1979; Koch et al., 1982; Yeung et al., 1983). 4-Cyano-2-oxybutyraldehyde semicarbazone was also recovered when nitrofurazone was administered to the germfree rat, the recovery of this metabolite being only slightly less than that recovered from the conventional rat (Yeung et al., 1983). Similarly, metabolites characteristic of misonidazole reduction were recovered from the germfree rat, although they were recovered in lower yields than from the conventional rat (Yeung et al., 1983). In contrast, when the 5-nitroimidazole, metronidazole, was administered to the germfree rat, no metabolites indicative of nitro group reduction were detected (Yeung et al., 1983).

The mechanism believed to be responsible for reduction of the nitroheterocyclic compounds may explain why nitroflurazone and misonidazole are reduced, whereas metronidazole is not. ESR studies indicate that the first step in the reduction of nitroheterocyclic compounds is the formation of the radical anion (Peterson et al., 1979; Perez-Reyes et al., 1980). Thus nitrofurazone and misonidazole, with one electron reduction potentials respectively of -257 mV and -389 mV, should be reduced to the radical anion more readily than metronidazole, whose one electron potential is -486 mV. If formation of the radical anion is the rate limiting step in the sequence of reactions for reduction of the nitro group, then it would be expected that nitroheterocyclic compounds with relatively less negative one electron reduction potentials are reduced most readily in the tissues.

The difference in the fates of the different nitroheterocyclic compounds in the tissues might also be attributed to the fates of their radical anions. According to ESR studies

(Peterson et al., 1979; Perez-Reyes et al., 1980) the second step in the reduction of a nitroheterocyclic compound is the disproportionation of the radical anion to yield the nitroso functionality. If oxygen is present, however, the radical anion may react with it to form superoxide in a reaction which also restores the parent nitroheterocyclic compound. This reaction, which has been termed a "futile cycle", occurs more rapidly in compounds that have a more negative one electron reduction potential (Wardman and Clarke, 1976). It is possible, therefore, that the radical anions of all 3 nitroheterocyclic compounds are formed in the tissues but tend to have different fates because of the presence of oxygen. Thus, in the presence of oxygen the radical anion of metronidazole is more likely to enter a futile cycle, whereas the other nitroheterocyclic compounds, whose radical anions are less reactive with oxygen, are more likely to disproportionate to yield the nitroso functionality.

The failure to find acetamide and N-(2-hydroxyethyl)oxamic acid when metronidazole is administered to the germfree rat may therefore be explained by either or both of two phenomena. On the one hand, it may be due to a negligible formation of metronidazole's radical anion. On the other hand, it may result from rapid reaction of metronidazole's radical anion with oxygen which prevents its further reduction to yield significant quantities of the nitroso functionality. Unfortunately, there are presently no techniques available to distinguish between these two possibilities.

The question of whether or not the nitroso functionality actually forms in the tissue under physiological conditions may be an important one which bears on the carcinogenic and mutagenic potential of such drugs as metronidazole. The nitroso functionality of a 5-nitroimidazole has now been synthesized and shown to have the lability and biological properties (Ehlhardt et al., 1988a) postulated for the active form of these drugs (Chrystal et al., 1980). Thus, the nitroso functionality of a 5-nitroimidazole has been found to be orders of magnitude more mutagenic, cytotoxic and bactericidal than the 5-nitroimidazole from which it is derived and in addition has the lability in biological media that has been postulated for the biologically active form of the 5-nitroimidazoles (Ehlhardt et al., 1988a and b).

<u>Gnotobiotic animals in establishing the role of the flora in mediating toxicity</u>. Gnotobiotic animals have on occasion also been used to establish a link between a xenobiotic compound's metabolic transformation and its toxicity. Thus, nitrobenzene causes methaemoglobinemia in conventional rats,

but not in germfree rats (Reddy et al., 1976), apparently because the reduction of nitrobenzene occurs almost exclusively in the flora. Such evidence suggests that the flora's obligate role in the toxic reaction of nitrobenzene can be correlated with their obligate role in reducing the nitro group.

The germfree rat has also been used to demonstrate a correlation between the flora's role in reducing 2,4-dinitrotoluene (Rickert et al., 1981) and its role in liver toxicity. When rats associated with a defined bacterial flora are administered 2,4-dinitrotoluene, the rate of unscheduled DNA synthesis is increased in their hepatocytes (87% in repair) compared to that of the hepatocytes of germfree rats subjected to the same dose of 2,4-dinitrotoluene (14% in repair) (Mirsalis et al., 1982). It must be emphasized that the flora are not necessarily obligatory for all manifestations of toxicity caused by compounds as the result of nitro group reduction. Thus, toxicity can arise as the result of nitro group reduction in the tissues themselves, as exemplified by compounds containing a nitro group that act directly on red cells to cause methaemoglobinemia (Facchini and Griffiths, 1981). Clearly, then, a case-by-case analysis is required to determine whether a compound is reduced in the tissues or in the flora. The gnotobiotic rat is useful in providing data for such an analysis and hence in sorting out the mechanism responsible for the toxicity of compounds that may be metabolized both by the flora and the tissues of their host.

Azo bond reduction

The azo bond is another functional group whose reduction has important consequences for drug metabolism and where the site of reduction has been clarified by use of the germfree animal. It is now recognized that the earliest sulpha drugs, prontosil and neoprontosil, were actually pro-drugs, which after reduction of their azo bonds yielded the bacteriostatic agent, sulphanilamide. As was the case with the reduction of 4-nitrobenzoic acid, it was possible to demonstrate azo bond cleavage of prontosil in anaerobically incubated liver enzyme systems (Fouts et al., 1957). However, as in the case of the reduction of 4-nitrobenzoic acid, a role for the flora in azo bond reduction was suggested when the release of sulphanilamide from prontosil or neoprontosil was decreased by the concomitant administration of antibiotics (Gingell et al., 1971).

A more timely example of the importance of azo bond cleavage concerns the metabolism of sulphasalazine, which for 40 years has been an important drug for the treatment and prophylaxis of ulcerative colitis. Sulphasalazine consists of sulphapyridine in azo linkage to 5-aminosalicylate and also turns out to be a prodrug, the active molecule in this case being 5-aminosalicylate. Our current understanding of how sulphasalazine works was based on the observation that the azo bond of sulphasalazine was cleaved in conventional rats but not in germfree rats (Peppercorn and Goldman, 1972a; Schroeder and Gustafsson, 1973), and therefore that the flora were obligatory for this reaction. It was of particular interest to find that the 2 major metabolites of sulphasalazine had different fates. Thus, sulphapyridine apparently was readily absorbed from the gastrointestinal tract and, after some further metabolism, was excreted in the urine. 5-Aminosalicylate, on the other hand, was retained within the gastrointestinal tract and eliminated in the faeces. It was suggested, therefore, that sulphasalazine might serve merely as a vehicle for the delivery of 5-aminosalicylate to the diseased colon (Goldman and Peppercorn, 1973), an hypothesis that seemed particularly attractive at the time, because the salicylates had just been shown to inhibit the synthesis of prostaglandins, which were believed to be involved in the inflammatory reaction.

The suggested activity of 5-aminosalicylate was confirmed by a clinical study in which it was demonstrated that the rectal administration of 5-aminosalicylate was effective for the treatment of ulcerative colitis of the distal colon, whereas the administration of sulphapyridine had no discernible effect. These observations encouraged a search for other drugs that might prove more effective in delivering 5-aminosalicylate to the colon in the hope that they would increase the effectiveness of sulphasalazine both to treat ulcerative colitis and to prevent its recurrence. Furthermore, it was felt that the safety of sulphasalazine might be improved by developing delivery systems for 5-aminosalicylate that eliminated sulphapyridine, which appeared to be responsible for many of the adverse reactions of sulphasalazine (Goldman, 1982).

Dehydroxylation at the 4-position of a catechol

The recognition that the flora are responsible for the dehydroxylation at the 4-position of such catechols as L-dopa has also been aided by experiments in gnotobiotic rats. Thus, 3-hydroxyphenylacetic acid, which was found in the urine when conventional rats received either L-dopa or dopamine, was not found when these drugs were administered to germfree rats (Goldin et al., 1973). Similar studies implicate the flora in the dehydroxylation at the 4-position of other catechols (Peppercorn and Goldman, 1972b).

GLYCOSIDASES

Many drugs are inactivated by conjugation with glucuronic acid and then eliminated in either the urine or the bile (Smith, 1973). If eliminated in the bile, these conjugates make contact with the intestinal flora and as a result may undergo deconjugation to restore the active drug (Scheline, 1973). When the polar glucuronic acid group is removed, the parent drug may then be reabsorbed from the gastrointestinal tract. The flora therefore play a key role in this process, which has been termed the enterohepatic circulation.

Glycosidases of the flora may also play a critical role in the mechanism of toxicity of certain glycosides. A case in point is the pathogenesis of liver toxicity and cancer caused by the plant glycoside, cycasin. The actual mediator of cycasin toxicity is not cycasin itself, but rather its aglycone, methylazoxymethanol. The elucidation of the obligatory role of the flora in releasing methylazoxymethanol provided one of the first examples of the way that gnotobiotic animals can help to determine a toxic mechanism (Laqueur and Spatz, 1975). A critical experiment in establishing the obligatory role of the flora in cycasin toxicity was a comparison of the response of germfree and conventional rats to the administration of cycasin. On a diet containing 0.2% cycasin, conventional rats gained weight poorly and were all dead within 3 weeks. At autopsy their livers showed severe diffuse centrilobular haemorrhagic necrosis. Germfree rats on this diet, however, remained healthy and had normal livers. Furthermore, 97% of the cycasin intake was recovered in the excreta of the germfree rats, whereas the recovery in conventional rats was only 26%. When germfree rats were associated with bacteria capable of hydrolyzing cycasin, however, the rats gained the capacity to metabolize cycasin and also began to exhibit the manifestations of its toxicity. It seemed clear, therefore, that both the metabolism of cycasin and its toxicity occurred only in rats with a flora and that glycosidases of the flora were necessary to convert the glycoside cycasin to its toxic aglycone, methylazoxymethanol (Laqueur and Spatz, 1975).

Amygdalin is a cyanide-containing glycoside whose toxicity also depends on the enzymic activity found in the flora. Thus an oral dose of amygdalin (600 mg/kg) caused conventional rats to become lethargic and often to have convulsions and die. Rats affected in this way had high concentrations of cyanide in their blood (2.6-4.5 μg/ml).

Although the metabolic sequence for the degradation of amygdalin has not been characterized, it apparently depends on the glycosidases of the flora (Carter et al., 1980).

A NOTE OF CAUTION ABOUT INTERPRETING STUDIES IN GERMFREE ANIMALS

The studies described above indicate the value of gnotobiotic rats in studies of the metabolism of xenobiotic compounds and how conclusions may be drawn by comparing metabolic reactions in conventional and germfree rats. It must be recognized, however, that the result of such comparisons, which implicate the flora in a metabolic reaction, should be confirmed *in vitro*, because germfree and conventional rats may have quantitative differences in metabolism as the result of indirect effects of the flora on the conventional rat. Such differences can arise from differences in anatomy and physiology between germfree and conventional rats.

One obvious difference is the morphology of the gastrointestinal tract. A large caecum with "paper thin" walls is one of the hallmarks of the germfree rat that distinguishes it from its conventional counterpart. Such anatomical differences raise the possibility that the absorption of compounds from the intestinal tract may differ in germfree and conventional rats. The result may be that xenobiotic compounds have different access to mammalian enzymes. Furthermore, the activity of the hepatic cytochrome P-450 system toward some substrates may be lower in the germfree rat than in the conventional rat (Short and Davis, 1969). These considerations indicate that an increased excretion of a metabolite by the conventional rat cannot necessarily be attributed to the metabolic activity of the flora.

The question of how the germfree state may affect the distribution and metabolism of xenobiotic compounds of compounds by mammalian enzymes has not been subjected to any systematic study. Nevertheless, some studies show that a drug and its metabolites may be distributed differently in the excreta of germfree and conventional rats (see, for example, Koch et al., 1979). One elegant study has defined such differences by comparing the fate of warfarin in germfree and conventional rats in terms of the familiar one-compartment pharmacokinetic model. These studies indicate that warfarin has a shorter half-life in germfree animals, although the volume of distribution is the same in both germfree and conventional rats (Remmel et al., 1981). The authors of this study suggested that a primary effect of the flora might be responsible for this difference, namely an interruption of the enterohepatic cycle of warfarin as the result of the absence of bacterial glucuronidases in the germfree state. It is also possible, however, that pharmacokinetic differences between germfree and conventional rats may result from such secondary effects of the flora as those mentioned above.

The assessment of the role of gut flora in a metabolic reaction by examining the effect of the administration of antibiotics has limitations. The rationale of this approach is that antibiotics suppress the bacteria of the gut and hence act to diminish the metabolic reactions that are catalyzed by these bacteria. There are, however, several reaons why the use of antibiotics may be misleading. Firstly, the flora are composed of approximately 450 different species of bacteria and so the degree to which the metabolically active ones are suppressed by an antibiotic regimen may be uncertain. Therefore, failure to suppress a reaction when an antibiotic is administered does not necessarily exclude a role for the flora in the reaction sequence. Furthermore, the administration of an antibiotic often causes diarrhoea which, by speeding up intestinal transit time, may change a compound's absorption and hence its access to the mammalian enzymes that may be involved in its metabolism. Therefore more than one interpretation may be possible when an antibiotic is observed to suppress the formation of a metabolite of a xenobiotic compound. Some of the complications of the use of antibiotics to suppress the flora are suggested by the unpredictable effect that antibiotics have on the pharmacokinetics of warfarin (Remmel et al., 1981).

DERIVING CONCLUSIONS FOR HUMANKIND FROM STUDIES OF THE FLORA IN THE RAT

A further cautionary note should be mentioned about the interpretation of studies of drug metabolism and its consequences in the gnotobiotic rat. As a rule we study a compound's metabolism in rats because we are interested in the implications of such results for the compound's human pharmacology and toxicity. In considering the human implications of data obtained in the rat, however, one must consider the anatomical and physiological differences between the two species that may affect one's conclusions. There are several important differences in the way the flora are distributed in the human and the rat gastrointestinal tract (Rowland, 1986) and hence in how the flora may metabolize a compound in the two species. Firstly, the rat has a forestomach which harbors a fairly high concentration of yeast and lactobacilli. Human beings lack a forestomach and because of high gastric acidity will have an upper gastrointestinal tract that is relatively free of microorganisms. Thus compounds are apt to make immediate contact with flora when ingested by the rat, but when ingested by human subjects will generally not contact the flora until they reach the lower ileum. Under these circumstances a compound that has good bioavailability may be absorbed high in the human gastrointestinal tract and thereby avoid contact with the flora unless it re-enters the gastrointestinal tract either in the bile or as the result of diffusion across the gut wall (Facchini and Griffiths, 1980). The rat also has a prominent caecum which provides xenobiotic compounds with considerable exposure to flora. In spite of these caveats, it is quite clear that data obtained in gnotobiotic rats have contributed a great deal to an understanding of the in vivo metabolism of xenobiotic compounds and have served to show how such metabolism may explain the pharmacology or toxicity of several compounds.

REFERENCES

Bakke, J.E. and Gustaffson, J.-Å., 1986, Role of intestinal flora in metabolism of agrochemicals conjugated with glutathione. Xenobiotica, 16, 1047-1056.

Carter, J.H., McLafferty, M.A. and Goldman, P., 1980, The role of the gastrointestinal microflora in amygdalin (laetrile)-induced cyanide toxicity. Biochemical Pharmacology, 29, 301-304.

Chrystal, E.J.T., Koch, R.L., McLafferty, M.A. and Goldman, P., 1980, The relationship between metronidazole metabolism and bactericidal activity. Antimicrobial Agents Chemotherapy, 18, 566-573.

Ehlhardt, W.J., Beaulieu, B.B., Jr. and Goldman, P., 1988a, Nitrosoimidazoles: Highly bactericidal analogues of 5-nitroimidazole drugs. Journal of Medicinal Chemistry, 31, 323-329.

Ehlhardt, W.J., Beaulieu, B.B., Jr. and Goldman, P., 1988b, Mammalian cell toxicity and bacterial mutagenicity of nictosoimidazoles. Biochemical Pharmacology, in press.

Facchini, V. and Griffiths, L.A., 1980, The metabolic fate of nitromide in the rat. I. Metabolism and excretion. Xenobiotica, 10, 289-297.

Facchini, V. and Griffiths, L.A., 1981, The involvement of the gastro-intestinal microflora in nitro-compound-induced methaemoglobinaemia in rats and its relationship to nitrogroup reduction. Biochemical Pharmacology, 30, 931-935.

Fouts, J.R., Kamm, J.J. and Brodie, B.B., 1957, Enzymatic reduction of prontosil and other azo dyes. Journal of Pharmacology and Experimental Therapeutics, 120, 291-300.

Gardner, D.M. and Renwick, A.G., 1978, The reduction of nitrobenzoic acids in the rat. Xenobiotica, 8, 679-690.

Gingell, R., Bridges, J.W. and Williams, R.T., 1971, The role of the gut flora in the metabolism of prontosil and neoprontosil in the rat. Xenobiotica, 1, 143-156.

Goldin, B.R., Peppercorn, M.A. and Goldman, P., 1973, Contribution of host and intestinal microflora in the metabolism of L-dopa by the rat. Journal of Pharmacology and Experimental Therapeutics, 186, 160-166.

Goldman, P., 1978, Biochemical pharmacology of the intestinal flora. Annual Review of Pharmacology and Toxicology, 18, 532-539.

Goldman, P., 1982, Will there be a next generation of sulfasalazine? Gastroenterology, 83, 1138-1141.

Goldman, P. and Peppercorn, M.A., 1973, Salicylazosulfapyridine in clinical practice. Gastroenterology, 65, 166-169.

Jonen, H.G., 1980, Reductive and oxidative metabolism of nitrofurantoin in rat liver. Naunyn-Schmeideberg's Archives of Pharmacology, 315, 167-175.

Koch, R.L., Chrystal, E.J.T., Beaulieu, B.B. Jr. and Goldman, P., 1979, Acetamide: a metabolite of metronidazole formed by the intestinal flora. Biochemical Pharmacology, 28, 3611-3615.

Koch, R.L., Rose, C., Rich, T.A. and Goldman, P., 1982, Comparative misonidazole metabolism in anaerobic bacteria and hypoxic Chinese hanster lung fibroblast. Biochemical Pharmacology, 31, 411-414.

Laqueur, G.L. and Spatz, M., 1975, Oncogenicity of cycasin and methylazoxymethanol. GANN, 17, 189-204.

McLafferty, M.A. and Goldman, P., 1981, Germfree rats. In Methods in Enzymology, Vol. 77, edited by W.B. Jakoby (New York: Academic Press), pp. 34-43.

Mirsalis, J.C., Hamm, R.E., Sherrill, J.M. and Butterworth, B.E., 1982, Role of gut flora in the genotoxicity of dinitrotoluene. Nature, 295, 322-323.

Peppercorn, M.A. and Goldman, P., 1972a, The role of intestinal bacteria in the metabolism of salicylazosulfapyridine. Journal of Pharmacology and Experimental Therapeutics, 181, 555-562.

Peppercorn, M.A. and Goldman, P., 1972b, Caffeic acid metabolism by gnotobiotic rats and their intestinal bacteria. Proceedings of the National Academy of Sciences USA, 69, 1413-1415.

Perez-Reyes, E., Kalyanaraman, B. and Mason, R.P., 1980, The reductive metabolism of metronidazole and ronidazole by aerobic liver microsomes. Molecular Pharmacology, 17, 329-244.

Peterson, F.J., Mason, R.P., Hovsepian, J. and Holtzman, J., 1979, Oxygen-sensitive and -insensitive nitroreduction by Escherichia coli and rat hepatic microsomes. Journal of Biological Chemistry, 254, 4009-4014.

Reddy, B.G., Pohl, L.R. and Krishna, G., 1976, The requirment of the gut flora in nitrobenzene-induced methemoglobinemia in rats. Biochemical Pharmacology, 25, 1119-1122.

Remmel, R.P., Pohl, L.R. and Elmer, G.W., 1981, Influence of the intestinal microflora on the elimination of warfarin in the rat. Drug Metabolism and Disposition, 9, 410-414.

Rickert, D.E., Long, R.M., Krakowa, S. and Dent, J.G., 1981, Metabolism and excretion of 2,4-[^{14}C]dinitrotoluene in conventional and Axenic Fischer-344 rats. Toxicology and Applied Pharmacology, 59, 574-579.

Rowland, I.R., 1986, Reduction by the gut microflora of animals and man. Biochemical Pharmacology, 35, 27-32.

Scheline, R.R., 1973, Metabolism of foreign compounds by gastrointestinal microorganisms. Pharmacological Reviews, 25, 451-532.

Schroeder, H. and Gustafsson, B.E., 1973, Azo reduction of salicylazosulfapyridine in germfree and conventional rats. Xenobiotica, 3, 225-231.

Short, C.R. and Davis, L.E., 1969, A comparison of hepatic drug-metabolizing enzyme activity in the germfree and conventional rat. Biochemical Pharmacology, 18, 945-947.

Smith, R.L., 1973, The Excretory Function of Bile: The elimination of drugs and toxic substances in bile (New York: Halsted Press).

Wardman, R. and Clarke, E.E., 1976, Oxygen inhibition of nitroreductase: Electron transfer from nitro radical-anions to oxygen. Biochemical and Biophysical Research Communications, 69, 942-949.

Wheeler, L.A., Soderberg, F.B. and Goldman, P., 1975, The relationship between nitro group reduction and the intestinal microflora. Journal of Pharmacology and Experimental Therapeutics, 194, 135-144.

Yeung, T.-C., Sudlow, G., Koch, R.L. and Goldman, P., 1983, Reduction of nitroheterocyclic compounds by mammalian tissues in vivo. Biochemical Pharmacology, 32, 2249-2253.

Zachariah, P. and Juchau, M.R., 1974, The role of the gut flora in the reduction of aromatic nitro-groups. Drug Metabolism and Disposition, 2, 74-78.

NMR AND THE STUDY OF XENOBIOTIC METABOLISM: AN INTRODUCTION

T.A. Moore and P.G. Morris

Department of Biochemistry,
University of Cambridge, Tennis Court Road,
Cambridge, CB2 1QW, UK

INTRODUCTION

The phenomenon of nuclear magnetic resonance (NMR) was successfully demonstrated for the first time in 1945 by Bloch at Stanford (Bloch et al., 1946) and independently by Purcell at MIT (Purcell et al., 1946). Its subsequent development into perhaps the most powerful analytical procedure available to the synthetic chemist was reflected in the joint award to its primary instigators of the 1952 Nobel prize for physics.

Atomic nuclei, unless they have both even atomic and even mass numbers (for example, ^{12}C, ^{16}O and ^{32}S), will have a net spin by virtue of the spins of the protons and neutrons which constitute them. As nuclei are positively charged, there will be a magnetic moment associated with the spin. The combination of spin angular momentum and magnetic moment causes the nuclei to precess in a magnetic field in much the same way that a gyroscope precesses in the earth's gravitational field. The frequency of precession is the resonant or Larmor frequency and is directly proportional to the magnetic field B. Thus: $\omega = \gamma B$. The proportionality constant γ or magnetogyric ratio, differs widely for the different nuclei, ensuring that there is little chance of heteronuclear spectral overlap. Only certain orientations of the spins and therefore, energy states, are allowed with respect to the magnetic field. Nuclei with a spin of 1/2 have two allowed orientations, those with spin 3/2 have four and so on. Irradiation with electromagnetic radiation at the Larmor frequency induces transitions between these energy states; this is the phenomenon of nuclear magnetic resonance. For practically attainable magnetic fields (up to 14T or about 300,000 times the earth's field), the radiation is at the low frequency end of the electromagnetic spectrum, typically in the radiofrequency region. Consequently it is non-ionizing. That is to say, there is no danger of ejecting an electron and thereby disrupting a molecular bond. This is one reason why NMR is attractive for in vivo biochemical studies - it will not damage samples and indeed, it is used for clinical diagnosis in situ. However, the low energy nature of NMR represents a major disadvantage in another respect, the method has very low sensitivity compared with traditional analytical methods. Its use is therefore restricted to metabolites present at high intracellular concentrations (normally $> 10\mu M$).

In addition to the large difference in resonance frequency due to the different nuclear magnetogyric ratios, much smaller differences exist between nuclei of the same atomic species located in different molecular environments. These differences arise because it is the field at an individual nucleus which determines the resonance frequency and many nucleii experience slightly different fields because of the shielding effects of their electronic environments. This is expressed in the following form: Bnucleus = Bapplied$(1-\sigma)$ where σ is the shielding constant. This is the basis of the NMR spectrum. Nuclei in different functional groups are distinguished by their different electronic environments. The differences in resonance frequency are known as chemical shifts which are normally measured in parts per million from some arbitrarily chosen reference.

CHOOSING THE NUCLEUS

For nuclei other than the simple proton or deuteron the magnitude of the nuclear spin, quantized in units of $\hbar/2$, is governed by three simple rules, namely: (i) if the mass number A is odd, the spin is half-integral; (ii) if A is even but the atomic number Z is odd, the spin is integral; (iii) if both A and Z are even the spin is zero. This means that most elements have at least one naturally-occurring isotope which possesses a non-zero nuclear spin and is therefore observable by NMR. Table 1 lists nuclear properties for those NMR observable isotopes which are potentially of biomedical interest. Note that the common isotopes of carbon (^{12}C) oxygen (^{16}O) and sulphur (^{32}S) are ruled out. Fortunately, however, these elements do have other isotopes with nuclear spin, namely ^{13}C, ^{17}O, and ^{33}S. There are practical difficulties in using certain nucleii to localize or trace compounds involved in cellular metabolism which are beyond the scope of this article, however, certain points are worth summarising.

Firstly, it is important to note that it is the material present in free solution which responds; the presence of an

TABLE 1: NMR SENSITIVITIES OF NUCLEI OF BIOMEDICAL INTEREST

Nucleus	Spin	γ (MHz/T)	Natural abundance (%)	Rel. sensitivity at constant field	Typical human physiological concentration of the element[1]
^{1}H	1/2	268	99.98	1	100 M
^{2}H	1	41	0.0015	2.4×10^{-6}	(10 mM)
^{3}H	1/2	285	0	—	(10 mM)
^{13}C	1/2	67	1.108	2.5×10^{-4}	10 mM
^{14}N	1	19	99.63	1.9×10^{-3}	10 mM
^{15}N	1/2	-27	0.37	6.8×10^{-6}	10 mM
^{17}O	5/2	-36	0.037	1.9×10^{-5}	50 M
^{19}F	1/2	252	100	8.5×10^{-1}	(10 mM)
^{23}Na	3/2	71	100	1.3×10^{-1}	80 mM
^{25}Mg	5/2	-16	10.13	5.5×10^{-4}	1 mM
^{31}P	1/2	108	100	8.3×10^{-2}	10 mM
^{33}S	3/2	21	0.76	3.3×10^{-5}	0.3 mM
^{35}Cl	3/2	26	75.53	6.3×10^{-3}	20 mM
^{39}K	3/2	12	93.1	1×10^{-3}	45 mM
^{43}Ca	7/2	-18	0.145	1.8×10^{-5}	0.5 mM

[1] Figures in brackets refer to artificially introduced material.

$\omega = \gamma \, B$

element in large quantity does not guarantee its NMR visibility. For example, most of the naturally-occurring ^{19}F in the body is present as fluorapatite in teeth. Similarly, much of the calcium is present in bone as crystalline hydroxyapatite ($Ca_{10}(PO_4)Ca(OH)_2$) or is bound to proteins rendering it solid-like and invisible as far as conventional studies are concerned.

Those nuclei with spins greater than one-half will have nuclear quadrupole moments which can lead to splitting of the resonance in motionally restricted environments, e.g. deuterated membrane lipids, or line-broadening due to quadrupolar relaxation effects. The latter effect can be severe and has led, for example, to the widespread use of ^{15}N (spin 1/2) rather than the more abundant ^{14}N (spin 1) for nitrogen NMR studies. Only four nuclei have been used extensively in biological investigations, ^{1}H, ^{13}C, ^{19}F and ^{31}P. Of these, ^{1}H is the generally most attractive because of its inherent high sensitivity and universality. What holds back the development is the spectral complexity and the dynamic range problem; thus it is often necessary to observe metabolites at concentrations of < 1 mM in the presence of 100 M water. Sophisticated solvent suppression and spin echo techniques are overcoming this problem although a second solvent suppression problem arises from the broad -CH_2-resonances of mobile lipids present at concentrations of up to 1 M. This region often includes resonances of interest to biologists.

^{31}P NMR spectroscopy has been extensively used to follow the changes in the levels of the energetically important phosphorylated metabolites. The sensitivity of the ^{31}P nucleus is again attractive but resonances are sometimes broadened by the presence of paramagnetic ions or, at higher field strengths, by chemical shift anisotropy effects.

^{19}F is increasingly being incorporated into NMR indicators designed to measure the intracellular concentrations and fluxes of ions such as Na^+, Ca^{++} and Mg^{++} (see Morris, 1988 for a review). It is a particularly attractive nucleus as there are no naturally occurring fluorine-containing compounds to complicate the spectra, and the sensitivity is second only to ^{1}H. It is possible to follow the metabolic fate of a drug in an animal using fluorinated derivates and in vivo NMR techniques. ^{19}F gives intense NMR signals with a wide chemical shift range and many fluorinated drugs are in clinical use. Later in this article we describe the use of ^{19}F NMR to monitor the metabolism of 5-fluorouracil in tumours and in the liver.

Potentially, the most attractive nucleus to biochemists is ^{13}C which unfortunately only has a natural abundance of

1.1%. Two approaches may be taken to overcome this
difficulty. Firstly, biological material may be enriched by
feeding ^{13}C enriched precursor compounds and following their
metabolic fate. For an excellent example see the perfusion
studies of Cohen (1983). The second approach is to
concentrate the sample greatly, either by maintaining
concentrated 'sludges' of cells in special apparatus or by
concentrating extracts as we describe later. Despite the
inherent difficulties of ^{13}C NMR, it does possess the
advantage of showing the states of all other carbon
containing metabolites whilst the compound in question is
being monitored. This is rarely possible using other
analytical techniques and is often of prime importance to
those studying the effects of xenobiotic metabolites.

LOCALIZATION METHODS

If the objective is to screen urine or plasma samples or
to analyze tissue extracts, then conventional NMR methods
can be used and the sample is simply placed in a standard 5
or 10 mm NMR tube. If, on the other hand, the intention is
to study metabolism in situ, then some form of spatial
localization will be required. A list detailing methods
appropriate for different biological samples is given in
Table 2.

TABLE 2 - Localization methods for NMR spectroscopy

---

1. None/Conventional Probe
   Appropriate for cell suspensions or systems where a
   single tissue type predominates.
2. Surgical Intervention
   (a) Remove organ of interest and use perfusion techniques
       to study in conventional probe.
   (b) Expose organ and place coil around or over it.
   (c) Catheter coils e.g. for cardiac studies.
3. Surface Coils
   (a) On their own.
   (b) In combination with other selective techniques.
4. Volume Selective Techniques
   (a) Field profiling methods e.g. FONAR, TMR.
   (b) Volume selective excitation methods e.g. ISIS.
5. Chemical Shift Imaging (CSI)
   (a) 4D CSI (3 spatial dimensions + 1 chemical shift).
   (b) 3D CSI (requires slice selection).

NMR is an inherently insensitive technique and the sample therefore needs to be in its most 'concentrated' form: in the case of cellular systems this means high cell densities, typically in the range $10^7 - 10^{12}$ cells/ml. Three main problems have to be overcome: (i) the cells must be kept in uniform suspension throughout the NMR sensitive volume; (ii) they must be suitably perfused so that nutrients are delivered and waste products eliminated; (iii) in the case of aerobic cells adequate oxygenation must also be ensured. It is of course important that a medium is chosen appropriate to the cell type under study and that adequate pH buffering is provided.

If there is no sample limitation (large numbers of cells are readily available) it is possible to circulate the entire cell suspension. Provided this is done sufficiently rapidly that the contents of the sensitive volume are exchanged on a time scale short compared to the T_1s of the species under study, sensitivity advantages also accrue. This type of approach might be considered a possibility for on-line monitoring of large-scale ferments used in the biotechnology industries, for example. Usually, however, cells are not available in such abundance and it is necessary to circulate the medium through the cell suspension. The problem is then one of constraining the cells. Several ingenious solutions have been advanced including the use of kidney dialysis fibre bundles, perfusion through artificial capillaries (Guillino and Knazek, 1979), growth on microcarrier beads (Ugurbil et al., 1981) and coextrusion of the cell suspension with low gelling temperature agarose solution. In this case, the small diameter (0.5 mm) of the threads ensures proper perfusion. Diffusion of nutrients to the centre of the thread typically requires ~1 min (Knop et al., 1984).

Isolated organs are well-suited to NMR study. Indeed a simple ^{31}P NMR study of excised muscle was the experiment which first drew the attention of the scientific world to the possibilities of in vivo NMR (Hoult et al., 1974). Such crude methods are generally not very useful since, deprived of the normal provision of oxygen and nutrients through the vascular system, the organ quickly perishes. Thus it is normal to perfuse the isolated tissue. (See Ross, 1972, for a comprehensive account of organ perfusion techniques). The heart lends itself to a particularly simple arrangement involving cannulation of the aorta and retrograde circulation of oxygenated medium according to the method of Langendorff (Morris et al., 1985). Figure 1 shows a ^{31}P NMR spectrum recorded under control conditions from an isolated ferret heart perfused in this way. Peak assignments are

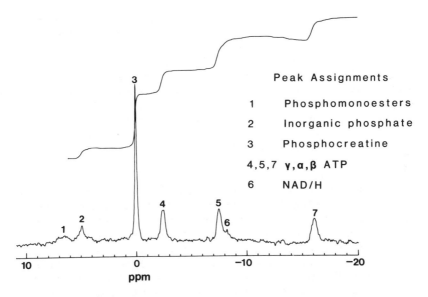

Figure 1. ^{31}P NMR spectrum of an isolated ferret heart, Langendorff-perfused at 30°C. Peak assignments are given in the Figure and the integral (upper trace) is directly proportional to the concentration of ^{31}P metabolites.

given on the figure. Since the data were accumulated under fully relaxed conditions, the integral (shown above the spectrum) is directly proportional to concentration. Of the mammalian organs, other than the heart, the liver is perhaps the next simplest to perfuse, involving only portal venous and venous cannulation, though it is usually considered necessary to perfuse with blood. It is also possible to perfuse the liver in situ. NMR studies of many other perfused organs have also been reported (see Morris 1988, for a fuller review). Although they are extremely valuable, nevertheless it is important to show that the in vitro observations can also be demonstrated in vivo. Ideally the animal should be monitored in as natural a state as possible, with no surgical intervention. Such experiments require NMR localization techniques to select the region of interest.

Surface coils (Ackerman et al., 1980) offer a simple, potentially non-invasive, means of studying metabolism in brain, muscle and liver. The ^{19}F studies described below were conducted using this approach. Alternatively, chronically implanted coils can be used. For in vivo heart studies, localization is more of a problem and both open chest procedures, with the heart placed in a solenoidal coil, and implanted coils have been described. An interesting elliptical catheter received coil has also been suggested (Kantor et al., 1984). There is little mystery surrounding the construction and practical use of a surface coil. In its simplest form it consists of a single loop of copper wire which is placed over the organ of interest. To a first approximation, it will receive signals from a hemispherical region lying immediately beneath it. Used with care, the surface coil is extremely effective. Much of the popularity and success it enjoys derives from the excellent sensitivity it affords. This is in part due to the fact that the noise originates only from that volume of tissue which contributes signal. This is in contrast to the situation with larger coils where noise originates from the entire volume even though the signal may derive from a small region within it.

If it is required to look at deeper lying tissues using a surface coil, then greater control is needed. Table 3 details some of the methods which can be used to improve the spatial selectivity of the surface coil. These include techniques such as DRESS which make use of the slice-selective schemes developed for NMR imaging (see Morris, 1986, for details). Such schemes, applied in each of three spatial dimensions, also form the basis for the successful volume-selective methods such as ISIS (Ordidge et al., 1986).

TABLE 3 - Surface coil improved spatial localization

---

1. DEPTH or COMPOSITE PULSES.
   Select region according to B_1 amplitude.
2. In combination with field profiling methods.
3. In combination with linear field gradient and selective pulse for plane definition, DRESS. Multiplanar version SLITDRESS.
4. Rotating frame Fourier imaging. Utilizes the approximately linear axial dependence of B_1.

It is also possible to 'go the whole way' and use three- or four-dimensional chemical shift imaging (CSI) methods. These involve an evolution in the presence of two or three mutually orthogonal magnetic field gradients whose amplitudes are systematically varied for the different records from which the image is reconstructed. Although such schemes have been considered esoteric and of little practical value, they represent one of the most efficient means of acquiring spectra. Recently they have come of age with the first publication of ^{31}P CSI images of the human liver (Bailes et al., 1988).

5-FLUOROURACIL METABOLISM MONITORED IN VIVO BY ^{19}F NMR

There are few examples where it has proved possible to monitor the distribution, metabolism and elimination of a drug at its site of action. The classical pharmacokinetic approach is at best indirect and requires extreme care in its interpretation. ^{19}F NMR applied in vivo offers a new approach to this problem. Several important drugs are fluorine-containing and, for those that are not, fluorination remains a possibility. Amongst the more important of the fluorine-containing drugs is 5-fluorouracil (5FU), which has been widely used in the treatment of disseminated human tumours such as those of the breast, ovary and gastro-intestinal tract. Figure 2 illustrates the metabolic pathways in which 5FU participates; anabolic horizontally, catabolic vertically. There are two possible modes of action: (i) 5FU can replace uracil in RNA, and (ii) it can block DNA synthesis through its metabolism to fluorodeoxyuridine monophosphate (FdUMP), a potent inhibitor of dTMP synthetase. Livers of adult female C57 mice were examined in situ by ^{19}F NMR using a flat one-turn 7 mm-diameter surface coil (Stevens et al., 1984). 5FU was injected into a catheterized jugular vein. At a dose of 30 mg/kg there is no evidence for any accumulation of 5FU. Only the intermediate and final elimination products, dihydrofluorouracil and fluoro-β-alanine respectively, are observed. Their time courses are shown in Figure 3a. At a higher dose of 180 mg/kg, 5FU is observed and the time course of its elimination can be followed directly (b). Note that as it is converted into fluoro-β-alanine, the concentration of the breakdown intermediate, dihydrofluorouracil, is maintained roughly constant. Similar studies were also conducted in C57 mice with Lewis lung carcinomas implanted by subcutaneous injection of 2.5×10^4 viable tumour cells. The surface coil was placed directly over the tumours after they had grown to a mean diameter of 6-7 mm. The resulting ^{19}F NMR time courses are shown in Figures 3c and 3d for the same low and high doses respectively as used in the liver studies of tumour-free animals. Even at the low dose, 5FU is readily seen in the tumour and its elimination can be followed over a period of some 2.5 h. Clearly there is potential here for improved patient management based on an ability to monitor the level of drug actually at its site of action. At the higher dose, not only does 5FU accumulate to a greater

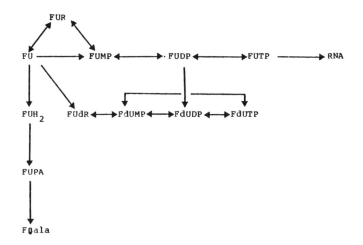

Figure 2. The pathway of 5FU metabolism. 5FU (fluorouracil); FUR (fluorouracil ribose); FUMP, FUDP, FUTP (fluorouridine mono, di and triphosphate); FUdR (fluorouracil deoxyribose); FdUMP, FdUDP, FdUTP (fluorodeoxyuridine mono, di, triphosphate); FUH$_2$ (dihydrofluorouracil); FUPA (fluoro-β-ureidopropionic acid); Fβala (fluoro-β-alanine). (From Stevens, et al., 1984, with permission).

extent, but also it is possible to identify the anabolic product FdUMP responsible for inhibition of DNA synthesis. In these experiments the minimum concentration of 5FU which could be detected was 100 μM. The absolute detection limit is currently in the region of 10 μM. Above this limit ^{19}F NMR offers a unique opportunity to monitor the drug metabolism <u>in situ</u>.

NATURAL ABUNDANCE ^{13}C NMR STUDIES

The use of fluorinated derivatives is clearly helpful in tracing the metabolic fate of foreign compounds but changes in cellular metabolites in response to these compounds cannot be assessed. Additionally, it may be difficult or impossible to synthesize a fluorinated derivative which will be metabolized in a similar way to the non-fluorinated compound. The same sort of problems apply to the use of

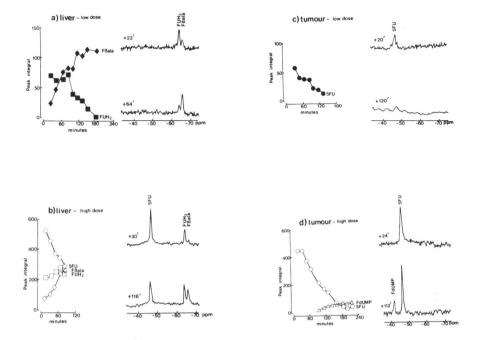

Figure 3. Uptake of 5-fluorouracil (5FU) into liver and implants of Lewis lung tumour in C57 mice monitored by ^{19}F NMR spectroscopy after the i.v. injection of 30 mg/kg 5FU (a,c) and 180 mg/kg 5FU (b, d) into the jugular vein. Peak assignments have been obtained from aqueous tissue extracts doped with tracer amounts of the suspected compounds. Times refer to time after injection of 5FU. Peak A (Δ) arises primarily from the conversion of 5FU to the major anabolic product FdUMP and probably contains contributions from the corresponding di- and tri-phosphates. (When prepared from FdUMP by the action of nucleoside monophosphate kinase both gave resonances shifted only 1 ppm upfield from FdUMP when dissolved in 200 μM TEA buffer pH 7.6.). Peak C (●, 0) is 5FU. Peak D (■,□) probably represents the first ring cleavage product of the 5FU catabolic pathway (FUH$_2$). Peak E (◆,◇) corresponds to fluoro-β-alanine (Fβala) the end product of 5FU metabolism. Chemical shift values have been derived from the assignment of 5FU and Fβala which provide non-titratable internal references. All shift values are given relative to a pure solution of 6-fluorotryptophan. (From Stevens et al., 1984, with permission).

enriched ^{13}C compounds. Ideally we would wish to have instrumentation available which would have the sensitivity to cope successfully with natural abundance ^{13}C studies. However, sensitivity is not the only problem associated with natural abundance ^{13}C spectroscopy; the identification and resolution of so many resonances requires painstaking studies. It is possible to assess the extent of the problems and the advantages that are likely to accrue, by looking at concentrated cellular extracts. Isolation of cellular extracts at well-defined times can be used to gain a picture of temporally separated events despite the very long acquisition times required. We have been developing such a technique which we hope to apply to experimental situations and although this study is not an *in vivo* one, it does show the difficulties and potential advantages of using natural abundance ^{13}C. We chose to use pig lymphocytes as our model as fresh pig mesenteric lymph nodes were readily available from a local slaughterhouse and the lymphocyte has well characterised metabolism. All spectra were obtained using a Bruker 400 MHz wide bore NMR machine operating at 100 MHz for ^{13}C. ^{13}C spectra were recorded using a gated programme with broadband decoupling of the ^1H resonances during signal acquisition only.

We have attempted to fulfil three criteria in our experiments: (i) To capture the 'cell state' when harvesting, and to maintain it over the time required for NMR data acquisition. The latter is the more difficult to achieve as it is common for samples to be in the NMR machine at room temperature for 20 h. Sodium azide was added to all aqueous extracts to prevent bacterial action and data were gathered in blocks of 4000 scans so that the first and last blocks could be compared. So far no changes with time have been detected. (ii) To avoid harsh treatments unless they had well-defined chemical actions. (iii) To produce a minimal number of well-defined extracts. The experiments concentrated the cell contents by approximately one order of magnitude to give a final volume for NMR study of approximately 3 ml. Each preparation was derived from the mesenteric nodes from 7 pigs and the extracts from 7 preparations were combined. Collagenase and DNAse were used in the cell preparations and the cells were quiesced for 2.5 h in 100 mM phosphate buffer. Phosphate buffer was used in these initial studies to prevent spectral contamination by buffer concentrated in the final samples. Microscopic examination of cells in the presence of eosin showed no sign of lysis over a period of 4 h in this buffer. Cells were harvested by rapid centrifugation and freeze clamped in liquid nitrogen. The frozen tissue was freeze dried,

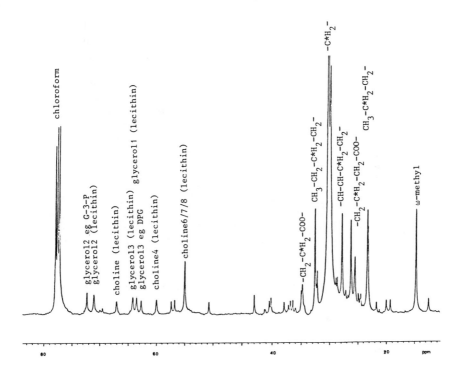

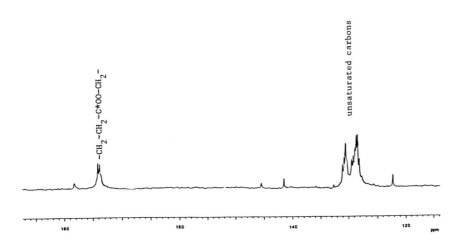

Figure 4. Natural abundance ^{13}C NMR spectrum of chloroform:methanol extract of pig lymphocytes. (a) 0-100 ppm, and (b) 100-200 ppm regions.

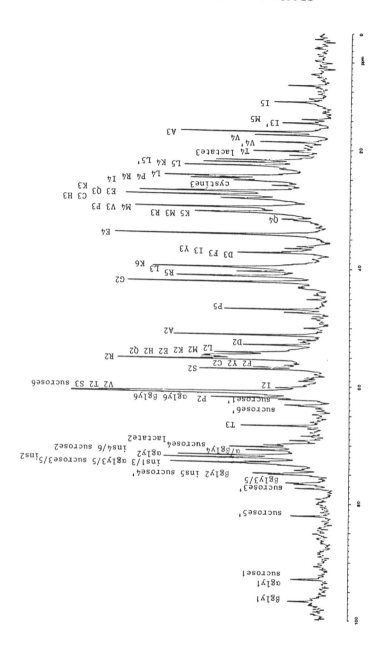

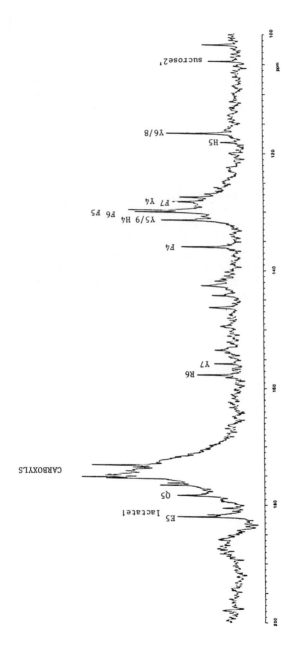

Figure 5. ^{13}C NMR spectra of aqueous extract of solid residue resulting from treatment of pig lymphocytes with chloroform:methanol. (a) 0-100 ppm and (b) 100-200 ppm regions

resuspended in chloroform:methanol (3:1) and agitated over a period of 30 min. The solid residue derived from the supernatant was resuspended in deuterated chloroform. The ^{13}C spectrum and tentative identifications from this extract are shown in Figures 4a and 4b. Several points are worth noting from these figures. Many of the resonances correspond to those from lecithin, in particular the doublets at approximately 174 and 35 ppm are characteristic of the α- and β-carbons of the 1 and 2 substituted acyl chains. Other substituted glycerides appear to be present but there is no evidence of phosphatidyl inositol compounds. Chloroform appears as a triplet caused by spin coupling with deuterium.

The dried solid residue after chloroform:methanol treatment was resuspended in distilled water and again agitated over a 30 min period. The resulting supernatant was freeze dried and resuspended in deuterium oxide. Residual quiescence buffer is extracted at this step so the solution remains at the pH of quiescence. The ^{13}C spectrum of this extract is shown in Figures 5a and 5b. Approximately 150 resonances are well resolved, most of which arise from amino acids. These are labelled using the single letter amino acid code followed by the carbon number. Lactate is clearly resolved at approximately 21 and 181 ppm, and indicates that the lactate level is acceptably low. In contrast we have prepared and run similar extracts from the lymph nodes before separation of the cells or quiescence, and found very high lactate levels. This indicates that our cells have returned to an aerobic state during quiescence. The sucrose resonances between 70 and 105 ppm are present as contamination from the tissue disruption buffer used with collagenase. Very clear resonances of myoinositol can be seen between 70 and 87 ppm but there is no evidence for any phosphorylated inositol compounds.

The insoluble material was resuspended with pronase at pH 7.4 and incubated at 37°C for 30 min. The resulting supernatant was dried and resuspended in deuterium oxide. The ^{13}C spectrum of this digest contained large numbers of broad resonances all of which corespond to those from a crude polymerised calf thymus DNA preparation which we ran separately. It is probable that all the resonances seen in the region 65 - 168 ppm are DNA, apart from those labelled otherwise. It is not clear whether the hydrolysis of protein and the appearance of DNA has any physiological significance.

Resonances derived from motionally restricted environments are broadened. So signals from cellular lipid environments or macromolecules such as DNA are much reduced in in vivo studies. The resonances from smaller soluble molecules associated with such environments are likewise broadened.

In vivo natural abundance ^{13}C studies are likely to be most useful for slowly changing soluble metabolites. However, extraction procedures can be used to take "snapshots" of the cell state and can be further used for ^{31}P or even ^{1}H studies to gain a fairly comprehensive view of the cellular state. These may be the only NMR methods to assess accurately immobile xenobiotic lipids.

CONCLUSIONS

The possibilities afforded by in vivo NMR, and by NMR studies of extracts, already form part of the biochemist's armoury. Many medical applications have been explored in the literature but the full clinical potential will not be realised for many years to come.

Acknowledgements. We are grateful to Shell Research Ltd., U.K. for a research grant in support of the natural abundance ^{13}C studies and for a postdoctoral fellowship for TAM.

REFERENCES

Ackerman, J.J.H., Grove, T.H., Wong, G., Gadian, D.G. and Radda, G.K., 1980, Mapping of Metabolites in Whole Animals by ^{31}P NMR using Surface Coils. Nature, 283, 167-170.

Bailes, D.R., Bryant, D.J., Case, H.A., Collins, A.G., Cox, I.J., Hall, A.S., Harman, R.R., Khenia, S., McArthur, P., Ross, B.D. and Young, I.R., 1988, In vivo Implementation of Three-Dimensional Phase-Encoded Spectroscopy with a Correction for Field Inhomogeneity. Journal of Magnetic Resonance, 77, 460-470.

Bloch, F., Hansen, W.W. and Packard, M., 1946, The Nuclear Induction Experiment. Physical Review, 70, 474-485.

Cohen, S.M., 1983, Simultaneous ^{13}C and ^{31}P NMR Studies of Perfused Rat Liver. Journal of Biological Chemistry, 258, 14294-14308.

Guillino, P.M. and Knazek, R.A., 1979, Tissue Culture on Artificial Capillaries. Methods in Enzymology, 58, 178-184.

Hoult, D.I., Busby, S.J.W., Gadian, D.G., Radda, G.K., Richards, R.E. and Seeley, P.J., 1974, Observation of Tissue Metabolites using ^{31}P Nuclear Magnetic Resonance. Nature, 252, 285-287.

Kantor, H.L., Briggs, R.W. and Balaban, R.S., 1984, In vivo ^{31}P Nuclear Magnetic Measurements in Canine Heart using a Catheter Coil. Circulation Research, 55, 261-266.

Knop, R.H., Chen, C.-W., Mitchell, J.B., Russo, A., McPherson, S. and Cohen, J.S., 1984, Metabolic Studies of Mammalian Cells using a Continuous Perfusion Technique. Biochemica et Biophysica Acta, 804, 275-284.

Morris, P.G., Allen, D.G. and Orchard, C.H., 1985, High-Time-Resolution ^{31}P NMR Studies of the Perfused Ferret Heart. Advances in Myocardiology, 5, 27-38.

Morris, P.G., 1986, NMR Imaging in Medicine and Biology (Oxford: University Press).

Morris, P.G., 1988, NMR Spectroscopy in Living Systems. In Annual Reports of NMR Spectroscopy edited by G.A. Webb (London:Academic Press) 20, 1-60.

Ordidge, R.J., Connelly, A. and Lohman, J.A., 1986, Image-Selected in vivo Spectroscopy (ISIS). A New Technique for Spatially Selective NMR Spectroscopy. Journal of Magnetic Resonance, 66, 283-294.

Purcell, E.M., Torrey, H.C. and Pound, R.V., 1946, Resonance Absorption by Nuclear Magnetic Moments in a Solid. Physical Review, 69, 37.

Ross, B.D., 1972, Perfusion Techniques in Biochemistry (Oxford: Clarendon Press).

Stevens, A.N., Morris, P.G., Iles, R.A., Sheldon, P.W. and Griffiths, J.R., 1984, 5-Fluorouracil Metabolism Monitored in vivo by ^{19}F NMR. British Journal of Cancer, 50, 113-117.

Ugurbil, K., Guernsey, D.L., Brown, T.R., Glynn, P., Tobkes, N. and Edelman, I.S., 1981, ^{31}P NMR Studies of Intact Anchorage-Dependent Mouse Embryo Fibroblasts. Proceedings of the National Academy of Sciences (USA). 78, 4843-4847.

THE USE OF AUTORADIOGRAPHY AS A TOOL TO STUDY
XENOBIOTIC METABOLISM

Ingvar Brandt and Eva B. Brittebo

Department of Pharmacology and Toxicology,
Faculty of Veterinary Medicine,
The Swedish University of Agricultural Sciences,
Uppsala Biomedical Center, Box 573, S-751 23
Uppsala, Sweden

INTRODUCTION
The presence of a labelled substance in a histological section can be localized by means of a radiation-sensitive photographic emulsion. This procedure is termed autoradiography. Depending on sample-preparation technique, the kind of radioisotope, film emulsion, etc., autoradiograms can be given different levels of resolution ranging from the ultrastructural level to the whole-body level. Basically, autoradiography registers the presence of a β- or positron-emitting isotope in the tissue-section, and the recorded blackening of the film does not per se give information about the chemical identity of the labelled compound. Therefore, the potential of autoradiography for metabolism studies has not always been fully appreciated. However, there is ample evidence that autoradiographic experiments, properly designed, may be very useful in investigating tissue-selective metabolic processes. The aim of the present review is to show that autoradiography, at the whole-body and light-microscopic levels, may be used as an efficient tool to study intermediary xenobiotic metabolism.

AUTORADIOGRAPHIC PROCEDURES
Some procedures used to study xenobiotic metabolism in rodents will be very briefly described. For more comprehensive information about autoradiography, see reference literature (e.g. Rogers, 1979; Ullberg, 1977).

Tape-section autoradiography
Tape-section autoradiography (also often referred to as whole-body autoradiography) was originally developed by Ullberg (1954) in connection with studies on the distribution of ^{35}S-labelled benzyl penicillin in the body.

The specimen (a piece of tissue or a whole animal) is rapidly frozen in an aqueous gel of carboxymethyl cellulose. Thin tissue sections (10-20 μm) are cut and collected onto tape in a cryostat microtome. The tissue-sections are freeze-dried; in some cases they are heated (ca. 50°C) in order to evaporate all volatile radioactivity. In addition, tissue-sections may be stepwise extracted with a series of organic solvents in order to study the distribution of irreversibly bound radioactivity. The sections are then apposed to X-ray film. After exposure, the film is removed from the sections and developed. The sections may be stained. Diffusion of a lipophilic radiolabelled compound in the sections may occasionally reduce the resolution of autoradiograms; this is avoided by doing all manipulations at -20°C (Appelgren, 1967).

The technique of tape-section autoradiography has been successively refined over the years. Stable cryostat microtomes have been constructed that allow sectioning of animals as large as pregnant cynomolgus monkey (Ullberg, 1977). Recently, computerized image analysis systems have been developed, allowing quantification of radioactivity in delicate anatomical structures (d'Argy, 1986).

Light-microscope autoradiography

Microautoradiography involves a permanent combination of a thin histological section (1-5 μm) and the photographic emulsion. A tissue is fixed by perfusion in situ or immersed in buffered formaldehyde or glutaraldehyde. After fixation, tissue-pieces are dehydrated with graded series of H_2O-ethanol solutions (and sometimes with xylene) and embedded in methacrylate resin or paraffin. The embedded tissue is sectioned, and the paraffin sections are deparaffinized in xylene and a series of H_2O-ethanol solutions before being combined with the photographic emulsion. The section is generally dipped into a liquid film emulsion but may also be covered with stripping film. Staining of the section may be done either before apposition of film or after development of the autoradiogram, depending on the staining procedure used. In order to assure efficient extraction of unbound radioactivity, the tissue-sections may be subjected to additional extraction procedures. Hence, the resulting autoradiograms will represent irreversibly bound radioactivity only. For the accurate localization of the extractable or diffusible radioactivity, different dry-mounting procedures have been described, involving apposition of a freeze-section onto a glass slide covered with a photographic emulsion. Such autoradiograms are generally stained after development of the film emulsion. Methods using freeze-dried tape-sections have also been described.

SITES OF METABOLITE ACCUMULATION AND BINDING

At the whole-body level, tape-section autoradiography may be used to identify uptake of radioactivity in tiny anatomical structures, which cannot be easily dissected and measured. The resolution of this technique enables the identification of tissues such as surface epithelia, subepithelial glands, germinal centra of lymph nodes and components of endocrine glands. Labelling of different secretions and fluids in the body may also be demonstrated. The autoradiogram in Figure 1 shows a selective localization of radioactivity in the uterine fluid and lung of a pregnant mouse injected intravenously with a trichlorobiphenyl. The localization of metabolites in the target tissues is sometimes observed almost instantaneously after administration of the labelled xenobiotic, while in other cases the transformation to tissue-binding metabolites is a slow process. A series of autoradiograms obtained using animals killed at different time-points after administration of the labelled substance will indicate whether metabolites are formed and selectively localized. After some experience in interpreting autoradiography, the investigator will be able to draw conclusions about the metabolic fate of the compound studied. Based on the autoradiographic information, proper metabolite identification experiments can be designed. The result of such a study is depicted in Figure. 1.

DETERMINATION OF NON-VOLATILE AND IRREVERSIBLY BOUND METABOLITES

Generally, autoradiograms obtained from freeze-dried tissue-sections represent the combined radiation from the parent compound and its metabolites. When the parent compound is volatile, heating of tape-sections before apposition to X-ray film will allow determination of non-volatile metabolites only. A volatile parent compound can be registered (in combination with metabolites) by low-temperature autoradiography (Bergman, 1979).

In order to determine sites of irreversibly bound metabolites, thin solvent-extracted tape-sections are used. The extraction solvents must be chosen so that they do not dissolve the tape adhesive. Suitable solvents include heptane, butanol, ethanol, methanol and water. The efficiency of the extraction procedure may differ depending on the physico-chemical properties of the labelled xenobiotic and its metabolites. Usually, the radioactivity

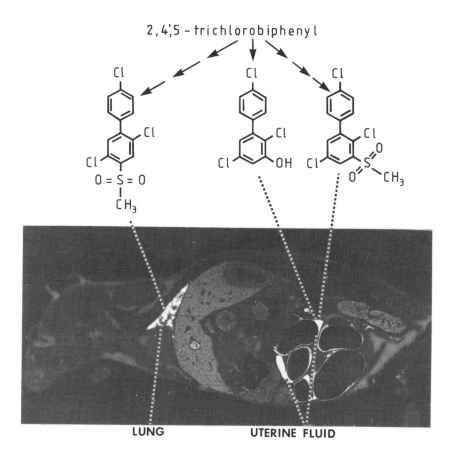

Figure 1. Tape-section autoradiogram from a pregnant mouse injected intravenously with ^{14}C-labelled 2,4',5-trichlorobiphenyl on gestation day 11 and killed 3 days later. A high concentration of radioactivity is seen in the lung and the uterine fluid. The metabolites identified in extracts of lung tissue and uterine fluid are depicted. (From Brandt et al., 1982a, with permission).

is completely extracted from most tissues, whereas the labelling remains in a few tissues. The extraction procedure may be monitored by measuring the radioactivity in the solvents.

Although the resolution of tape-section autoradiography is impressive, microautoradiography is useful to confirm and

detail the information about sites of irreversible binding.
When a ^{3}H-labelled substance is under investigation, the
resolution of microautoradiograms allows determination of
individual target cells, cell regions and larger organelles
such as the nucleus. Figure 2 shows the localization of
irreversibly bound metabolites in the nasal mucosa of a rat
given ^{3}H-labelled phenacetin (Brittebo, 1987). Due to the
higher radiation energy of ^{14}C and ^{35}S, the
microautoradiographic resolution obtained when these
isotopes are used is lower than that when ^{3}H is used.
However, different cell-layers in stratified epithelia, the
different zones in the adrenal cortex, etc. can still be
differentiated.

TISSUE-SPECIFIC METABOLISM TO REACTIVE INTERMEDIATES

When considering the formation of reactive intermediates,
it seems likely that they bind at or very near their sites
of formation. If the intermediate is sufficiently stable,
migration and binding to remote targets is a possibility,
however. Conceivably, this migration could result in a high
residue concentration in a localized area away from the site
of activation; however, in general it seems unlikely that a
reactive metabolite would be enriched and selectively bound
in a cell or tissue where it was not formed.

The sites of metabolic activation of a xenobiotic to a
reactive metabolite can be examined by autoradiography in
vitro. Tissue-explants containing the relevant cell-types
are incubated in a physiological medium together with the
labelled xenobiotic. The incubated explants are fixed,
extracted and subjected to tape-section and/or light-
microscope autoradiography. If the observed cellular
binding-sites are identical after in vivo- and in vitro-
exposure, it may be concluded that binding occurs mostly in
the vicinity of the activation site. Hence, the combination
of autoradiography in vitro and in vivo provides a means
for the verification of activation-sites in vivo. By using
this approach, the surface epithelia in the respiratory and
upper alimentary tracts have been shown to transform a
number of halogenated hydrocarbons and N-nitrosamines into
reactive products (Brittebo and Tjälve, 1983; Kowalski et
al., 1985). Other in vitro experiments confirmed that the
surface epithelia have great ability to metabolize these
compounds.

Occasionally, the localization of metabolites is not
identical after in vivo- and in vitro-exposure. For
instance, when the nasal carcinogen phenacetin was
intravenously injected into rats, an irreversible binding of
metabolites was observed predominantly in the Bowman's

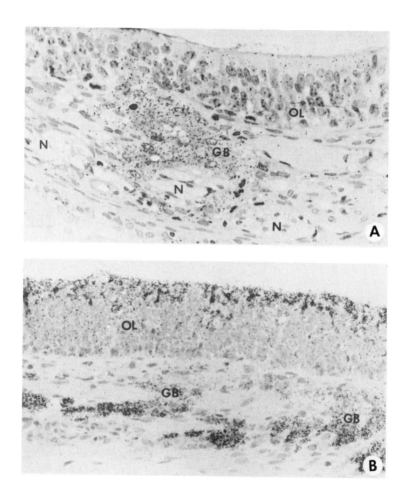

Figure 2. Microautoradiograms (methacrylate sections) of (A) the olfactory mucosa (nasal septum) from a rat intravenously injected with a trace-dose of ring-^{3}H-labelled phenacetin, and (B) the olfactory mucosa (septum) incubated with (^{3}H)-phenacetin in vitro. The rats had been pretreated with phorone in order to deplete the nasal glutathione pool. In (A), tissue-bound metabolites (represented by black silver grains) are present in the glands of Bowman (GB), whereas in (B), tissue-bound metabolites are seen also in the supporting cells of the olfactory epithelium (OL). N = nerve bundles. (From Brittebo, 1987, with permission).

glands of the olfactory mucosa (Figure 2A); however when phenacetin was incubated with nasal septa in vitro, both the Bowman's glands and the supporting cells of the olfactory epithelium were labelled (Figure 2B). Obviously, both celltypes contained the activating enzyme. However, the diffusion of the parent compound to the supporting cells in vivo is probably poor as compared to the diffusion in the better perfused subepithelial glands of Bowman (Brittebo, 1987).

THE USE OF METABOLIC INHIBITORS AND INDUCERS

The enzymes involved in the formation and inactivation of tissue-binding metabolites may be studied by means of compounds that modulate phase I and II reactions. The involvement of cytochrome P-450 can be indicated by autoradiography of animals pretreated with a suitable inhibitor. In Figure 3, metyrapone is shown to inhibit the in vivo-localization of non-volatile metabolites of the lung-toxic compound chlorobenzene to the bronchial epithelium, while the labelling of the liver remains the same (Brittebo and Brandt, 1984). In vitro-autoradiography of tissue-explants incubated together with the inhibitor and the labelled xenobiotic may provide information about activation in situ. Likewise, specific cytochrome P-450 inducers are useful to obtain information about the activating enzyme. It has been shown, for instance, that the cytochrome P-448 inducer β-naphthoflavone induces binding of the mutagenic tryptophan-pyrolysis product Trp-PI very efficiently in the lung tissue in vivo (Brandt et al., 1983). Reactive metabolites that are inactivated by glutathione (GSH) conjugation often give rise to an increased tissue-binding when the GSH-pool of the cell has been depleted. Such threshold phenomena can be studied conveniently with autoradiography, employing different doses of the labelled xenobiotic, or the use of a GSH-depleting agent. A high dose of phenacetin gives rise to labelling in the olfactory mucosa, while a low dose does not. Following depletion of the nasal GSH-pool with phorone, an irreversible binding of metabolites in the olfactory mucosa is also observed after a low dose of phenacetin (Figure 4; Brittebo, 1987).

METABOLISM OF FOETAL TISSUES - QUANTITATION BY IMAGE ANALYSIS

Autoradiography is well suited to the study of tissuespecific metabolite binding in delicate foetal tissuestructures. The labelled xenobiotic can be administered to the dam, injected into the foetuses in utero or exposed to foetal tissue-explants in vitro. Generally, the capacity of

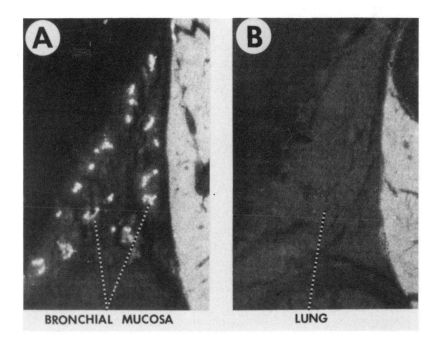

Figure 3. Tape-section autoradiograms showing the localization of non-volatile ^{14}C-labelled metabolites in the lung region after intravenous injection of ^{14}C-chlorobenzene in a control mouse (A) and a metyrapone-pretreated mouse (B). ^{14}C-Metabolites are selectively localized in the bronchial epithelium of the control mouse, whereas no metabolites are present in the bronchial epithelium of the mouse treated with the cytochrome P-450 inhibitor. (From Brittebo and Brandt, 1984, with permission).

the rodent foetus to metabolize and bind xenobiotics is absent or low, as compared to adult tissues. In certain cases, however, the late gestational foetus is capable of activating the xenobiotic in a tissue-specific manner similar to that of the adult. As shown in Figure 5, there is a preferential labelling of the olfactory mucosa when rat foetuses are injected in utero with phenacetin. In vitro experiments confirmed that the foetal nasal mucosa has a marked ability to dealkylate phenacetin; the activity exceeds considerably that of the foetal liver and increases with foetal age (Brittebo and Åhlman, 1984).

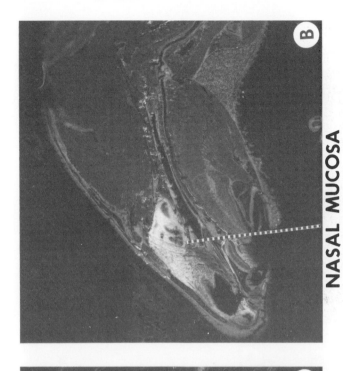

Figure 4. Tape-section autoradiograms from the head region of rats intravenously injected with a trace-dose of ^{3}H-labelled phenacetin. Rat (B) had been pretreated with the glutathione-depleting agent phorone, while rat (A) served as control. In the control rat, the radioactivity in the nasal mucosa is very low, while in the nasal mucosa from glutathione-depleted rat there is a marked uptake of radioactivity (metabolites). (From Brittebo, 1987, with permission).

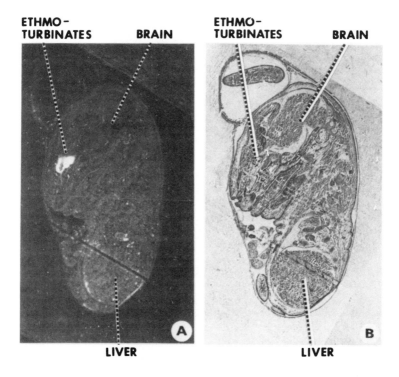

Figure 5. (A) Detail of a tape-section autoradiogram from the uterus of a pregnant rat (gestation day 22). The foetus was injected with ^{14}C-ethyl-labelled phenacetin in the neck region 1 h before freezing. (B) Corresponding tape-section stained with hematoxylin-eosin. There is a selective uptake of radioactivity in the mucosa of the ethmoturbinates of the foetus (representing olfactory mucosa). In the other foetal tissues there is a low and homogenously distributed radioactivity. (From Brittebo and Åhlman, 1984, with permission).

Several halogenated hydrocarbons and N-nitrosamines are activated and irreversibly bound in the surface epithelia of the respiratory and/or upper alimentary tract in adult animals. An ability of foetal surface epithelia to activate these compounds in vivo and in vitro has also been demonstrated by various autoradiographic procedures and in vitro experiments. By means of computer-assisted image analysis it was shown that the level of irreversibly bound metabolites in the foetal forestomach and oral epithelium was equal to, and 3 times higher, respectively, than that of the maternal liver after administration of the carcinogenic 1,2-dibromoethane to the dam (Figure 6; Table 1; Kowalski et al., 1986). In certain cases, maternal metabolites penetrate the placenta and become bound in a foetal tissue. Examples of such metabolites are PCB methylsulphones, which are formed during entero-hepatic circulation involving metabolic transformations both in the maternal liver and intestinal microflora (see below).

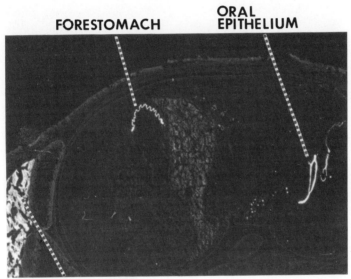

Figure 6. Detail of a tape-section autoradiogram showing the distribution of irreversibly bound (non-extractable) metabolites in the foetal tissues 90 min. after intravenous injection of ^{14}C-labelled 1,2-dibromoethane into a pregnant mouse (gestation day 17). A high level of tissue-bound metabolites is seen in the epithelia of the foetal forestomach and oral cavity. (From Kowalski et al., 1986, with permission).

TRANSLOCATION OF BILIARY METABOLITES BETWEEN TISSUES
The biliary elimination of radioactivity may be followed by tape-section autoradiography subsequent to parenteral administration of a radiolabelled xenobiotic. Biliary

TABLE 1 - Quantitation of irreversibly bound metabolites by computer-assisted image analysis. Pregnant mice (gestation day 17) were injected with ^{14}C-labelled 1,2-dibromoethane and killed 4 h later. Tape-sections were cut and solvent-extracted before autoradiographic exposure. The levels of bound metabolites were expressed as percentage of the maternal liver-level in the same section.

	Tissue-bound metabolites	
Tissue	Foetus	Dam
	(%)	(%)
Blood	n.d.[a]	19
Thymus	5	5
Liver	24	100
Oral epithelium	305	238
Oesophagus	-[b]	329
Forestomach epithelium	100	-
Nasal mucosa	105	-
Bronchi	57	442

[a]n.d. = not detectable; [b]- = not determined.
(From Kowalski et al., 1986, with permission).

excretion of radioactivity is often observed in late gestational rodent foetuses. Mercapturic acid pathway metabolites excreted in the bile may be transformed by the intestinal microflora to compounds that are reabsorbed and selectively taken up or further metabolized in various tissues. Examples are methylsulphone metabolites of PCB (Bakke et al., 1983) (Figure 7) and DDT. The DDT metabolite, 3-methylsulphonyl-DDE, is activated specifically in the adrenal zona fasciculata, to yield a highly toxic reactive metabolite which is irreversibly bound (Lund et al., 1988). Certain PCB methylsulphone metabolites are selectively taken up and reversibly bound, e.g. in the lung and kidney (Figure 1; Brandt and Bergman, 1987).

Metabolic sequences involving entero-hepatic circulation may be investigated by autoradiographic analysis. Germfree animals may be particularly useful to determine the role of the intestinal microflora. Autoradiographical examination of the lung region of germfree and conventional mice given a trichlorobiphenyl was used to establish the role of the microflora in the formation of PCB methylsulphone metabolites (Figure 7, Brandt et al., 1982b).

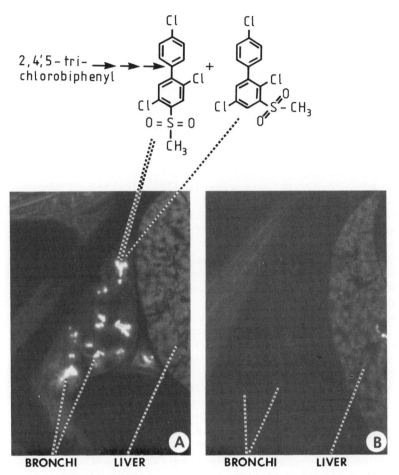

Figure 7. Tape-section autoradiograms of the lung region of a conventional (A) and a germfree (B) mouse one day after injection of ^{14}C-labelled 2,4',5-trichlorobiphenyl. In mouse (A), a high accumulation of radioactivity is seen in the bronchi, while in mouse (B) no specific labelling of the bronchial mucosa occurred. A major and a minor metabolite identified in extracts of lung tissue are depicted. (From Brandt et al., 1982b, with permission).

TRANSLOCATION OF METABOLITES WITHIN A TISSUE

Whole-body autoradiography was used to demonstrate that the PCB methylsulphone metabolite 4,4'-bis-(methylsulphonyl)-2,2'-5,5'-tetrachlorobiphenyl (($MeSO_2$)$_2$-TCB) was accumulated in a dose-dependent manner in the rodent lung (Brandt et al., 1985). When fixed tissue-sections from $(MeSO_2)_2$-TCB-injected mice and rats were extracted in ethanol and xylene, microautoradiography showed that no radioactivity was present in any lung cell. This result showed that $(MeSO_2)_2$-PCB was reversibly associated to the lung tissue. When fixed tissue-sections were extracted with ethanol (but not with xylene), 20-30% of the initial radioactivity remained in the lungs. Microautoradiography of these sections demonstrated a high and selective labelling in the non-ciliated bronchiolar (Clara) cells, and in goblet-cells at higher levels of the airways. Finally, microautoradiography of freeze-sections from lung tissue was performed in order to determine the localization of the ethanol-extractable radioactivity. These autoradiograms showed that the radioactivity was present to a large extent in the airway lumen, from where it could subsequently be isolated with lung lavage and demonstrated to be associated with a specific PCB methylsulphone-binding protein (Figure 8; Brandt et al., 1985). This protein is present in high concentration in rat and mouse lung cytosol (Lund et al., 1985).

Based on these experiments, the following conclusions were drawn: The PCB methylsulphones are initially bound to a specific protein residing in Clara and goblet cells. The protein-sulphone complex is subsequently secreted into the airway lumen and spread over the entire surface lining. It also seems likely that the protein-sulphone complex is transported with the mucociliary escalator to the pharynx and swallowed. In the gut, the sulphone may be released, reabsorbed and recirculated to the lung. Such an enteropulmonary circulation of PCB methylsulphones could contribute to the exceptionally long retention times in the rodent lung observed by tape-section autoradiography (generally several months).

NEW SITES OF DRUG METABOLISM

As indicated above, one advantage of autoradiography is the possibility of determining drug metabolism in defined cell populations and delicate anatomical structures. In fact, several previously unforeseen sites of drug metabolism were first observed by autoradiography. The high drug-metabolizing ability of the bronchiolar Clara cells was first indicated in microautoradiographic in vivo studies of the pulmonary toxin 4-ipomeanol (Boyd, 1977). The high metabolic ability of the nasal mucosa was demonstrated in studies combining in vivo autoradiography with in vitro experiments with some N-nitrosamines (Brittebo and Tjälve, 1983; Brittebo et al., 1986). The nasal and tracheo-

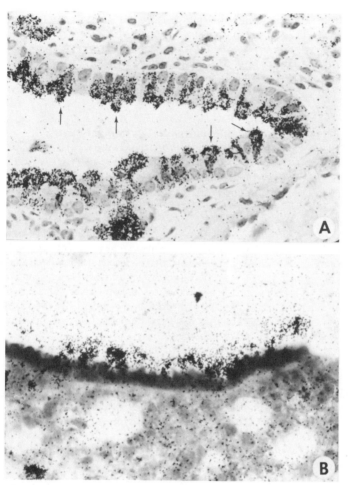

Figure 8. Microautoradiograms from lung tissue of mice 6 days after injection of ^{3}H-labelled 4,4'-bis(methylsulphonyl)-2,2',5,5'-tetrachlorobiphenyl. Autoradiogram (A) was prepared from a methacrylate section (1 μm) from ethanol-extracted tissue (containing 20-30% of the initial radioactivity). The labelling (black silver grains) is confined to the Clara cells (arrows) of a terminal bronchiole, while the adjacent (ciliated) cells are devoid of radioactivity. Autoradiogram (B), which was prepared from a thaw-mounted freeze-section (5 μm; no radioactivity extracted), shows that the radioactivity is present also in the bronchiolar lumen. The PCB methylsulphone is primarily taken up in the Clara cells, from where it is secreted into the bronchiolar lumen, bound to a specific protein. (From Brandt et al., 1985, with permission).

bronchial mucosa are now recognized as tissues with a marked drug-metabolizing activity.

We have recently shown by the use of in vivo and in vitro autoradiography that the cervico-vaginal epithelium is very active in metabolizing short-chained organic halides during certain phases of the oestrous cycle (Brittebo et al., 1987).

PITFALLS IN THE INTERPRETATION OF AUTORADIOGRAMS

The proper evaluation of autoradiograms requires a detailed knowledge about the anatomy and histology of the experimental animal used. Apart from difficulties in identifying tiny anatomical structures, a variety of technical problems may occur, depending on the particular technique used. Different types of artefacts may also arise. For instance, an intense labelling of reticuloendothelial cells in the lung, liver, bone marrow and spleen may be observed when the xenobiotic is dissolved in a lipid emulsion and injected intravenously. Likewise, i.v. injection of a lipophilic compound may result in labelled embolies which are trapped in the lung capillaries. The latter artefact may easily be misinterpreted as an uptake into the bronchial epithelium.

Of particular relevance when studying intermediary drug metabolism is the problem of differentiating between covalent metabolite binding and the incorporation of radioactivity into natural products. When a labelled xenobiotic is dealkylated to yield ^{14}C-labelled formaldehyde, acetaldehyde etc., a typical whole-body autoradiogram will be observed, characterized by a high uptake of radioactivity in tissues with a high cell proliferation and/or protein synthesis (Johansson and Tjälve, 1978; Johansson-Brittebo and Tjälve, 1979). These tissues include the bone marrow, thymus, lymph nodes, small intestinal epithelium, exocrine pancreas and salivary glands. Hence, when intense labelling is observed in these tissues it should be considered whether metabolic incorporation of one- and two-carbon fragments into natural products has occurred. This can easily be checked by, for example, measuring the exhalation of $^{14}CO_2$ from the animals. In fact, the degradation of a ^{14}C-labelled compound to $^{14}CO_2$ can be predicted when the characteristic autoradiographic distribution pattern is observed. This was done in our studies on the degradation of chlorinated paraffins in mice (Darnerud et al., 1982).

Another way of differentiating between covalent binding and incorporation into natural products is to study the distribution of radioactivity at very short post-injection times (1-5 min.), since the characteristic tissue-labelling resulting from incorporation into natural products generally needs more time to develop. Another possibility is to change the position of the ^{14}C-labelled atom or introduce another radioisotope into the molecule.

At very long post-injection times ($\geq$ 12 days), a selective localization of radioactivity may occasionally be observed in the adrenal cortex and in the central nervous system of adult rodents. This pattern is seen after administration of xenobiotics which are metabolized to ^{14}C-labelled two-carbon fragments (Figure 9). The two-carbon fragments are utilized

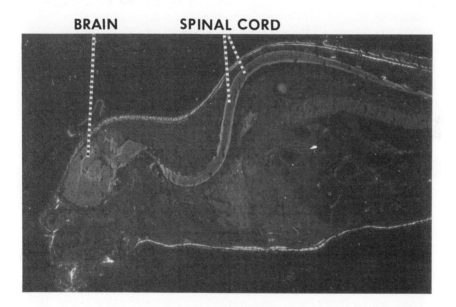

Figure 9. Tape-section autoradiogram of a mouse 30 days after injection of ^{14}C-ethyl-labelled 4,4'- bis(ethylsulphonyl)-2,2',5,5'-tetrachlorobiphenyl. The autoradiogram demonstrates a retention of radioactivity in the central nervous system. This distribution pattern is characteristic for compounds that are degraded to ^{14}C-labelled two-carbon fragments. The fragments are incorporated metabolically into cholesterol and other endogenous compounds, which are built into the myelin.

for the biosynthesis of cholesterol and other lipids, which are subsequently used or stored in the adrenal cortex and brain (Darnerud and Brandt, 1985; Brandt et al., 1984). Notably, whole-body autoradiography of animals given ^{14}C-labelled cholesterol shows a similar localization of radioactivity (Appelgren, 1967).

Acknowledgements. We thank all collaborators, particularly Drs. Å. Bergman and C.A. Wachtmeister, for stimulating cooperation. Ms Anki Adlercreutz provided expert secretarial assistance. Economic support was given by the Swedish Environment Protection Board, the National Council for Forestry and Agricultural Research (I. Brandt), the Swedish Work Environment Fund and the Swedish Medical Research Council (E.B. Brittebo).

REFERENCES

Appelgren, L.-E., 1967, Sites of steroid hormone formation: Autoradiographic studies using labelled precursors. Acta Physiologica Scandinavica, Suppl. 301, 1-108.

d'Argy, R., 1986, Image analysis in whole-body autoradiography (WBA). Uppsala Journal of Medical Science, 91, Special issue, 257-262.

Bakke, J.E., Bergman, Å.L., Brandt, I., Darnerud, P. and Struble, C., Metabolism of the mercapturic acid of 2,4',5-trichlorobiphenyl in rats and mice. Xenobiotica, 13, 597-605.

Bergman, K., 1979, Whole-body autoradiography and allied tracer techniques in distribution and elimination studies of some organic solvents. Scandinavian Journal of Work, Environment and Health, 5, Suppl. 1, 1-263.

Boyd, M.R., 1977, Evidence for the Clara cell as a site of cytochrome P450-dependent mixed-function oxidase activity in lung. Nature, 269, 713-715.

Brandt, I., Darnerud, P.O., Bergman, Å. and Larsson, Y., 1982a, Metabolism of 2,4',5-trichlorobiphenyl: Enrichment of hydroxylated and methyl sulphone metabolites in the uterine luminal fluid of pregnant mice. Chemico-Biological Interactions, 40, 45-56.

Brandt, I., Klasson-Wehler, E., Rafter, J. and Bergman, Å., 1982b, Metabolism of 2,4',5-trichlorobiphenyl: Tissue concentrations of methylsulphonyl-2,4',5-trichlorobiphenyl in germfree and conventional mice. Toxicology Letters, 12, 273-280.

Brandt, I., Gustafsson, J.-Å. and Rafter, J., 1983, Distribution of the carcinogenic tryptophan pyrolysis product Trp-P-1 in control, 9-hydroxyellipticine and β-naphthoflavone pretreated mice. Carcinogensis, 4, 1291-1296.

Brandt, I., Darnerud, P.O. and Bergman, Å., 1984, Degradation to $^{14}CO_2$ of 4,4'-bis[(2-^{14}C)-ethylsulphonyl]-2,2',

5,5'-tetrachlorobiphenyl. Acta Pharmacologica et Toxicologica, 55, 429-430.

Brandt, I., Lund, J., Bergman, Å., Klasson-Wehler, E., Poellinger L. and Gustafsson, J.-A., 1985, Target cells for the polychlorinated biphenyl metabolite 4,4'-bis (methylsulphonyl)-2,2',5,5'-tetrachlorobiphenyl in lung and kidney. Drug Metabolism and Disposition, 13, 490-496.

Brandt, I. and Bergman, Å., 1987, PCB methyl sulphones and related compounds: Identification of target cells and tissues in different species. Chemosphere, 16, 1671-1676.

Brittebo, E.B. and Åhlman, M., 1984, Nasal mucosa from rat fetuses and neonates metabolizes the nasal carcinogen phenacetin. Toxicology Letters, 23, 279-285.

Brittebo, E. and Brandt, I., 1984, Metabolism of chlorobenzene in the mucosa of the murine respiratory tract. Lung, 162, 79-88.

Brittebo, E.B. and Tjälve, H., 1983, Metabolism of N-nitrosamines by the nasal mucosa. In: Nasal Tumors in Animals and Man, Vol. III, edited by G. Reznik and S.F. Stinson (Boca Raton, Florida: CRC Press, Inc.), pp. 233-250.

Brittebo, E.B., Castonguay, A., Rafter, J.J., Kowalski, B., Åhlman, M. and Brandt, I., 1986, Metabolism of xenobiotics and steroid hormones in the nasal mucosa. In: Toxicology of the Nasal Passages, edited by C.S. Barrow (Washington: Hemisphere Publishing Corporation), pp. 211-234.

Brittebo, E.B., 1987, Metabolic activation of phenacetin in rat nasal mucosa: Dose-dependent binding to the glands of Bowman. Cancer Research, 47, 1449-1456.

Brittebo, E.B., Kowalski, B. and Brandt, I., 1987, Binding of the aliphatic halides 1,2-dibromoethane and chloroform in the rodent vaginal epithelium. Pharmacology and Toxicology, 60, 294-298.

Darnerud, P.O., Biessmann, A. and Brandt, I., 1982, Metabolic fate of chlorinated paraffins: Degree of chlorination of [^{14}C]-chlorododecanes in relation to degradation and excretion in mice. Archives of Toxicology, 50, 217-226.

Darnerud, P.O. and Brandt, I., 1985, Pitfalls in the interpretation of whole-body autoradiograms: Long-time retention in brain and adrenal cortex caused by metabolic incorporation of ^{14}C from various labelled xenobiotics. Acta Pharmacologica et Toxicologica, 56, 55-62.

Johansson, E.B. and Tjälve, H., 1978, The distribution of (^{14}C)dimethylnitrosamine in mice. Autoradiographic

studies in mice with inhibited and noninhibited dimethyl-
nitrosamine metabolism and a comparison with the
distribution of (^{14}CO)formaldehyde. Toxicology and
Applied Pharmacology, 45, 565-575.

Johansson-Brittebo, E. and Tjälve, H., 1979, Studies on the
tissue-disposition and fate of N-(^{14}C)ethyl-N-nitrosourea
in mice. Toxicology, 13, 275-285.

Kowalski, B., Brittebo, E.B. and Brandt, I., 1985,
Epithelial binding of 1,2-dibromoethane in the respiratory
and upper alimentary tracts of mice and rats. Cancer
Research, 45, 2616-2625.

Kowalski, B., Brittebo, E.B., d'Argy, R, Sperber, G.O. and
Brandt, I., 1986, Fetal epithelial binding of 1,2-di-
bromoethane in mice. Carcinogenesis, 7, 1709-1714.

Lund, J., Brandt, I., Poellinger, L., Bergman, Å., Klasson-
Wehler, E. and Gustafsson, J.-Å., 1985, Target cells for
the polychlorinated biphenyl metabolite 4,4'-bis(methylsulfonyl)-2,2',5,5'-tetrachlorobiphenyl. Characterization of high affinity binding in rat and mouse lung
cytosol. Molecular Pharmacology, 27, 314-323.

Lund, V.-O., Bergman, Å. and Brandt, I., 1988, Metabolic
activation and toxicity of a DDT-metabolite, 3-methyl-
sulphonyl-DDE, in the adrenal *zona fasciculata* in mice.
Chemico-Biological Interactions, 65, 25-40.

Rogers, A.W., 1979, Techniques of Autoradiography, Third
completely revised edition (Elsevier/North Holland
Biomedical Press), 1-429.

Ullberg, S., 1954, Studies on the distribution and fate
of S^{35}-labelled benzylpenicillin in the body. Acta
Radiologica, Suppl. 118, 1-110.

Ullberg, S., 1977, The technique of whole body auto-
radiography. Cryosectioning of large specimens.
Science Tools, The LKB Instrument Journal, Special
Issue, 2-29.

IN SITU PERFUSION AND COLLECTION TECHNIQUES FOR STUDYING XENOBIOTIC METABOLISM IN ANIMALS

K.L. Davison

Metabolism and Radiation Research Laboratory,
Agricultural Research Service,
United States Department of Agriculture,
Fargo, North Dakota 58105, USA

INTRODUCTION
Animal models are used for studying the metabolic fate of various compounds, including xenobiotics. During metabolism, some xenobiotics are cycled among various tissues before being eliminated in faeces or urine. A well known example of this is enterohepatic circulation where compounds are absorbed from the intestine, carried to the liver in blood and returned to the intestine in bile. Thus, techniques are needed for studying the contribution of specific organs and tissues to the metabolism of xenobiotics.

This paper presents techniques used to perfuse directly livers and kidneys of various animals, and to perfuse the gastrointestinal tract. If interested in performing any of the techniques described herein, the reader first may wish to read one of the following books: Lambert, 1965; Petty, 1982; Ross, 1972; or Waynforth, 1980. While books and reviews are excellent sources of information regarding experimental surgical methods, seemingly trivial details which enhance performance of the surgical preparations are often omitted. Thus, I have included details which I believe are beneficial to the surgical procedures or to performance of the preparations.

NEPHRIC PERFUSIONS
Anaesthesia
I have perfused kidneys in calves, chickens, pigs and rats. All perfusions were with anaesthetized animals and lasted two to four hours. Calves, pigs and rats were anaesthetized with halothane ($CF_3CHBrCl$) by using a closed circuit anaesthesia machine with oxygen as the carrier gas. On completion of the surgery, anaesthesia was maintained with sodium pentobarbital (Pentobarbital Sodium Solution,

Fort Dodge Laboratories, Inc., Fort Dodge, IA). All animals were given the pentobarbital, to effect, two or more times during the experiments. Pigs and calves were given pentobarbital intrajugularly and rats were given pentobarbital intramuscularly. The total dose varied from 60 mg/kg of body weight for calves to 90 mg/kg for rats. Chickens were anaesthetized with a solution of diallylbarbituric acid (Wideman and Braun, 1982), which will be described in more detail later.

The experimental subjects were fasted overnight (about 16 hours). Guidelines of the laboratory Animal Care Committee were followed. The surgeon wore plastic gloves and used clean instruments, but asceptic procedures were not followed. On completion of the surgery, the animals were covered with towels and provided heat from an infra-red lamp to maintain body temperature.

Rats

Three to four-month-old male Sprague Dawley rats weighing 350 to 400 g were used. The rats were anaesthetized and placed on their backs, hair was clipped on their neck and abdomen, and an antiseptic solution was applied to the clipped areas.

A 12 mm slit was made in the skin over the area of the jugular vein, the external jugular was located by blunt dissection with micro dissecting forceps and occluded with a 5-0 silk ligature. A PE-10 cannula (polyethylene tubing, 0.28 mm i.d. X 0.61 mm o.d.; Clay Adams, Parsippany, NJ) filled with physiological saline solution (0.15 M NaCl) was inserted into the jugular vein, caudal to the occlusion, and anchored with a 5-0 silk ligature. Infusion of physiological saline by means of a syringe pump began immediately, and the skin was closed with 2-0 silk suture.

A midline incision extending from the xiphoid cartilage to just anterior to the penis was made in the abdomen. The abdominal walls were retracted with a baby Balfour retractor, its weight supported by wooden blocks, and the intestinal mass was reflected to the right (directions are given with respect to the animal's body) and covered with saline-dampened gauze. The left ureter, located on the ventral surface of the lumbar muscle mass, was dissected free of other tissue with the aid of micro forceps, beginning just anterior to the bladder and continuing about 5 mm toward the kidney. A 5-0 silk ligature was used to occlude the ureter. Anterior to the ligature, using microdissecting scissors, a cut was made part way through the ureter and a PE-10 cannula was inserted and anchored with a 5-0 silk ligature. The free end of this cannula was

brought outside the body through 14 ga stainless steel
hypodermic needle tubing inserted at the left of the anus
near the base of the tail; the 14 ga steel tubing was then
removed. The intestinal mass was then reflected to the left
and the right ureter cannulated in like manner. The bile
duct, located in pancreatic tissue between the duodenum and
liver, was dissected free of surrounding tissue and
cannulated with PE-10 tubing in the same manner as the
ureters. The free ends of the cannulae from the right
ureter and bile duct were brought outside the body,
together, at the base of the tail to the right of the anus.

The right renal artery was cannulated because it was more
conveniently located. The cannula (Fine et al., 1974,
Figure 1) was made of PE-10 tubing and 30 ga hypodermic
needle stock (stainless steel hypodermic needle tubing), 11
mm long. The PE-10 tubing was slipped over the blunt end of
the needle stock a distance of 3-4 mm and a 5-0 silk
ligature was tied securely around it. The ends of the
ligature were cut about 3 mm long. About 3 mm from the
point, the bevelled end of the needle stock was bent to $90°$
(bevel toward the inside of the bend). The cannula was
filled with physiological saline. The right renal artery
usually lies immediately dorsal to the right renal vein and
posterior vena and is not visible through the ventral
incision. With cotton tipped swabs, the vena cava was
teased to the right, exposing the renal artery and aorta
(Figure 1). A 5-0 silk suture was placed around the renal
artery adjacent to the aorta, but not tied (Beuzeville,
1968), so that by pulling on the ends of the suture the
artery was stretched and its blood flow stopped, and
by releasing tension on the suture blood flow resumed
immediately. With gentle tension on the suture, the artery
was cleaned of surrounding tissue for a distance of 2 to 3
mm, the needle on the renal cannula was grasped with micro
needle holders and inserted into the artery to the bend in
the needle. A syringe attached to the free end of the
cannula prevented blood from flowing back into the cannula.
Saline perfusion was started immediately at 0.1 ml/min and
(because blood flow decreases in traumatized arteries)
continued for about 30 min before the surgical preparation
was used for an experiment.

Curved micro needle holders without catches and with tips
1 mm wide (for better visibility) were used for inserting
the cannula into the artery. A drop of cyanoacrylate glue
(Super Glue Corp, Hollis, NY) applied where the needle
entered the artery anchored the needle, and cyanoacrylate
glue applied to the ends of the silk sutures previously
placed around the cannula anchored the cannula to nearby

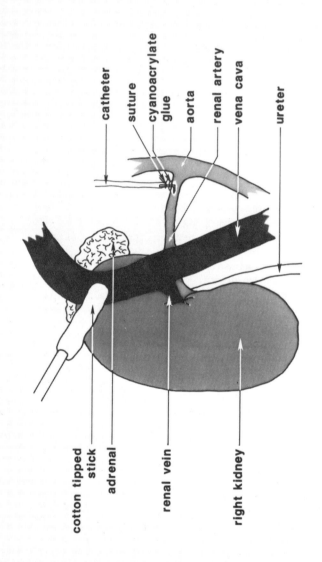

Figure 1. Cannulation of the renal artery of the rat (redrawn from Smits et al., 1983)

adventitious tissue. The free end of the cannula was then brought outside the body cavity at any place in the ventral incision that would minimize tension and the abdomen was closed with a continuous 2-0 silk suture.

Pigs

Young pigs weighing 12 to 25 kg were used. Standard size surgical instruments were used. Cannulation procedures for the jugular vein, bile duct and ureters were essentially the same as for rats, except for cannula size. PE-160 cannulae (1.14 mm i.d. x 1.57 mm o.d.) were used for the jugular vein (the vein is about 3 cm beneath the skin in this size of pig) and PE-240 cannulae (1.67 mm i.d. x 2.42 mm o.d.) were used for the bile duct and ureters. The cannulae were anchored with 2-0 silk ligatures. Free ends of the cannulae from the bile duct and ureters were brought outside the body with the aid of a Trocar and cannula inserted through the posterior abdominal wall.

Cannulae for the renal artery were made from PE-50 tubing (0.58 mm i.d. x 0.965 mm o.d.; Herd and Barger , 1964; Figure 2). A piece of tubing was heated ($\sim$ 130°) in an oil bath and stretched. The tubing was then cut at the smallest point in the stretched area and threaded through the lumen of 23 ga thin-wall hypodermic needle stock, 17 mm long. When the stretched end of the polyethylene cannula protruded through the pointed end of the needle stock, it was grasped and pulled until the cannula wedged firmly in the needle. While under slight tension, the excess polyethylene was then cut off flush with the bevelled needle point. Finally, the needle stock was bent to about 3/8 circle (Figure 2). A 5-0 cardiovascular suture (atraumatic needle attached) was tied securely about 2 cm above the needle stock. The cannula was filled with saline and attached to a syringe.

The right renal artery is more accessible from the ventral approach than the left renal artery. Reflection of the visceral mass to the left is necessary but difficult. The right artery was located by palpating around the easily visible renal vein. The artery lies dorsal to the vein, but may lie slightly posterior to slightly anterior to the vein. The artery was isolated by blunt dissection and lifted slightly with a blunt metal hook. The needle end of the arterial cannula was grasped with a needle holder and inserted toward the kidney into the lumen of the artery and back out again, as in making a suture (Figure 2). The cannula was pulled through the artery to the level of the attached suture, the suture was anchored to the musculature of the artery, and the leading tip of the cannula was then removed flush with the surface of the artery. Slight reverse tension then pulled the tip of the cannula back into the arterial lumen (the musculature of the artery seals the puncture and blood immediately flows into the cannula). The syringe on the opposite end of the cannula was attached to a syringe pump and saline was pumped into the renal artery at 0.3 ml/min. The ventral incision was closed around the arterial cannula with 1-0 or No. 1 silk in a continuous suture. Saline was pumped into the renal artery for about 30 min before the pigs were used for xenobiotic perfusion studies.

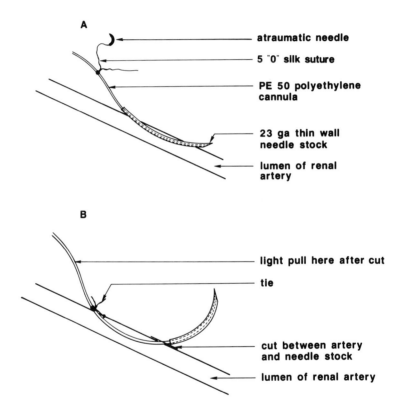

Figure 2. Cannula and procedure for cannulating the renal artery of calves and pigs

Calf

Two-week-old calves weighing about 50 kg were used. Surgery was identical to that for the pigs, except that the left renal artery was used because the location of the rumen made it more accessible. The cannula sizes were PE-240 for the jugular vein and bile duct, and PE-190 (1.19 mm i.d. x 1.70 mm o.d.) for the ureters.

Chickens

White Leghorn roosters, 3- to 6-months-old and weighing 1.1 to 1.9 kg, were used. The experiments were conducted with anaesthetized, surgically modified chickens in the supine position. Test xenobiotics were administered on a per chicken basis in avian saline (0.11 M NaCl) perfused at about 0.33 ml/min for 2 hours.

The chickens were anaesthetized with diallylbarbituric acid (Wideman and Braun, 1982). The anaesthetic was prepared by dissolving 40 g of ethyl urea, 40 g of urethane and 5 g of diallylbarbituric acid in 60 ml of distilled water at 80-90°C, diluting to 200 ml with hot distilled water, and then storing at room temperature without refrigeration or further dilution. Multiple 1 ml injections were made into various sites in the breast muscle to provide a total of 3 ml of anaesthetic/kg of body weight. Anaesthesia to a surgical plane takes about 30 min. Procaine (2% solution) was injected along the site of the abdominal incision before making the incision.

The abdominal cavity was opened continuous right to left flank so that the viscera could be reflected to expose the kidneys. A ligature was placed around the colon to prevent defaecation. The common iliac vein was partly freed from the surface of the left kidney by blunt dissection (Figure 3) and occluded with an aneurysm clip (1X8 mm, Roboz Surgical Instrument Co., Inc., Washington, DC) placed near the bifurcation of the external iliac vein. The caudal branch of the left renal portal vein was freed from the surface of the kidney and from the ureter and occluded with an aneurysm clip placed near the interiliac anastomosis (Odlind, 1978). The venous supply to the right kidney was then occluded in an identical manner. The aneurysm clips on both renal portal veins were then examined to assure that they were not obstructing the ureters which lie immediately ventral to these veins, and the abdomen was closed with metal wound clips. The superficial metatarsal vein of one leg was exposed and a PE-50 cannula filled with avian saline was inserted and anchored.

Blood from the superficial metatarsal vein ultimately flows into the external iliac vein. With the modifications described, blood from the external iliac vein is forced through the kidney parenchyma to the renal vein. From the renal vein, blood goes via the posterior vena cava to the

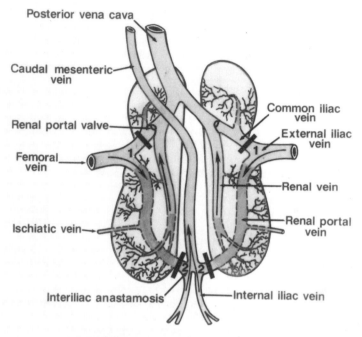

1 & 2 : Clip placement for forced portal perfusion

Figure 3. Renal portal system of chickens (used with permission, Akester, 1967; also see Odlind, 1978, and Wideman and Braun, 1982)

heart. The kidney ipsilateral to the cannulated superficial metatarsal vein was then perfused with saline at 0.3 to 0.4 ml/min as the preparation proceeded. Test xenobiotic was also perfused through this cannula.

The cloaca, including proctodaeum and urodium, was everted exposing the orifice of the ureters (Wideman and Braun, 1982). Faecal material was swabbed away, a tissue paper plug was placed in the coprodium and latex cannulae (2 mm i.d. X 4 mm o.d. X 6 cm long) were glued over the orifice of each ureter with cyanoacrylate glue. Cyanoacrylate glue was then applied over the entire everted cloaca and paper plug to prevent seepage of tissue fluid and to maintain the cloacal eversion. The chicken was inclined slightly, head up, and test tubes were placed beneath the latex cannulae to collect urine. To stimulate urine flow, a PE-50 cannula was inserted into a brachial (wing) vein and either a 2.5% aqueous solution of mannitol or avian saline was perfused at 0.2 to 0.4 ml/min.

Anaesthesia was maintained throughout the experiments. Urine flow was observed for about 15 min after the preparation was completed, then test substrates were perfused for about 2 hours. The chickens were euthanized by injecting sodium pentobarbitol (26% w/v aqueous solution) into a brachial vein soon after perfusion of the xenobiotic was completed.

HEPATIC PERFUSIONS
Rats

Through a ventral, midline incision, the superior mesenteric vein (Figure 4) was located, isolated and lifted slightly with a blunt metal hook. The vein was punctured

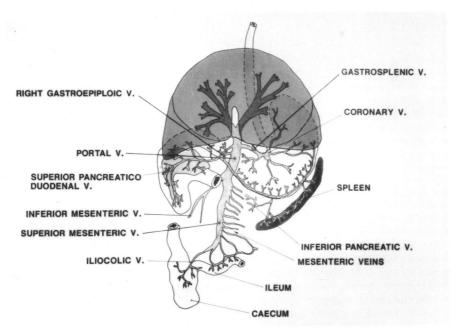

Figure 4. Hepatic venous system of the rat (used with permission, Chiasson, 1980).

about 3 cm from the liver with a small hypodermic needle (~23 ga) and a PE-10 cannula with bevelled tip was inserted through the puncture about 2 cm toward the liver. A drop of cyanoacrylate glue sealed the vein. A 5-0 silk ligature around the cannula and glued to adventitious tissue anchored the cannula. PE-10 cannulae were placed in the bile duct and ureters as previously described and the incision was closed. Xenobiotics in physiological saline were perfused at 0.1 ml/min with a syringe pump.

For our experiments, the rats remained anaesthetized. However, the rats could be allowed to recover from the anaesthesia, provided that they were restrained in a suitable device such as a Bollman type cage, and a ligature around both the cannula and the superior mesenteric vein would provide a more secure anchor for the infusion cannula. The reader is referred to three other very good methods for in vivo cannulation of the rat portal vein (Gallo-Torres and Ludorf, 1974).

Pigs

The hepatic portal system of the pig was perfused through either the splenic or the ileal veins. These veins were approached through a ventral, midline incision. They were isolated by blunt dissection, partially transected with micro dissecting scissors, and cannulated with PE-50 tubing. The cannula was anchored with 5-0 silk ligatures. Cannulae were inserted into the bile duct and ureters as described for renal perfusions. Xenobiotics were perfused in saline at 0.3 ml/min.

Chickens

The caudal mesenteric vein was used (Figure 3) for hepatic portal perfusion of chickens. With the chicken on its back, an incision was made in the right abdomen beginning at the ribs and continuing posteriorly to within a few millimeters of the midline. Since the caudal mesenteric vein lies dorsal to the colon, the colon was lifted with forceps. A few millimeters of the vein were isolated by blunt dissection, and the vein was partially transected with micro dissecting scissors. A PE-50 cannula was inserted, with the tip directed anteriorly, and anchored with 5-0 silk ligatures. Avian saline was perfused at 0.3 ml/min.

In chickens, two ducts carry bile to the duodenum. The cystic duct lies anterior to the hepatic duct and carries bile from the gall-bladder. The hepatic duct carries bile directly from the liver. Both ducts originate from the same sinuses in the liver. An abdominal incision is made beginning on the right side at the ribs and following along the dorsal edge of the breast muscle. This exposes the edge of the ventral lobe of the liver. The duodenum touches and lies dorsal to this edge of the liver. The duodenum was grasped with serrated toothed forceps and pulled gently into the incision (the duodenum is firmly attached and cannot be pulled very far without tearing). Both bile ducts appeared adjacently and on the medial side of the duodenum. 2-0 silk ligatures were used to occlude both the cystic and hepatic ducts. Gentle tension was applied to the hepatic duct by pulling on the ligature while a few millimeters of the duct were isolated by blunt dissection. The duct was partially transected with micro dissecting scissors and a PE-50 cannula was inserted toward the liver. The cannula was anchored by a second 2-0 silk ligature around the hepatic duct.

The portal and bile cannulae were exteriorized (passed through a 14 ga hypodermic needle in the abdominal wall) and the incision was closed with 2-0 silk in a continuous suture. Urine was collected as described previously.

GASTROINTESTINAL PERFUSIONS
All species

Windmueller and Spaeth (1981, 1984) describe an excellent, and quite elaborate procedure for perfusing the small intestine of the rat. Their method allows access to the intestinal lumen, arterial and venous blood, and lymph.

I have perfused substrates directly into the stomach and small intestine of rats through PE-10 or PE-50 cannulae placed in these organs. The cannulae were inserted into the GI tract through a puncture made with a small-gauge hypodermic needle and anchored to the serosa and musculature of the GI tract with 6-0 silk suture with attached atraumatic needle. A small bend or hook on the end of the cannulae (made by heating them in hot water) may make them easier to align with the GI tract, depending on the site selected for exiting the cannulae from the abdominal cavity. Bile was collected as described earlier in this paper. The rats were placed in restraining cages, allowed to recover from the anaesthetic, and dosed 24 hours later. Substrate dissolved in saline was perfused into the lumen of the GI tract at 0.1 ml/min with a syringe pump.

Dr. Craig Struble in our laboratory modified the procedure of Doluisio et al. (1969) to study absorption _in situ_ of radiolabelled xenobiotics from intestinal segments of chickens, pigs and rats. An experimental preparation similar to that used by Dr. Struble has been published for calves (Murray et al., 1987, Figure 5).

Struble maintained anaesthesia with halothane in all species throughout the one hour absorption studies. Through a midline incision, cannulae were placed in the intestinal

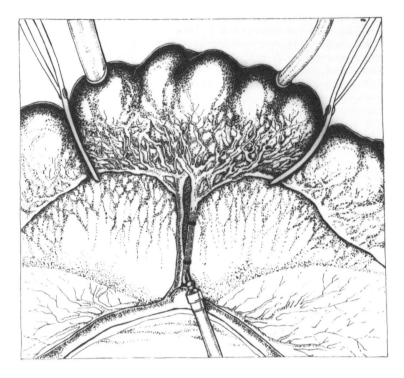

Figure 5. Drawing of an isolated intestinal segment of the calf for perfusion experiments (used with permission, Murray et al., 1987)

segment to be perfused and syringe barrels were attached to the cannulae. The test solution was added through the syringe barrels. After the test solution had funnelled into the intestinal segment, air was used to move it back and forth in the intestinal segment. Air moved the test solution into the opposite syringe barrel for sampling at selected time intervals. The bile duct, jugular vein, portal vein (for rats, by the method of Yorgey et al., 1986) and urinary bladder were also cannulated.

Five ml of test solution was used in rats. The length of intestinal segment to be perfused was determined by inserting the first cannula in the duodenum, near the pylorus, then introducing 5 ml of solution. The second cannula was inserted at the end of the space filled with the 5 ml, which was about one half of the small intestine.

Fifteen ml of test solution was used in chickens. The perfused intestinal segment started at the pylorus and extended to within about 5 cm of the ilioceacal junction. Rapid absorption of water by the chicken intestine necessitated the addition of 1.0 ml of saline solution in the intestinal segment. Fifty ml of test solution was used in pigs, and large glass cannulae replaced the syringe barrels and polyethylene cannula. The perfused intestinal segment started at the pylorus and extended to a point about 10 cm past the duodenocolic fold. Sampling was via teflon tubing inserted through the cannulae into the intestinal lumen.

GENERAL DISCUSSION

Some metabolic sequences investigated with the aid of the perfusion techniques reported herein are included by Dr.J.E.Bakke in his chapter: "Metabolites derived from glutathione conjugation". Radiolabel balance data are given here to demonstrate use of the techniques and interpretation of the results.

The recovery of radiolabel from various substrates perfused into kidneys is shown in Table 1. Propachlor (2-chloro-$\underline{N}$-isopropylacetanilide) or its glutathione conjugate (Propachlor-GSH) was used as a reference compound because its metabolic pathway is known and its metabolites can be synthesized. With perfusion directed to only one kidney, and the urine collected separately from both kidneys, one can determine the contribution of the perfused kidney to first pass metabolism and clearance of the xenobiotic. Infused material that does not clear the perfused kidney on the first pass is transported in the blood and then has equal opportunity for excretion by either kidney. The first pass metabolism and clearance by a kidney is estimated by the difference between the amount of radiolabel eliminated in urine by the perfused and nonperfused kidneys.

Rats given ^{14}C-propachlor-GSH eliminated 54% of the ^{14}C in urine from the perfused kidney (Table 1) and 13% in urine from the nonperfused kidney. Only small amounts of ^{14}C remained in the tissue of either kidney. These results are interpreted to mean that 41% of the perfused dose was eliminated in urine on the first pass through the kidney, i.e. 54 - 13 = 41. Infused material that did not clear on the first pass either remained in the tissues, was eliminated in bile, or was eliminated by both kidneys. In this case, probably 26% of the ^{14}C was returned in the blood to the kidneys and eliminated in the urine (i.e. 2 times 13%), and 4% of the ^{14}C remained in the liver, 23% was eliminated in bile and 6% remained in the rest of the rat carcass.

TABLE 1 - Recovery of radiolabel from substrates perfused directly into kidneys in situ (2-3 hr recovery, % of dose)

Substrate	Dose (μmol)	Species	Kidney output (urine)		Kidney tissue		Liver	Bile	Carcass	Total
			Perfused	Nonperfused	Perfused	Nonperfused				
14C-Propachlor	1.02	Rat(3)[a]	24	19	4	1.1	7	27	10	94
14C-Propachlor-GSH	1.02	Rat(4)	54	13	0.3	0.1	4	23	6	100
14C-Propachlor-S-homocysteine	1.52	Rat(3)	49	13	1.4	0.7	9	17	9	101
14C-Hydroxydichlobenil-GSH	1.97	Rat(2)	32	38	6	0.8	1.7	12	11	103
14C-Dichlorovinylcysteine	0.91	Rat(3)	41	12	6	2	6	14	16	98
14C-Dichlorovinylcysteine	232.0	Calf(1)	16	4	27	3	15	2	27	102
14C-Propachlor-GSH	12.3	Chicken(3)	72	28				nil		100
14C-Hydroxydichlobenil-GSH	3.6	Chicken(2)	60	30						90
35S-Benzothiazole-GSH	0.29	Chicken(3)	66	38	~1	~1				104
14C-Tetrachloromethyl-sulphinylphenyl-GSH	0.85	Chicken(3)	59	36	0.3	0.3	1.9	7		104
14C-Tetrahydrotrihydroxy-naphthalene-S-cysteine	4.87	Chicken(1)	43	36	0.5	0.5	1.4	4	19	107
14C-Hydroxydichlobenil-GSH	3.5	Pig(2)	67	25	0.6	0.3				93
14C-Propachlor-GSH	17.1	Pig(2)	90	10						100
35S-Benzothiazole-GSH	0.34	Pig(1)	84	17				nil		101
14C-Tetrachloromethyl-sulphinylphenyl-GSH	0.71	Pig(3)	83	9	1	0.2		0.2		94

The substrates were: 2-chloro-N-isopropylacetanilide (propachlor); 2-S(glutathionyl)-N-isopropylacetanilide (propachlor-GSH); 2-S(homocysteinyl)-N-isopropylacetanilide (propachlor-S-homocysteine); 2-S(glutathionyl)-3-hydroxy-6-chlorobenzonitrile (hydroxydichlobenil-GSH); 2-S(cysteinyl)-1,2-dichloroethylene (dichlorovinylcysteine); 2-S(glutathionyl)benzothiazole (benzothiazole-GSH); S(glutathionyl)tetrachloromethylsulphinylbenzene (tetrachloromethylsulphinylphenyl-GSH); and S(cysteinyl)tetrahydrotrihydroxynaphthalene (tetrahydrotrihydroxynaphthalene-S-cysteine).

[a]Number of animals

Radiolabel elimination and retention patterns of rats given the homocysteine conjugate of ^{14}C-propachlor were similar to those for rats given ^{14}C-propachlor-GSH; but radiolabel elimination and retention patterns of rats given ^{14}C-hydroxydichlobenil-GSH differed markedly from those for rats given ^{14}C-propachlor-GSH. Elimination of ^{14}C in urine from the nonperfused kidney of rats given hydroxydichlobenil was slightly greater than that from the perfused kidney, suggesting that first pass elimination of hydroxydichlobenil or its metabolites was essentially nil. Retention of ^{14}C in the perfused kidney and carcass of rats perfused with hydroxydichlobenil-GSH was greater than for rats perfused with propachlor-GSH, and elimination of ^{14}C in bile was less. The identity of the metabolites in the urine was not determined for these hydroxydichlobenil-dosed rats, but the data suggest that some of the metabolism occurred in tissues other than the kidney before the metabolites are eliminated in the urine.

Dichlorovinylcysteine (DCVC) is nephrotoxic in the calf, more so than in the rat. Based on percentage of the dose, the perfused calf kidney retained 4 times more ^{14}C from DCVC than the rat kidneys (Table 1) and eliminated 60% less ^{14}C in the urine. The calf also retained more ^{14}C from DCVC in the liver and carcass than the rat and eliminated less in the bile. The data suggest that ^{14}C from DCVC is bound in the calf kidney, but more radiolabelled DCVC is needed before we can proceed further with research on this compound.

As one scans the remainder of Table 1, other differences among compounds and species become apparent. From the data collected, it appears that the kidneys of pigs are more efficient in metabolizing or clearing xenobiotics than kidneys of chickens and rats, and bile appears to be a more important route for eliminating xenobiotics, or their metabolites, in rats than in chickens and pigs.

Recovery of radiolabel from some substrates perfused into the GI tract or into the hepatic portal system is shown in Table 2. The tissue distribution and routes of elimination of ^{14}C for rats given ^{14}C-propachlor via the stomach or duodenum were essentially identical, but different from those for rats dosed via the hepatic portal system. Rats dosed via the GI tract eliminated less ^{14}C in urine than rats dosed via the portal system and more in bile, and had less ^{14}C in their kidneys, liver and carcass. Other comparisons which may be of interest to the reader are available in the table.

TABLE 2 - Recovery of radiolabel from substrates perfused into the hepatic portal vein or the gastrointestinal tract (3 hr recovery, % of dose)

Substrate/Route Perfused	Dose (μmol)	Species	Urine	Faeces	Bile	Liver	Kidneys	GI Tract	Carcass	Total
14C-Propachlor/Stomach	0.15	Rat(4)[a]	22	trace	26	9	0.5	29	4	91
/Duodenum	0.15	Rat(5)	22	trace	27	8	0.7	31	5	94
/Portal	0.15	Rat(3)	55	0	13	11	2	3	9	93
14C-Dichlobenil/Oral[b]	8.48	Rat(5)	43	0.2	44	0.7	0.1	2	2	92
14C-HO-DCBN-GSH/Portal	4.43	Rat(2)	62	2.3	14	0.8	0.5		2	80
	4.92	Chicken(3)	78	0	15	1.2	0.2			94
	4.92	Pig(2)	82	0	5	1.1	1.1			89
14C-HO-DCBN-Cysgly/Portal	0.72	Rat(2)	62	0.1	13	4.1	0.2	2	8	90

The substrates were: 2-chloro-N-isopropylacetanilide (propachlor); 2,6-dichlorobenzonitrile (dichlobenil); 2-S(glutathionyl)-3-hydroxy-6-chlorobenzonitrile (HO-DCBN-GSH); and 2-S(cysteinylglycyl)-3-hydroxy-6-chlorobenzonitrile (HO-DCBN-Cysgly).

[a]Number of animals.

[b]A bolus dose dissolved in ethanol was given orally and excreta were collected for 24 hr.

GENERAL COMMENTS ON USE OF SURGICAL PREPARATIONS

Phenol red (phenosulphonphthalein, sodium salt) is quite useful as a quick visual test for kidney function. With chickens, it is essential to inject a small quantity of phenol red dissolved in water or saline to establish positively that the venous system of the kidney is properly occluded for the forced perfusion. The appearance of phenol red in urine from the perfused kidney in about 90 seconds and in urine from the nonperfused kidney in about 3 minutes indicates a successful preparation. Phenol red should also appear in greater concentration in urine from the perfused kidney in a successful preparation. Urinary output (volume) obviously affects the timing. Phenol red injected at the end of the perfusions in any of the species mentioned verifies that the preparation remained patent and that kidney function remained.

Latex or silicone cannulae were unsatisfactory for this research because they retained (adsorbed) some of the radiolabelled substrates. Polyethylene cannulae were satisfactory. Perfusion rates greater than 0.1 ml/min caused excessive hydration of the rat. Excessive hydration did not occur at perfusion rates used in the other species.

Saline was a suitable carrier (solvent) for all metabolites of xenobiotics tested. However, it was not a suitable solvent for the parent xenobiotics. Saline containing 10% ethanol would dissolve propachlor in the quantities needed for perfusion (20 to 40 ml of perfusate), but this amount of ethanol inhibited urine production. An aqueous solution containing 4% dextran (~ 70,000 daltons) proved to be a suitable carrier for propachlor. A suitable aqueous solution for perfusing dichlobenil was not found. Small volumes (~ 0.5 ml) of ethanol and ethyl acetate have routinely been used as solvents when dosing was by oral or gastrointestinal routes in rats, but these solvents are not suitable for direct perfusion into kidneys or the hepatic portal system.

In conclusion, techniques for _in situ_ perfusion of various substrates into intestines, kidneys and livers of calves, chickens, pigs and rats have been presented. Metabolites produced during the first pass of the substrates through the tissues can be collected in blood, bile and urine for isolation and identification. These techniques are suitable for studies of the intermediary metabolism of any compound, natural or synthetic, provided that identifying characteristics are present on the resultant metabolites which link them to their parent compound.

REFERENCES

Akester, A.R., 1967, Renal portal shunts in the kidney of the domestic fowl. Journal of Anatomy, 101, 569-594.

Beuzeville, C., 1968, Catheterization of renal artery in rats. Proceedings of the Society for Experimental Biology and Medicine, 129, 932-936.

Chiasson, R.B., 1980, Laboratory Anatomy of the White Rat, 4th ed. (Dubuque, IA: Wm. C. Brown Co.).

Doluisio, J.T., Billups, N.F., Dittert, L.W., Sugita, E.T. and Swintosky, J.V., 1969, Drug absorption I: An in situ rat gut technique yielding realistic absorption rates. Journal of Pharmaceutical Sciences, 58, 1196-1200.

Fine, L.G., Lee, H., Goldsmith, D., Weber, H. and Blaufox, M.D., 1974, Effects of catheterization of renal artery on renal function in the rat. Journal of Applied Physiology, 37, 930-933.

Gallo-Torres, H.E. and Ludorf, J., 1974, Techniques for the in vivo catheterization of the portal vein in the rat. Proceedings of the Society for Experimental Biology and Medicine, 145, 249-254.

Herd, J.A. and Barger, A.C., 1964, A simplified technique for chronic catheterization of blood vessels. Journal of Applied Physiology, 19, 791-792.

Lambert, R., 1965, Surgery of the Digestive System in the Rat (Springfield, IL: Charles C. Thomas).

Murray, R.A., Nocek, J.E., Schwab, C.G., Hylton, W.E. and Bozak, C.K. 1987, Description and validation of an in situ autoperfusion method to determine nutrient absorption and metabolism in bovine small intestine. Journal of Animal Science, 65, 841-860.

Odlind, B., 1978, Blood flow in the renal portal system of the intact hen. A study of the venous system using microspheres. Acta Physiologica Scandinavia, 102, 342-356.

Petty, C., 1982, Research Techniques in the Rat (Springfield, IL: Charles C. Thomas).

Ross, B.D., 1972, Perfusion Techniques in Biochemistry (Oxford: Clarendon Press).

Smits, J.F.M., Kasbergen, C.M., van Essen, H., Kleinjans, J.C. and Struyker-Bouldier, A.J., 1983, Chronic local infusion into the renal artery of unrestrained rats. American Journal of Physiology, 244, H304-H307.

Waynforth, H.B., 1980, Experimental and Surgical Technique in the Rat (London: Academic Press Inc., Ltd.).

Wideman, R.F. Jr. and Braun, E.J., 1982, Ureteral urine collection from anesthetized domestic fowl. Laboratory Animal Science 32, 298-301.

Windmueller, H.G. and Spaeth, A.E., 1981, Vascular autoperfusion of rat small intestine *in situ*. Methods in Enzymology, 77, 120-129.

Windmueller, H.G. and Spaeth, A.E., 1984, Vascular perfusion of rat small intestine for permeation and metabolism studies. In *Handbook of Experimental Pharmacology*, Vol. 70/I, edited by T.Z. Csaky (Berlin: Springer-Verlag), pp. 113-156.

Yorgey, K.A., Pritchard, J.F., Renzi, N.L. and Dvorchik, B.H., 1986, Evaluation of drug absorption and presystemic metabolism using an in situ intestinal preparation. Journal of Pharmaceutical Sciences, 75, 869-872.

FOOTNOTE: No warranties are herein implied by the U.S. Department of Agriculture.

SUBSTRATE DELIVERY AS A CRITICAL ELEMENT IN THE
STUDY OF INTERMEDIARY METABOLITES OF LIPOPHILIC
XENOBIOTICS IN VITRO

Tsutomu Nakatsugawa, Julita Timoszyk, and John M. Becker

State University of New York,
College of Environmental Science and Forestry,
Syracuse, NY 13210, USA

INTRODUCTION

A major purpose of using in vitro systems in biotransformation studies is to understand toxicological events occurring in vivo. Much of our understanding of xenobiotic metabolism in vivo, especially of intermediates, has been gleaned from in vitro studies. As previously reviewed (Menzer, 1979; Nakatsugawa and Tsuda, 1983; Nakatsugawa and Becker, 1987), cell-free systems such as microsomes and homogenates provide building blocks of information for the understanding of biotransformation: e.g. optimum reaction conditions, cofactor requirements, characteristics of individual enzymes, parallel and serial pathways, generation and fate of reactive metabolites and relative enzyme titers among various tissues. Current trends toward the use of isolated cells and organ perfusion systems in the study of biotransformation are driven by the anticipation that a more integral system would aid in the synthesis of information gained through cell-free systems and ultimately in the understanding of toxic expression at the cellular, organ and whole organism levels. It is easy to conclude that these artificial systems reproduce events occurring in the intact animal when one finds a common metabolic pattern between the in vitro and in vivo studies. Such results, however, could be produced coincidentally under quite different enzymological circumstances. While certain intermediary metabolites can only be detected in vitro, the occurrence of metabolites which are not found in vivo could also reflect the unnatural conditions under which the in vitro assays were performed. The level of transient metabolites, especially genotoxic metabolites, in a cell in vivo is of critical concern (Oesch, 1987), but can be erroneously estimated by conventional in vitro assay methods.

We must be constantly aware of limitations in extrapolating in vitro results to in vivo situations. As was already pointed out (Nakatsugawa and Becker, 1987), the burden of proof that an enzyme reaction does indeed proceed in vivo as observed in vitro rests with the investigator who designs the experiment. Considerations which are important in understanding the significance of reactive intermediary metabolites observed in vitro include the reaction rate of the enzyme producing the metabolite, the location of the cell in which the metabolite is generated relative to the site of toxic action, the lability of the metabolite within the cell and in the blood, and the residence time of the reactive species in the producing and target cells. Levels of important cofactors such as Ca^{2+}, ATP, NADPH and GSH, the rate of diffusion of the xenobiotic substrate and metabolites, and concentration of the substrate actually available to the enzyme in vivo may greatly influence the enzyme rate and need to be considered in designing in vitro experiments. Researchers often alter these conditions when conducting assays in vitro in order to "optimize" the system. Optimum conditions for the highest reaction rate, however, may not reflect situations in vivo. For example, saturation of enzyme systems (tissue homogenates or microsomes) with substrate to achieve a maximal rate may cause otherwise inactive enzymes to become involved in biotransformation so that the "normal" metabolic pathways are altered.

Toxicological consequences of biotransformation in vivo depend upon both structural as well as chemical environments of the enzyme in question. By "going in vitro", we delete or modify a number of elements which are important in metabolism and toxicity in vivo, including vascular supplies as well as hormonal and neural control. To provide closer simulation, we may examine those elements and design our in vitro experiments to incorporate as many in vivo-like parameters as technically feasible. Elimination of vascular supplies may affect enzyme reaction rates because in cellular systems or tissue fragments such as isolated hepatocytes, kidney tubules and isolated intestinal sacs, suspension media may not exchange over the cell surface at the rate prevailing in vivo. This could alter residence time of both the substrate and metabolites in a cell. In addition, the loss of vasculature is usually accompanied by replacement of blood with artificial media both in organ perfusion and in cell suspensions. The composition of the medium is more often dictated by the maintenance of cell viability than by other considerations. Similarly in cell-free systems, the intracellular milieu is substituted by an artificial medium. All these factors add to the difficulty of _in_ _vitro_ simulation of the production, transport and interaction of a reactive intermediate in the animal _in_ _vivo_. For instance, the production of a neurotoxic intermediate of phosphorothioate insecticides by liver microsomes will only demonstrate that the enzymatic mechanism exists for this biotransformation. _In_ _vivo_ physiological factors may prevent the metabolite release or exposure to the target macromolecule. Parathion activation and paraoxon degradation in the liver have been compared by many investigators, but it is usually very difficult to translate enzyme rates measured _in_ _vitro_ to situations _in_ _vivo_. The actual level of paraoxon available in the hepatocytes as it is produced from parathion has never been determined. The activation of parathion to paraoxon within the liver _in_ _vivo_ does not mean that paraoxon will exist systemically and affect the nerves.

One of the most neglected basic elements that pervades the entire range of _in_ _vitro_ experimentation is how lipophilic xenobiotic substrates are introduced into the system. _In_ _vivo_, xenobiotics and drugs are gradually absorbed into the body and are "buffered" from the enzyme systems by macromolecular binding so that delivery is a moderated process. The extent of binding influences distribution, metabolism and toxicity (Barre et al., 1988). Our own studies have given us opportunities to realize some technical pitfalls of _in_ _vitro_ systems and make improvements, first with cell-free systems and then with rat hepatocyte suspensions and the perfused liver _in_ _situ_. We present our understanding of this problem with reference to hepatic metabolism in the hope that the lessons learned will also help in the use of other _in_ _vitro_ systems.

SOLUBILITY OF A LIPOPHILIC SUBSTRATE

Researchers have always struggled with the low water solubility of many xenobiotic substrates. For instance, in an early study of DDT dehydrochlorinase, Lipke and Kearns (1960) noted that the low water solubility of DDT severely limited the activity of this soluble enzyme. They solved this problem by providing a sufficient reserve of substrate by using egg yolk lipoprotein. The current practice of many investigators is to stir the lipophilic xenobiotic substrate into the medium in a small volume of water-miscible co-solvent such as ethanol, acetone, dimethylformamide, acetonitrile or dimethyl sulphoxide, sometimes with the help of an emulsifier such as Triton X-100. Typically, the lack of deleterious effects on the enzyme, cells or cell organelles under study is the major consideration in the selection of a solvent. A small amount of solvent, however, may increase the solubility of highly lipophilic xenobiotics in aqueous media and, therefore, affect their availability to enzymes. It should be noted also that in vivo levels of ethanol at 100 mg/dl and acetone above 2 mg/dl represent intoxicated and pathological situations, respectively (Tietz, 1976). Another common method of introducing xenobiotic substrates is to evaporate the solvent from a substrate stock solution and hope to dissolve the residue by vigorous mixing with the incubation medium. Both of these methods have limitations when dealing with lipophilic xenobiotics. It is noteworthy that an early report (Moss and Hathway, 1964) recognized the powerful solubizing effect of serum constituents on dieldrin and telodrin and their complex partitioning between blood cells and macromolecules and metabolic tissues. These interactions of lipophilic insecticides are critical in understanding metabolism in vivo based on in vitro data.

PARTITIONING OF A LIPOPHILIC SUBSTRATE

In addition to the problem of solubilization, lipophilic non-ionic xenobiotics present a special problem of partitioning in in vitro studies. While the fugacity of lipophilic chemicals to seek hydrophobic sites in aqueous systems may be obvious, most in vitro experiments are carried out with little regard to this feature. Problems stemming from this, however, were recognized early. Lewis et al. (1967) observed in their study of aldrin epoxidation by pig liver microsomes in vitro that the rate was little influenced by the overall concentration of the substrate, but was determined by the total amount of substrate added to a given amount of microsomes in an incubation mixture. The authors pointed out that the substrate concentration in or on microsomes affects the reaction rate, and noted that this consideration should apply to in vitro systems involving similarly lipophilic substrates. Surprisingly, under such conditions, a normal relationship fitting the Lineweaver-Burk equation was observed between the reaction rate and the concentration of aldrin. The reaction medium contained no protein except the enzyme source and aldrin was introduced in a small volume of ethanol (100 μl in 5ml). This study clearly indicated that essentially all substrate partitioned into the microsomes. A later study in our laboratory with rat liver microsomes (Stewart, 1979) confirmed this, and showed that with a given amount of the substrate, the reaction rate is not proportional to the enzyme concentration due to different available substrate concentrations. The latter fact was consistent with earlier observations by other investigators (Ghiasuddin and Menzer,

1976; Krieger and Wilkinson, 1969). A recent study (Rabovsky and Judy, 1987) described the non-linear relationship between liver and lung cytochrome P-450 activity and increasing microsomal concentration. These effects were alleviated by addition of heat-inactivated microsomes, phospholipid, or concentrating the assay preparation. All of these increase the lipophilicity of the reaction medium. Once cells are disrupted, the density of xenobiotic-binding components that normally surround biotransformation enzymes is drastically reduced since the reaction volume is far greater than the volume of the intact cells. Without some correction for this highly aqueous condition, enzymes can be exposed to a much higher concentration of xenobiotics than ever possible in vivo and give artifactual kinetic values.

Similar limitations of partitioning exist for cellular systems in vitro. Isolated hepatocytes and other cells in culture have been used increasingly in an effort to substitute these controlled systems for animals in vivo (Tyson, 1987). Protein-free salt medium is gradually being replaced by more in vivo-like preparations containing fetal bovine serum or BSA, cellular growth factors and carbogen (95:5% $O_2:CO_2$) gaseous exchange (Jauregui et al., 1986; Nakamura et al., 1986; Tyson, 1987). Our own studies have shown, however, that even with precautions regarding substrate buffering, hepatocyte suspensions do not yield in vivo-like metabolite profiles for parathion (Becker, J.M. and Nakatsugawa, T., poster 7D-11 presented at 6th IUPAC Congress of Pesticide Chemistry). A hepatocyte suspension in 3% BSA (1×10^6 cells/ml) was exposed to parathion at 1.7×10^{-7} M by continuous input and metabolites were identified. In contrast to the case in vivo, only two major metabolites were identified (paraoxon : diethylphosphorothioic acid = 3:1) and their proportions were unlike that in urine. It is apparent that much of the paraoxon produced partitioned out to the suspension medium and was not available for subsequent metabolism. In vivo, the volume of the hepatocytes is proportionally greater than that of the sinusoid which favours partitioning of intermediary metabolites within the cell and hence additional metabolism. In fact, at low doses in vivo neither parathion nor paraoxon emerges from the liver due to their total degradation (Tsuda et al., 1987). In order to simulate the in vivo biotransformation and disposition of parathion, one must be able to reproduce in vitro the intracellular concentration of both the parent chemical and metabolites attained in vivo, thereby allowing hepatocytes to function in a normal manner. With a freely diffusing intermediary metabolite like paraoxon, the large extracellular volume in vitro makes this difficult to accomplish.

SUBSTRATE DELIVERY IN VIVO

Discrepancies between typical in vitro systems and conditions in vivo become apparent if one traces the movement of an oral dose of a typical lipophilic xenobiotic from the gut lumen to the liver. Both solubility and partitioning are important factors to consider. Initially, the oral dose may, for example, be in food, oil or saline, given to a fed or fasted animal and administered in a bolus dose or over an extended time. Uptake from the gut can be variable but is usually moderated over some time. Several membrane barriers are involved in absorption, and the xenobiotic molecules typically cross them by passive diffusion of the free unbound form. The process of diffusion into a cell may be considered to consist of two partitioning steps, one between the extracellular medium and the membrane itself and the second between the membrane and intracellular phase. The chemical must then traverse the cell to the distal membrane which is in dynamic equilibrium with the vascular flow. Again diffusion occurs. Partitioning between the inner cell membrane and the blood is probably controlled by the degree of binding of the lipophilic chemical with the membrane and with the blood components.

Typical lipophilic components of rodent blood include albumins (3.4 - 4.3 g/dl), globulins (0.6 and 2.0 g/dl for mice and rats, respectively), lipoproteins and the cell membranes and contents of blood cells. Intestinal absorption of highly lipophilic xenobiotics may be limited by their solubility in the blood, which has a limited volume and carrying capacity. Kinetic studies of human serum albumin (Maliwal and Guthrie, 1981) indicated the presence of a few binding sites with association constants in the 10^4 to 10^6 M^{-1} range for several organic chemicals and insecticides, while more low affinity sites exist. The high affinity binding is primarily hydrophobic, since it is affected little by temperature but is inversely dependent on the lipophilicity of the chemical. The highly lipophilic insecticides, DDT and dieldrin, appeared to have 5 high affinity sites. Sultatos and coworkers (Sultatos et al., 1984) reported that albumin has only one binding site with a dissociation constant of about 3 and 11 $\times 10^{-6}$ M for chlorpyrifos and parathion, respectively. At high concentrations of parathion (10^{-5} to 10^{-4}M) the bound portion was about 95, 96, 94, 88, and 84% for blood and 4, 2, 1, and 0.5% solutions of BSA in buffer, respectively

(Sultatos and Minor, 1986). The albumin did have some catalytic activity for both insecticides, which is relevant to *in vitro* metabolism studies. The data which show differences in biotransformation and disposition between normal and analbuminaemic rats (Hirate et al., 1984; Inoue et al., 1985) emphasizes the importance of xenobiotic delivery in metabolism and toxicity studies.

Xenobiotics are carried by the blood from the intestine to the liver lobule within seconds. Lipophilic xenobiotics bound to blood macromolecules must dissociate into the free form for uptake into hepatocytes. Since the hepatic transit time for blood is only a few seconds, uptake by hepatocytes along the sinusoid is affected by how rapidly the reversal of xenobiotic-macromolecular binding occurs within this time frame. Free unbound molecules diffuse extremely rapidly into the lipophilic hepatocytes and reach biotransformation enzymes. However, the actual free concentration of chemical available to the hepatocyte at a given time may be but a small fraction of the total blood concentration. Within the hepatocyte, partitioning must involve various membranous organelles, such as endoplasmic reticulum, Golgi apparatus, and mitochondria, as well as cytosolic proteins, which may or may not metabolize the xenobiotic. Throughout these steps, the processes of diffusion and eventual biotransformation are buffered by a complex array of macromolecular binding of the lipophilic xenobiotic. Previously, little attention has been paid to the consequences of such buffering action on the concentration and gradual delivery of xenobiotics to biotransformation enzymes *in vitro*.

IN VITRO SUBSTRATE BUFFER

The phenomenon of "substrate buffering", which occurs *in vivo*, is thus critical in modeling *in vivo* events by *in vitro* assays. The use of such binding components has the added advantage of solubilizing a highly lipophilic substrate without an organic solvent or a detergent. The classical method of designing an enzyme assay is to find optimum conditions to maximize enzyme activity. Through the study of parathion oxidation, we have come to realize that the highest activity has little to do with activities *in vivo* and the optimum conditions may not resemble conditions prevailing *in vivo*. Few studies of xenobiotic biotransformation estimate actual *in vivo* reaction rates. We do not typically know the substrate concentration available to the enzyme *in vivo*. This is not to say that the classical optimization of the assay is not to be used, but that results from such experiments should be evaluated with these limitations in mind.

A good illustration of the importance of the substrate buffer principle for cell and cell-free systems is provided by our recent study of the classical microsomal suicidal reaction, parathion desulphuration (Nakatsugawa and Timoszyk, 1988a). Typically, half the enzyme activity in a microsomal system may be lost in 3 min (Morelli and Nakatsugawa, 1978) and a reconstituted system may lose its total activity within 5 min (Kamataki et al., 1976). The inactivation is not caused by depletion of cofactors or substrate, nor by the accumulation of metabolites. The oxidative reaction involving cytochrome P-450 yields a reactive sulphur metabolite which appears to inactivate this enzyme via covalent binding; however, not all of the macromolecular sulphur binding causes inactivation (Morelli and Nakatsugawa, 1978). The inactivation can be slowed down, though not abolished, by the addition of a thiol in vitro, suggesting an important protective role for glutathione. In vivo, little enzyme inactivation or sulphur binding occurs in the rat liver even at low toxic doses (parathion at 3 mg/kg, i.p.) (Morelli and Nakatsugawa, 1979). Obviously, therefore, these typical in vitro systems do not reflect events in vivo.

The expectation that a more integral isolated hepatocyte system would circumvent this problem was not realized since the reaction in the cell also slowed down quickly (Nakatsugawa et al., 1980). An important difference in the unsuccessful hepatocyte experiment as compared with metabolism in vivo is the absence of proteins that bind parathion outside hepatocytes. Under typical in vitro conditions, such as 0.5 million cells/ml and 1×10^{-5} M parathion in Waymouth's medium, the average intracellular parathion level reaches nearly 2×10^{-3} M (Nakatsugawa et al., 1980). These observations indicate that the inactivation was a result of a high parathion concentration which drove parathion oxidase to generate detrimental reactive intermediate species at a rate beyond the cell's defence capacity. Similar exaggerated partitioning from the aqueous to the lipoidal phase is likely to occur in most classical in vitro systems of lipophilic substrates (Lewis et al., 1967; Stewart, 1979) so the following concept should have broad implications. The suicidal reaction has merely provided a detectable means of improving the inadequacies of the in vitro methods.

We reasoned that the use of appropriate proteins to buffer the substrate would be a practical method of providing a sufficient reservoir of a low level substrate. When BSA was added to the medium to reduce intracellular parathion concentrations, the reaction rate was greatly reduced but the reaction proceeded much more linearly over 30 min when the substrate concentration was less than 1×10^{-6} M While 7% BSA was acceptable as the reaction medium, undiluted rat blood was even better with no reduction in the "parathion oxidase" rate over 30 min. With cell-free homogenates, 5% BSA was effective at parathion concentrations no greater than 10^{-6}M (Figure 1). In the presence of 5% BSA as the substrate buffer, the K_m values for parathion oxidase based on the rates over 20 min, with substrate concentrations of 2×10^{-6} M and less, were 2 to 4×10^{-6} M, an order of magnitude lower than those obtained with unbuffered systems. This may have significant implications because the difference between the low substrate levels used here and those in classical procedures is greater than apparent. In the presence of a substrate buffer, the actual concentration of free substrate is even less than its nominal average concentration in the incubation mixture. Therefore, when a lipophilic substrate is dissolved in a substrate-buffered system, the terms such as "apparent substrate concentration" and "apparent K_m" will always have to be used unless the actual level of unbound substrate available to the enzyme is defined. The substrate will still partition through hydrophobic binding between the ambient medium and the enzyme protein or the membranes with which the enzyme is associated.

There is a possibility that without substrate buffering, one may be measuring different isozymes than those operative in vivo. Numerous studies (Guengerich, 1977; Guengerich, 1988; Imaoka et al., 1987a; Daujat et al., 1987; Funae and Imaoka, 1987) have reported the existence of many isozymes of cytochromes P-450, having different kinetic characters and inducibilities, from liver and lung tissue of several species. Sex differences have been shown and may be related to toxicity differences between the sexes (Schenkman et al., 1987; Imaoka et al., 1987a, 1987b). Several P-450 systems (selectively induced for specific isozymes) with different turnover numbers and metabolite profiles have been identified in the rat liver for parathion oxidation (Guengerich, 1977). It is conceivable that we have, among the isozymes, one with low V_{max} and K_m and another with high kinetic constants. Then, at a very low substrate level, the former enzyme could have, say, 100x the activity of the latter whereas at a much higher substrate level, the opposite would be the case. The different K_m values between protein-free and substrate buffered methods, then, could represent two separate enzymes. The artifactually high concentration of a xenobiotic substrate could be almost

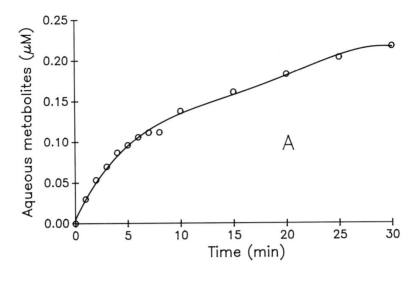
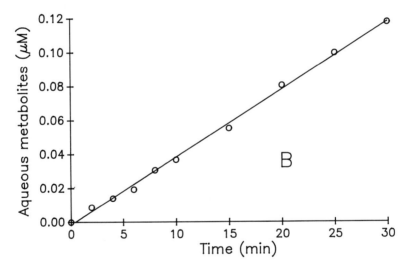

Figure 1. Time course of parathion oxidation by liver homogenate from male Sprague-Dawley rat (300 g). Reaction mixture contained 0.15 M KCl, 0.01 M phosphate buffer, pH 7.4, 0.25% w/v liver homogenate and A: 0.89×10^{-6} M [ethyl-^{3}H]parathion and B: 1.01×10^{-6} M parathion plus 5% bovine serum albumin. The reaction mixture was shaken with 4x volume of hexane:ether (1:1) for 20 min, the aqueous phase was washed with chloroform:ether (9:1) and 0.2 ml was sampled for scintillation counting.

completely metabolized by an isozyme which would have
negligible activity at substrate concentrations which could
exist in vivo. The metabolite profile determined in vitro
would thus be quite different from that in vivo. Of course,
actual cases can be even more complex with several isozymes
involved.

SUBSTRATE BUFFER IN THE PERFUSED LIVER
 The principle of "substrate buffering" is also significant
in the delivery of lipophilic xenobiotics to isolated whole
organs. Once an organ is isolated from the remainder of
the body as is the case in either in situ or isolated liver
perfusion, the choice of perfusate dictates what fraction of
the xenobiotic is readily diffusible into the cell. In the
case of liver, which consists of numerous lobules (or acini)
with hepatocytes being exposed to the influent fluid in a
serial manner (Weisiger et al., 1986), the degree of
xenobiotic binding can directly affect both the intralobular
region of initial uptake and the pattern of the xenobiotic
uptake along the sinusoid. As we have described
(Nakatsugawa et al., 1980; Tsuda et al., 1987; Tsuda et
al., 1988), the hepatic lobule acts almost as though it is a
reversed phase chromatographic column for those xenobiotics
that diffuse into and out of hepatocytes with extreme
rapidity. Under these circumstances, perfusates with
different xenobiotic-binding characteristics can influence
the region of hepatocytes within the lobule which are exposed
to the xenobiotic and actively involved in biotransformation.
This is particularly true for those chemicals which undergo
rapid metabolic degradation at the dose level in question,
which tends to create a very sharp concentration gradient
along the sinusoid. As an example, we have compared the
hepatic behavior of parathion, following pulse infusion into
the liver in situ, using protein-free Waymouth's medium and
several concentrations of autologous blood. A perfusate
with little lipophilic-binding capacity causes initial
absorption of the total dose within 1 or 2 rows of
periportal hepatocytes. The subsequent "chromatographic"
migration is delayed by the low solubility of parathion in
the perfusate. The hepatic residence time for a pulse dose
with protein-free perfusate is about 40 minutes whereas with
whole blood a hepatic residence time of about 4 minutes is
observed. Dilutions of blood as the perfusate give
intermediate results. Sultatos (1987) has also reported an
effect of BSA in the hepatic disposition of methyl parathion
in the mouse liver in situ. Perfusate supplemented with
increasing concentrations of BSA (0 to 4%) bound more of the
substrate and lengthened the halftime to reach steady state.

This demonstrates that equilibration times, hepatic extraction ratios (Sultatos and Minor, 1985) and elution of lipophilic metabolites (Sultatos et al., 1985) could become artifactual depending upon the choice of perfusate.

Reproducing the actual rate of biotransformation in the perfused liver requires consideration of an intralobular exposure gradient as a function of dose, residence time of the chemical and its metabolites and the activities of a multiplicity of enzymes. The effect of the perfusate on the extent of intralobular exposure to the chemical and the rate of migration through the lobule (for those undergoing chromatographic translobular migration) brings into play the implications of hepatocellular heterogeneity. If all the hepatocytes within the lobule were identical, the lack of substrate buffering would only cause a higher concentration of xenobiotic in the periportal hepatocyte. Hepatocellular heterogeneity, however, is more a rule than an exception for many biotransformation systems (Jungermann and Katz, 1982; James et al., 1981; Baron et al., 1978; Pang and Terrell, 1981; Smith et al., 1979), often with differential distribution of 2-3 fold (Bengtsson et al., 1987). Our own data obtained with hepatocytes sorted in a fluorescence-activated cell sorter from specific intralobular regions show that centrilobular hepatocytes are approximately twice as active as periportal cells in both paraoxon hydrolysis and paraoxon deethylation (Nakatsugawa and Timoszyk, 1988b). Microsomal glutathione S-transferase and glutathione reductase are evenly distributed intralobularly but cytoplasmic glutathione S-transferase activity is 1.6 times higher in the centrilobular hepatocytes (Kera et al., 1987). Glutathione peroxidase activity is about 2-fold higher in the periportal hepatocytes. The enzymes for glucuronidation and sulphation have higher activities in periportal hepatocytes (3 to 8-fold) (Araya et al., 1986). Due to the intralobular distribution of metabolic enzymes, the use of an inadequate perfusion medium can produce a metabolite profile that is quite different from that in vivo. Simply, the xenobiotic would be delivered to a different population of hepatocytes in situ than may be exposed in vivo. Hepatocellular heterogeneity of the intact liver is important especially when assessing the metabolism of chemicals which behave "chromatographically" within the liver, as we have previously described (Nakatsugawa et al., 1980; Nakatsugawa et al., 1983; Tsuda et al., 1987). Continuing research suggests that many pesticides and industrial chemicals having a log[octanol/water partition coefficient] (log P) in the range of 2-4 may be almost completely absorbed by periportal hepatocytes initially and show subsequent chromatographic translobular migration (Murakami et al., 1987; Tsuda et al., 1988). Regional hepatotoxicity, such as that of allyl alcohol (Belinsky et al., 1986; Badr et al., 1986), may be due to this behaviour. Typical isolated hepatocyte systems involving heterogeneous hepatocytes would only give weighted averages of enzyme activities (Quistorff et al., 1986; Quistorff, 1987). Broad implications of heterogeneity are thus apparent.

FUTURE RESEARCH NEEDS

In the absence of substrate-binding components such as albumin, Lewis et al. (1967) still obtained a "normal" relationship between the substrate concentration and the reaction rate. This demonstrates that apparent normality does not guarantee that the enzyme reaction is occurring under in vivo-like conditions. Obviously substrate concentrations in such reactions were several orders of magnitude higher than those in vivo (presumably attained within the endoplasmic reticulum or its surface). Since saturation was not reached even under conditions where essentially all the substrate molecules partition into microsomes, aldrin epoxidase seems to have an enormous K_m value. If so, under the in vivo conditions, would the rate be infinitesimally low? As was seen with parathion oxidation, multiplicity of enzymes may be involved. Re-examination of the biotransformation of other lipophilic xenobiotics in the light of the substrate buffer concept may be revealing. This is a worthwhile area for study in the future.

What is the right substrate buffer for isolated organs, cell systems, cell-free preparations and purified enzymes? We have shown that whole blood is the ideal substrate buffer for hepatocytes, but it cannot be used in many cases due to limited availability and its own enzymatic activity. Overall partitioning characteristics of blood can be approximated by comparing various perfusates for the lobular transit time of a pulse dose of parathion-like ("chromatographically" migrating) chemicals. This was done for parathion which showed a 4 min. transit time with 7% BSA, identical to that with autologous blood. Since many binding components in the blood are involved in vivo, it may be desirable to supplement perfusate with globulins and lipoproteins to improve the lipophilic makeup. This needs to be defined for each xenobiotic separately.

Simulation of intracellular substrate buffering is more difficult. One current limitation of in vitro work aimed at modeling in vivo biotransformation is the lack of information on the biochemical conditions surrounding the enzymes in vivo. In addition to enzymatic hepatocellular heterogeneity, pH or the levels of NADPH, GSH, etc. may also vary within the intralobular regions. In this context, efforts by Thurman and coworkers to investigate metabolic conditions in situ through the use of fiber optics is significant (Ji et al., 1981; Belinsky et al., 1984; Conway et al., 1985; Harris and Thurman, 1986). Such investigations may reveal the intracellular balance of various reactants such as NADPH/NADP, GSH/GSSG, bound vs. free calcium, etc., information required for the design of truly representative cell-free systems. In vitro studies of biotransformation systems cannot stand alone without information of the interplay of enzymes and xenobiotics in vivo.

Acknowledgements. The research described in this paper was supported by grant ES01019 from National Institute of Environmental Health Sciences, DHS and a grant from Sankyo Co., Japan.

REFERENCES

Araya, H., Mizuma, T., Horie, T., Hayashi, M. and Awazu, S., 1986, Heterogeneous distribution of the conjugation activity of acetaminophen and p-nitrophenol in isolated rat liver cells. Journal of Pharmacobio-Dynamics, 9, 218-222.

Badr, M.Z., Belinsky, S.A., Kauffman, F.C. and Thurman, R.G., 1986, Mechanism of hepatotoxicity to periportal regions of the liver lobule due to allyl alcohol: Role of oxygen and lipid peroxidation. Journal of Pharmacology and Experimental Therapeutics, 238, 1138-1142.

Baron, J., Redick, J.A. and Guengerich, F.P., 1978, Immunohistochemical localizations of cytochromes P-450 in rat liver. Life Sciences, 23, 2627-2631.

Barre, J., Urien, S., Albengres, E. and Tillement, J.P., 1988, Plasma and tissue binding as determinants of drug body distribution. Possible applications to toxicological studies. Xenobiotica, 18, 15-20.

Belinsky, S.A., Badr, M.Z., Kauffman, F.C. and Thurman, R.G., 1986, Mechanism of hepatotoxicity in periportal regions of the liver lobule due to allyl alcohol: Studies on thiols and energy status. Journal of Pharmacology and Experimental Therapeutics, 238, 1132-1137.

Belinsky, S.A., Matsumura, T., Kauffman, F.C. and Thurman, R.G., 1984, Rates of allyl alcohol metabolism in periportal and pericentral regions of the liver lobule. Molecular Pharmacology, 25, 158-164.

Bengtsson, G., Julkunen, A., Penttila, K.E. and Lindros, K.O., 1987, Effect of phenobarbital on the distribution of drug metabolizing enzymes between periportal and perivenous rat hepatocytes prepared by digitonin-collagenase liver perfusion. Journal of Pharmacology and Experimental Therapeutics, 240, 663-667.

Conway, J.G., Popp, J.A. and Thurman, R.G., 1985, Microcirculation in periportal and pericentral regions of lobule in perfused rat liver. American Journal of Physiology, 249, G449-G456.

Daujat, M., Pichard, L., Dalet, C., Larroque, C., Bonfils, C., Pompon, D., Li, D., Guzelian, P.S. and Maurelm, P., 1987, Expression of five forms of microsomal cytochrome P-450 in primary cultures of rabbit hepatocytes treated with various classes of inducers. Biochemical Pharmacology, 36, 3597-3606.

Funae, Y. and Imaoka, S., 1987, Purification and characterization of liver microsomal cytochrome P-450 from untreated rats. Biochimica Biophysica Acta, 926, 349-358.

Ghiasuddin, S.M. and Menzer, R.E., 1976, Microsomal epoxidation of aldrin to dieldrin in rats. Bulletin of Environmental Contamination and Toxicology, 15, 324-329.

Guengerich, F.P., 1977, Separation and purification of multiple forms of microsomal cytochrome P-450. Activities of different forms of cytochrome P-450 towards several compounds of environmental interest. Journal of Biological Chemistry, 252, 3970-3979.

Guengerich, F.P., 1988, Cytochrome P-450. Comparative Biochemical Physiology, 89C, 1-4.

Harris, C. and Thurman, R.G., 1986, A new method to study glutathione adduct formation in periportal and pericentral regions of the liver lobule by microreflectance spectrophotometry. Molecular Pharmacology, 29, 88-96.

Hirate, J., Horikoshi, I., Watanabe, J., Ozeki, S. and Nagase, S., 1984, Disposition of salicylic acid in analbuminemic rats. Journal of Pharmaceutics and Dynamics, 7, 929-934.

Imaoka, S., Kamataki, T. and Funae, Y., 1987a, Purification and characterization of six cytochromes P-450 from hepatic microsomes of immature female rats. Journal of Biochemistry, 102, 843-851.

Imaoka, S., Terano, Y. and Funae, Y., 1987b, Purification and characterization of two constitutive cytochromes P-450 (F-1 and F-2) from adult female rats: identification of P-450F-1 as the phenobarbital-inducible cytochrome

P-450 in male rat liver. Biochimica Biophysica Acta, 916, 358-367.

Inoue, M., Morino, Y. and Nagase, S., 1985, Transhepatic transport of taurocholic acid in normal and mutant analbuminemic rats. Biochimica et Biophysica Acta, 833, 211-216.

James, R., Desmond, P., Küpfer, A., Schenker, S. and Branch, R.A., 1981, The differential localization of various drug metabolizing systems within the rat liver lobule as determined by the hepatotoxins allyl alcohol, carbon tetrachloride and bromobenzene. Journal of Pharmacology and Experimental Therapeutics, 217, 127-132.

Jauregui, H.O., McMillan, P.N., Driscoll, J. and Naik, S., 1986, Attachment and long term survival of adult rat hepatocytes in primary monolayer cultures: comparison of different substrata and tissue culture media formulations. In Vitro Cellular and Developmental Biology, 22, 13-22.

Ji, S., Lemasters, J.J. and Thurman, R.G., 1981, A fluorometric method to measure sublobular rates of mixed-function oxidation in the hemoglobin-free perfused rat liver. Molecular Pharmacology, 19, 513-516.

Jungermann, K. and Katz, N., 1982, Functional hepatocellular heterogeneity. Hepatology, 2, 385-395.

Kamataki, T., Lin, M.C.M.L., Belcher, D.H. and Neal, R.A., 1976, Studies of the metabolism of parathion with an apparently homogeneous preparation of rabbit liver cytochrome P-450. Drug Metabolism and Disposition, 4, 180-189.

Kera, Y., Sippel, H.W., Penttila, K.E. and Lindros, K.O., 1987, Acinar distribution of glutathione-dependent detoxifying enzymes: Low glutathione peroxidase activity in perivenous hepatocytes. Biochemical Pharmacology, 36, 2003-2006.

Krieger, R.I. and Wilkinson, C.F., 1969, Microsomal mixed-function oxidases in insects-I: Localization and properties of an enzyme system effecting aldrin epoxidation in larvae of the southern armyworm (Prodenia eridania). Biochemical Pharmacology, 18, 1403-1415.

Lewis, S.E., Wilkinson, C.F. and Ray, J.W., 1967, The relationship between microsomal epoxidation and lipid peroxidation in houseflies and pig liver and the inhibitory effect of derivatives of 1,3-benzodioxole (methylenedioxybenzene). Biochemical Pharmacology, 16, 1195-1210.

Lipke, H. and Kearns, C.W., 1960, DDT-dehydrochlorinase III. Solubilization of insecticides by lipoprotein. Journal of Economic Entomology, 53, 31-35.

Maliwal, B.P. and Guthrie, F.E., 1981, Interaction of insecticides with human serum albumin. Molecular Pharmacology, 20, 138-144.

Menzer, R.E., 1979, Techniques for studying the metabolism of xenobiotics by intact animal cells, tissues and organs in vitro. In Xenobiotic Metabolism: In Vitro Methods, ACS Symposium Series 97, edited by G. D. Paulson, D.S. Frear and E.P. Marks, pp. 131-148.

Morelli, M.A. and Nakatsugawa, T., 1978, Inactivation in vitro of microsomal oxidases during parathion metabolism. Biochemical Pharmacology, 27, 293-299.

Morelli, M.A. and Nakatsugawa, T., 1979, Sulfur oxyacid production as a consequence of parathion desulfuration. Pesticicide Biochemical and Physiology, 10, 243-250.

Moss, J.A. and Hathway, D.E., 1964, Transport of organic compounds in the mammal: Partition of dieldrin and telodrin between the cellular components and soluble proteins of blood. Biochemical Journal, 91, 384-393.

Murakami, N., Uchida, M., Sugimoto, T. and Nakatsugawa, T., 1987, Relationship between translobular migration in perfused rat liver and hydrophobicity of dialkyl dithiolanylidenemalonates and related compounds. Xenobiotica, 17, 241-249.

Nakamura, T., Teramoto, H. and Ichihara, A., 1986, Purification and characterization of a growth factor from rat platelets for mature parenchymal hepatocytes in promary cultures. Proceedings of the National Academy of Sciences, U.S.A., 83, 6489-6493.

Nakatsugawa, T. and Becker, J.M., 1987, In vitro systems in the study of mechanisms of pesticide metabolism in animals. In Pesticide Science and Biotechnology, edited by R. Greenhalgh and T. R. Roberts (Oxford: Blackwell Scientific Publications) pp. 523-526.

Nakatsugawa, T., Bradford, W.L. and Usui, K., 1980, Hepatic disposition of parathion: Uptake by isolated hepatocytes and chromatographic translobular migration. Pesticide Biochemistry and Physiology, 14, 13-25.

Nakatsugawa, T. and Timoszyk, J., 1988a, Non-suicidal oxidation of parathion by cell-free and isolated hepatocyte systems in the presence of a substrate buffer. The Toxicologist, 8, 204.

Nakatsugawa, T. and Timoszyk, J., 1988b, Fluorescence labeling and sorting of hepatocyte subpopulations to determine the intralobular heterogeneity of paraoxon-metabolizing enzymes in DDE-treated and control rats. Pesticide Biochemistry and Physiology, 30, 113-124.

Nakatsugawa, T. and Tsuda, S., 1983, Metabolism studies with liver homogenate, hepatocyte suspension and perfused liver. In *IUPAC Pesticide Chemistry, Human Welfare and the Environment*, edited by J. Miyamoto and P.C. Kearney, Vol. 3, (Oxford: Pergamon Press) pp. 395-400.

Nakatsugawa, T., Tsuda, S. and Sherman, W.K., 1983, Chromatographic translobular migration of xenobiotics. In *IUPAC Pesticide Chemistry, Human Welfare and the Environment*, edited by J. Miyamoto and P.C. Kearney, Vol. 3, (Oxford: Pergamon Press) pp. 469-474.

Oesch, F., 1987, Significance of various enzymes in the control of reactive metabolites. Archives of Toxicology, 60, 174-178.

Pang, K.S. and Terrell, J.A., 1981, Retrograde perfusion to probe the heterogeneous distribution of hepatic drug metabolizing enzymes in rats. Journal of Pharmacology and Experimental Therapeutics, 216, 339-346.

Quistorff, B., 1987, Digitonin perfusion in the study of metabolic zonation of the rat liver: Potassium as an intracellular concentration reference. Biochemical Society Transactions, 15, 361-363.

Quistorff, B., Dich, J. and Grunnet, N., 1986, Periportal and perivenous hepatocytes retain their zonal characteristics in primary culture. Biochemical and Biophysical Research Communications, 139, 1055-1061.

Rabovsky, J. and Judy, D.J., 1987, The non-linear dependence of cytochrome P450 activity at low microsome concentration. Biochemistry International, 15, 1033-1041.

Schenkman, J.B., Favreau, L.V., Mole, J., Kreutzer, D.L. and Jansson, I.,1987, Fingerprinting rat liver microsomal cytochromes P-450 as a means of delineating sexually distinctive forms. Archives of Toxicology, 60, 43-51.

Smith, M.T., Loveridge, N., Wills, E.D. and Chayen, J., 1979, The distribution of glutathione in the rat liver lobule. Biochemical Journal, 182, 103-108.

Stewart, R.R., 1979, The epoxidation of aldrin in the intestine of the American cockroach Periplaneta americana (L.), Orthoptera: Blattidae -- A biochemical and histological study. Ph.D. Thesis, SUNY College of Environmental Science and Forestry, Syracuse, NY.

Sultatos, L.G., 1987, The role of the liver in mediating the acute toxicity of the pesticide methyl parathion in the mouse. Drug Metabolism and Disposition, 15, 613-617.

Sultatos, L.G. and Minor, L.D., 1985, Biotransformation of paraoxon and p-nitrophenol by isolated perfused mouse livers. Toxicology, 36, 159-169.

Sultatos, L.G. and Minor, L.D., 1986, Factors affecting the biotransformation of the pesticide parathion by the isolated perfused mouse liver. Drug Metabolism and Disposition, 14, 214-220.

Sultatos, L.G., Basker, K.M., Shao, M. and Murphy, S.D., 1984, The interaction of the phosphorothioate insecticides chlorpyrifos and parathion and their oxygen analogues with bovine serum albumin. Molecular Pharmacology, 26, 99-104.

Sultatos, L.G., Minor, L.D. and Murphy, S.D., 1985, Metabolic activation of phosphorothioate pesticides: role of the liver. Journal of Pharmacology and Experimental Therapeutics, 232, 624-628.

Tietz, Norbert W., editor, 1976, Fundamentals of Clinical Chemistry, (Philadelphia: W.B. Saunders Co.).

Tsuda, S., Rosenberg, A. and Nakatsugawa, T., 1988, Translobular uptake patterns of environmental toxicants in the rat liver. Bulletin of Environmental Contamination and Toxicology, 40, 410-417.

Tsuda, S., Sherman, W., Rosenberg, A., Timoszyk, J., Becker, J.M., Keadtisuke, S. and Nakatsugawa, T., 1987, Rapid periportal uptake and translobular migration of parathion with concurrent metabolism in the rat liver *in vivo*. Pesticide Biochemistry and Physiology, 28, 201-215.

Tyson, C.A., 1987, Correspondence of results from hepatocyte studies with in vivo response. Toxicology and Industrial Health, 3, 459-478.

Weisiger, R.A., Mendel, C.M. and Cavalieri, R.R., 1986, The hepatic sinusoid is not well-stirred: estimation of the degree of axial mixing by analysis of lobular concentration gradients formed during uptake of thyroxine by the perfused rat liver. Journal of Pharmaceutical Sciences, 75, 233-237.

STABLE ISOTOPES IN METABOLISM STUDIES:
USE OF ISOTOPE EFFECTS

Norio Kurihara

Radioisotope Research Center,
Kyoto University, Kyoto, Japan

INTRODUCTION

A xenobiotic metabolism study usually has one or more aims other than simply identifying metabolites and metabolic pathways. One aim is to obtain an indication of the role of metabolism in the biological activity of a compound, either by discovering the presence of a particular bioactivation reaction or by identifying various detoxication reactions. Another is to elucidate precise reaction mechanisms and reaction kinetics. For this extra dimension to metabolism studies, the use of isotopes has become almost indispensable. Many of the applications of isotopes are in tracer studies using radioisotopes, but stable isotopes have been used with increasing frequency in recent years, partly due to the rapid development and sophistication of mass spectrometric techniques and nuclear magnetic resonance spectroscopy (NMR). Table 1 presents general comparative data on radioisotopes and stable isotopes showing the differences in their application to xenobiotic metabolism studies.

For detection and quantification, radioisotopes are more convenient than stable isotopes because quick and automatic detection and counting of radioactivity are possible with very high sensitivity using liquid scintillation counting, for example. Stable isotopes offer much better performance than radioisotopes in some aspects. The use of a substrate labelled with stable isotopes at specific positions can provide much information on chemical structure by examining mass fragmentation patterns in mass spectra of its metabolites. Characteristic signals in NMR, in the case of carbon-13, in particular, can also help elucidate the structures of metabolites although substantial amounts of the metabolites are often needed for a good NMR spectrum. A

TABLE 1 - Stable isotopes and radioisotopes (SI and RI): characteristics in use for metabolic reaction studies

	Stable Isotopes	Radioisotopes
Information on structure	Abundant	None
Sensitivity in detection	High	Very high
Natural background	High	Low
Instrumentation	(Mass spectrometer)	(Liquid scintillation counter)
	Very expensive	Expensive
	Needs experience	Easy
	Unsealed sample (destructive, and susceptible to contamination)	Sealed sample (non-destructive, and not susceptible to contamination)
Labelled compounds		
Availability and cost	Unsatisfactory	Satisfactory
Labelling abundance	High (90-99.5%)	Very low
Facilities for safety	Unnecessary	Necessary & expensive

labelled compound is much more stable in its stable isotope-labelled form because it presents no radiolysis problem. Stable isotopes are much safer than radioisotopes for human and environmental studies and therefore usually more acceptable. Stable isotopes are also more convenient than radioisotopes when used in studies on isotope effect because labelled compounds having highly abundant labels with stable isotopes are more easily available, and because the measurement of the ratio of stable isotope-labelled and unlabelled compounds is done relatively easily by selected ion-monitoring with a modern mass spectrometer. The reasons for the increased use of stable isotopes in metabolism studies in recent years are summarized in Table 2.

ISOTOPE EFFECTS

When we use a labelled compound as a tracer in a metabolism study, we assume that the behaviour of the labelled compound is identical to that of its unlabelled analogue. This, however, is not the case. We can classify the identity and differences in their behaviours as shown in Table 3.

TABLE 2 - Basis for increased use of stable isotopes in metabolism studies

---

(1) Development of instruments and techniques:
Sensitive and selective analytical instruments, e.g. GC-MS; selected ion monitoring

(2) Increased availability of labelled compounds:
Compounds highly labelled at specific positions, e.g. 99% or more abundantly labelled ^{2}H(D)- and ^{13}C-compounds

---

This paper focuses on the important difference between an unlabelled compound and its labelled form, i.e. the isotope effect. This effect is noticed in particular with deuterium-labelled compounds.

The utility of labelled compounds in conventional tracer studies requires the near identical behaviour of the unlabelled compound and its labelled form. Such cases are shown in section (1) and (2.1) in Table 3. For studies utilising isotope effects we usually favour a compound labelled with an isotope giving a large mass ratio, e.g. a deuterium-labelled compound, as in section (2.2) of Table 3. Of course other stable isotopes also serve to give information on isotope effect that complements information on the deuterium isotope effect, but the much smaller magnitude of their isotope effects is barely measurable in complex biological reactions in most cases.

Deuterium-labelled compounds

Results of studies on <u>rate comparison</u> of deuterated and undeuterated compounds in a given metabolic reaction will show:

(i) whether or not the metabolic reaction is a <u>reaction involving bond-breaking</u> and/or -<u>forming</u> at the labelled position e.g. -C-H and -C-D,

(ii) whether or not it has a metabolic sequence involving a covalent bond-breaking and/or -forming step as a <u>rate-limiting step</u>, and

(iii) whether or not it has a <u>reaction mechanism</u> in which a rate-limiting step occurs at the labelled position, e.g. E2, irreversible E1cB mechanism, etc.

TABLE 3 - Behaviour of labelled compounds compared with non-labelled compounds; the isotope effect

---

1 <u>Physicochemical</u> Behaviour: Practically <u>Identical</u>.
  Boiling point, melting point, solubility, chromatographic behaviour etc.: very small difference, if any, e.g. boiling point of D_2O, 101.42°C, compared with 100°C of H_2O; and retention time of $[d_6]$methoxychlor 13.102 min, compared with 13.131 min for $[d_0]$methoxychlor in a capillary gas chromatograph.

2 <u>Covalent Bond</u> Cleavage and Formation: From practically identical to largely different.

  2.1 <u>Very Small Difference</u> (Observed Kinetic Isotope Effect ~ 1.0).
    2.1.1 <u>Intrinsically small</u>:
      (a) When bond cleavage (and formation) does not involve the labelled position, only <u>secondary</u> isotope effect can be observed.

$$\begin{array}{ccc} H & D \\ | & | \\ -C\!=\!C- \end{array} \longrightarrow \begin{array}{ccc} H & D \\ | & | \\ -C-C- \\ \diagdown O \diagup \end{array}$$

      (b) When bond cleavage (and formation) that involves the labelled position is not rate-limiting as in the E1 mechanism, isotope effect is small (~1.0).

$$\begin{array}{c} D \\ | \ | \\ -C-C- \\ | \ | \\ \ \ Cl \end{array} \xrightarrow{slow} \begin{array}{c} D \\ | \ | \\ -C-C- \\ | \ + \\ \ \ Cl^- \end{array} \xrightarrow{fast} \begin{array}{c} | \ | \\ -C\!=\!C- \\ \ \ D^+ \end{array}$$

      (c) When the mass ratio of isotopes is small; for example, the ^{13}C versus ^{12}C isotope effect is small (~1.0).

    2.1.2 <u>Apparently small</u>: When an observed overall reaction sequence involves slow reactions insensitive to isotopes, an isotope effect on a reaction in the sequence is <u>masked</u> (from largely to slightly).

  2.2 <u>Large Difference</u> (Observed Kinetic Isotope Effect >1.0) <u>Primary</u> isotope effect, <u>isotope-sensitive mechanism</u>, a large mass <u>ratio</u> (D vs. H etc.), and <u>not masked</u>.

With organic chemical reactions the situation is much simpler than with biochemical reactions. Observed kinetic isotope effects in an organic reaction directly elucidate the reaction mechanism, because in organic reactions we can usually observe the velocity of covalent bond transformation unperturbed by other slow steps such as enzyme-substrate complex-formation and dissociation in enzyme-catalyzed reactions.

In order to study kinetic isotope effects in an enzyme-catalyzed reaction, we often need another series of experiments using tritium together with deuterium in order to overcome heavy masking of the intrinsic isotope effect by slow steps in the overall reaction (to be discussed later). Some studies using tritium-labelled compounds are also described in this paper.

Isotope effect studies may supply some other types of information:

(i)   change in biological activity of a substrate and

(ii) perturbation of a network of metabolic pathways that may lead to the discovery or enhancement of a very minor metabolic pathway.

SIMPLE USE OF OBSERVED ISOTOPE EFFECTS ON REACTION RATES AND ON BIOLOGICAL ACTIVITIES

Intrinsic isotope effects are often masked in biological reactions by several slow steps with the result that the observed isotope effect may be very small (Northrop, 1977). However, we can use such an observed isotope effect value (or an apparent value) for the study of xenobiotic degradation pathways if the value is significantly larger than unity.

The first such example involves lindane metabolism. When this potent insecticide is hexadeuterated, its insecticidal activity is much enhanced as shown in Table 4 (Tanaka et al., 1981). Comparison in the penetration rates into insects and the electrophysiological activity on the insect nerve cord show no difference between labelled and unlabelled lindane. Biodegradation rates were also compared and here there is a remarkable difference. Figure 1 shows an example of the experiment. A one to one mixture of $[d_0]$- and $[d_6]$lindane was applied to house flies. After a specified time the remaining insecticide in the body was quantitated by monitoring selected ions with a GC-MS. There was a big difference in their recoveries from the insects at every time point. What this observed isotope effect on the biodegradation rate suggests is that C-H bond cleavages are

TABLE 4 - Insecticidal activity of lindane and [d_6]lindane

	LD$_{50}$ values (10^{-10}mol/insect)					
	Mosquito	House fly			Cockroach	
	NAIDM[a]	Toichi	3rd-Yum[b]		German	American
[d_0]Lindane	1.3	6.53	816	>2200	16.0	200
[d_6]Lindane	0.32	2.06	68	86.2	1.95	24.5
Ratio	4.06	3.17	11.8	>25	8.12	8.16

[a] NAIDM = National Association of Insecticide and Disinfectant Manufacturers susceptible strain.
[b] 3rd-Yum = 3rd-Yumenoshima strain

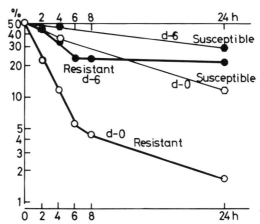

Figure 1. Percent decrease of [d_0]- and [d_6]lindane topically applied to two strains of house fly (resistant and susceptible)

involved in the first steps of lindane degradation and that these steps are the rate-limiting and detoxifying reactions of the insecticide. By using conventional techniques we identified various intermediary and terminal metabolites. The structures of these metabolites and the above isotope effect studies help us formulate a network of metabolic pathways for lindane in the house fly. These metabolic reaction pathways are cis-dehydrogenation, cis- and trans-dehydrochlorinations and oxygenations in the earlier phases, as shown in Figure 2. Glutathione conjugations are involved in the later steps.

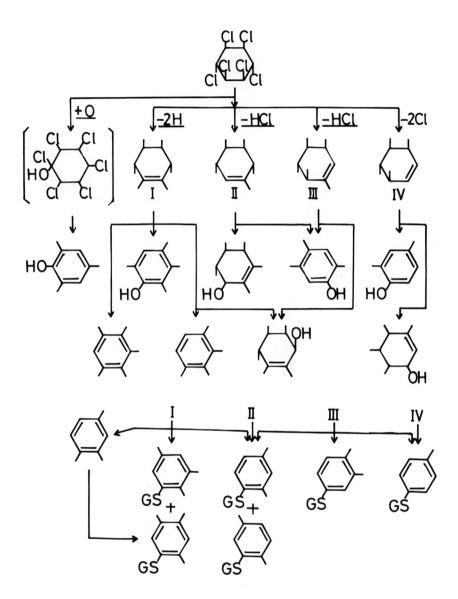

Figure 2. Metabolic pathways of lindane in house flies and in rats. Underlined reactions show a large magnitude of kinetic deuterium isotope effect. Most chlorine substituents are indicated only by short lines.

We observed metabolites very similar to the above in rats. We found mercapturic acids in their urine (Kurihara et al., 1979) as predicted, instead of simple glutathione conjugates that are found in the house flies. Isotope effects were also observed for metabolites from the rats. When a 1:1 mixture of $[d_0]$- and $[d_6]$lindane was administered to rats, we found in the urine a mixture of mercapturic acids with various proportions of undeuterated and deuterated phenyl ring. Table 5 shows a difference in the ratio between monochlorophenylmercapturic acids and dichloro- or trichloro-phenylmercapturic acids (Kurihara et al., 1980). Relative amounts of monochlorophenylmercapturic

TABLE 5 - Ratio of three chlorophenylmercapturic acid fractions[a] excreted from lindane ($[d_0]$ and $[d_6]$)-dosed rats during the first 48 hours after dosing

	Ratio	
	Dichloro/Monochloro	Trichloro/Monochloro
$[d_0]$lindane	3.96	3.08
$[d_6]$lindane	2.08	0.67
One to one mixture of $[d_0]$ and $[d_6]$	3.33	2.67

[a]See the structures in Figure 2 for the corresponding glutathione conjugates

acids increase when $[d_6]$lindane is metabolized. The dichloro- and trichloro-phenylmercapturic acid pathways have a larger magnitude of isotope effects than those for the monochloro-derivatives. A smaller isotope effect on the formation of monochlorophenylmercapturic acids can be ascribed to the first dechlorination step to tetrachlorocyclohexene followed by glutathione conjugation. The first dechlorination reaction must be slower than the later steps, and must show a smaller isotope effect: a secondary isotope effect, if any, because it is not a C-H bond cleaving reaction. For the formation of di- and trichlorophenylmercapturic acids, we must assume the occurrence of the first dehydrochlorination and dehydrogenation both of which would proceed slower showing a primary isotope effect because these involve C-H cleaving reactions. The results in Table 5 are consistent with the assumptions.

These experiments on lindane isotope effects have afforded important information about critical steps in the metabolic sequence. The earlier steps are all rate-limiting and detoxifying and the metabolic reaction rates are directly related to the biological activities of labelled and unlabelled lindane giving some very interesting isotope effects on insecticidal activity. The isotope effect study with rats in vivo gives evidence for the previously proposed pathways.

An elegant study on the bioactivation of chloroform to phosgene using deuterochloroform has been described by Pohl (1979) and illustrates another useful application of an isotope effect study by simply observing apparent isotope effect. Deuteration of the $C-CH_3$ group in antipyrine changes its main metabolic pathway to $\underline{N}$-demethylation; this is called metabolic switching (Horning et al., 1976). Metabolic formation of PAM from an antitumor agent, cyclophosphamide, decreases when the latter is deuterated at the 5-position. Its antitumor activity also decreases demonstrating that the active principle from cyclophosphamide is PAM (Spielman et al., 1980).

METHODS OF OBTAINING INTRINSIC ISOTOPE EFFECT VALUES

We can use isotope effect values to elucidate a more precise mechanism of a metabolic reaction but for such a precision we require an intrinsic isotope effect value. Even a simple enzymatic reaction often involves several slow steps other than the central catalytic step. When the formation and dissociation of the enzyme-substrate complex (k_1 and k_2) and the dissociation of the enzyme-product complex (k_4) are as slow as, or slower than, the catalytic step (k_3) in the simple model reactions below (also see Figure 3),

$$E + S \underset{k_2}{\overset{k_1}{\rightleftarrows}} ES \xrightarrow{k_3} EP \xrightarrow{k_4} E + P$$

the observed isotope effect values on V_{max} and V_{max}/K_m are (Northrop, 1977):

$V_H/V_D = [(k_{3H}/k_{3D}) + (k_{3H}/k_4)] / [1 + (k_{3H}/k_4)]$
$(V/K)_H/(V/K)_D = [(k_{3H}/k_{3D}) + (k_{3H}/k_2)] / [1 + (k_{3H}/k_2)]$

In these equations, k_2 is the rate constant for dissociation of the ES complex, k_3 is that of the catalytic step and k_4 is that of dissociation of the EP complex. When k_2 and k_4 values are small compared with k_{3H}, $(V/K)_H/(V/K)_D$ and V_H/V_D

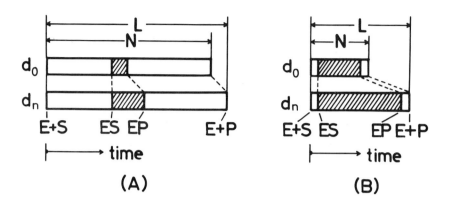

Figure 3. Heavy masking of an isotope effect value (A), and a relatively close approximation of an intrinsic isotope effect to observed value (B). The ratio of L/N (L=labelled, N=normal=unlabelled) is equal to v_H/v_D, i.e. observed isotope effect value. In (A), the slow steps (E)+(S) to (ES) and (EP) to (E)+(P) largely mask the intrinsic isotope effect, i.e. the effect on the rate from (ES) to (EP).

values become smaller than k_H/k_D values. I will use in this paper notation of $^D V$, $^D(V/K)$ and $^D k$ instead of V_H/V_D, $(V/K)_H/(V/K)_D$ and K^H/K^D, respectively, according to Northrop (1977).

The most orthodox method to obtain the value $^D k$ is to measure the magnitude of $^D(V/K)$ and $^T(V/K)$ in the same reaction condition. (A set of $^D V$ and $^T V$ should also give an intrinsic value, but in practice $^T V$ is not obtainable, because a carrier-free tritium-labelled compound is required to obtain a $^T V$ value). The rationale for this method is illustrated in the equations as shown below (Northrop 1977).

$$[^D(V/K) - 1]/[^T(V/K) - 1] = (^D k - 1)/(^T k - 1)$$
$$= (^D k - 1)/(^D k^{1.442} - 1)$$

Several examples have been published on the application of this technique to obtain intrinsic isotope effect values. For example, 7-ethoxycoumarin deethylation (Figure 4) was examined using rat liver microsomes and a reconstituted enzyme system containing purified cytochrome P-450 and cytochrome P-448 (Miwa et al., 1984). Reported values of $^D k$ on this oxidative deethylation are 12.8 and 14.0 for P-450 and P-448, respectively, and this group concluded that the reaction mechanism suggested by the above isotope effect

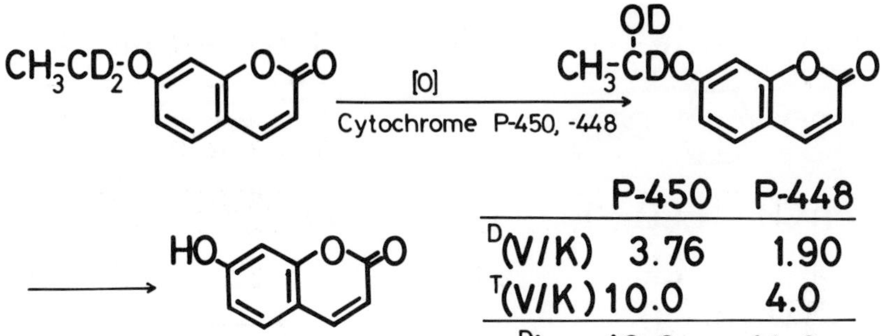

Figure 4. Oxidative deethylation of 7-ethoxycoumarin and its deuterium and tritium isotope effect values ($^D(V/K)$ and $^T(V/K)$). (Miwa et al., 1984).

	P-450	P-448
$^D(V/K)$	3.76	1.90
$^T(V/K)$	10.0	4.0
^Dk	12.8	14.0

values is consistent with the previously proposed homolytic cleavage of the C-H bond followed by recombination of an .OH radical with the above formed ⟩C. radical.

Another method has been proposed for obtaining intrinsic isotope effect values. The method is based on the intramolecular competition of two corresponding reacting groups in which one group is completely labelled with deuterium (Melander and Saunders, 1980). This is called "intramolecular isotope effect". Published examples (Figure 5) include metabolic reactions of [monomethoxy-d3]p-methoxyanisole (Foster et al., 1974), [1,1-d2]1,3-diphenylpropane (Hjelmeland et al., 1977), and tetradeutero-norbornane (Groves et al., 1978). In each example, pairs of similar reacting groups are found: methoxy-groups in p-methoxyanisole, 1- and 3-methylene groups in 1,3-diphenylpropane and methylene groups in norbornane. Intramolecular isotope effect values for these reactions are 10, 11±1 and 11.5±1, respectively, as large as the theoretical maximum deuterium isotope effect. Thus it will be safe to formulate reaction mechanisms with the help of such observed intramolecular effects.

This good approximation of intramolecular isotope effect values to the intrinsic values depends on a very rapid exchange of corresponding reacting groups in the same molecule at the site of enzyme catalysis (Figure 6). If the exchange does not proceed rapidly, the group-exchanging speed might mask the intrinsic value, giving a lower observed value (as happens in intermolecular isotope effect measurements).

Substrate	Ratio of rate v_H/v_D
H_3CO–⟨⟩–OCD_3	10
$Ph\text{-}CH_2CH_2CD_2\text{-}Ph$	11 ± 1
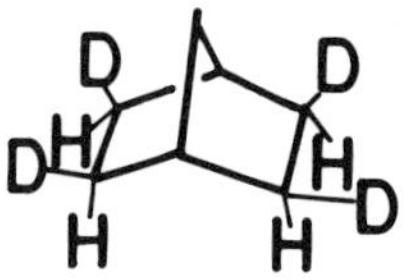 (norbornane with D/H)	11.5 ± 1

Foster et al. (1974), Hjelmeland et al. (1977), Groves et al. (1978).

Figure 5. Published examples of intramolecular isotope effects on enzyme-catalyzed oxygenation (or oxidative dealkylation) reactions.

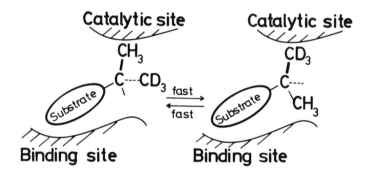

Figure 6. A frequently observed very fast exchange of corresponding reacting groups (CH_3 and CD_3 in this case) in the same molecule on the catalytic site of the enzyme.

We have measured oxidative demethylation rates of methoxychlor in rat liver microsomes with [hexadeuterated dimethoxy]methoxychlor ([d_6]methoxychlor) and [monomethoxy-d_3]methoxychlor ([d_3]methoxychlor) (Figure 7). First we compared the rate parameters of [d_0]- and [d_6]methoxychlor (Ichinose and Kurihara, 1986) using rat liver microsomes of untreated, phenobarbitone-treated and β-naphthoflavone-

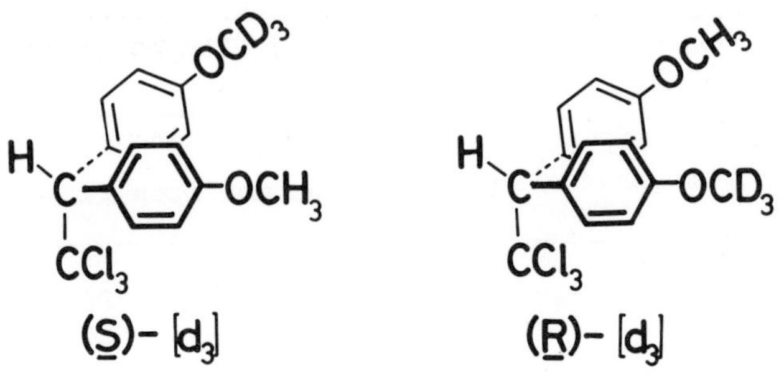

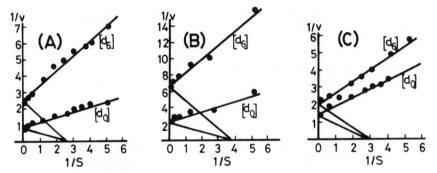

Figure 7. Structures of (S)- and (R)-[d3]methoxychlor

Figure 8. Double reciprocal plots for the oxidative demethylation of [d_0]- and [d_6]methoxychlor by rat liver microsomes. (A) Untreated rat, (B) β-naphthoflavone-treated rat and (C) phenobarbitone-treated rat. v = velocity (mol/mol P-450/min). S = substrate concentration (μM).

treated rats. The double reciprocal plots are shown in Figure 8 and the kinetic constants in Table 6. The observed deuterium isotope effects are considerably larger than unity, and the effects are certainly primary. However, these values do not reach the theoretical maximum of the deuterium isotope effect value. The intrinsic isotope effect value is essential to a detailed discussion on the reaction mechanism. Using the tritium isotope effect values together with the above values, we calculated intrinsic isotope effect values of these reactions. The values are as large as 15.2 and 19.2 for untreated and β-naphthoflavone-treated rat liver microsomes, respectively. The mechanism, therefore, is again consistent to the so-called radical-

TABLE 6 - Kinetic constants for oxidative demethylation of [d_0] and [d_6]methoxychlor by rat liver microsomes

	Liver microsomes from rats					
	Untreated[a]		NF-treated[b]		PB-treated[c]	
	[d_0]	[d_6]	[d_0]	[d_6]	[d_0]	[d_6]
V_{max}(min^{-1})	1.15	0.46	0.43	0.16	0.69	0.50
K_m (M)	0.36	0.34	0.27	0.28	0.36	0.33
V/K	3.19	1.35	1.59	0.57	1.92	1.52
D_V	2.53		2.72		1.38	
$D_{(V/K)}$	2.44		2.76		1.27	

[a] Observed tritium isotope effect: ^{T}(V/K) was 6.70. Thus, ^Dk=15.2.

[b] Observed tritium isotope effect: ^{T}(V/K) was 7.78. Thus, ^Dk=19.2. NF = β-Naphthoflavone.

[c] PB = Phenobarbitone.

formation and radical-recombination mechanism of cytochrome P-450-mediated oxygenation.

Next, we conducted intramolecular competition experiments with [d_3]methoxychlor (Ichinose and Kurihara, 1987). An introduction of one [d_3]methyl group into methoxychlor makes the methine-carbon a chiral centre. We thus obtained the S and R forms of [d_3]methoxychlor through optical resolution at an appropriate step in the synthetic route. A brief incubation of these chirally labelled compounds with rat liver microsomes produced a mono-demethylated product as the sole metabolite. The metabolite from [d_3]methoxychlor is composed of a d_3-compound and a d_0-compound, corresponding to the d_0-methyl elimination and the d_3-methyl elimination products, respectively. After pentafluoropropionylation, we measured the ratio of these two metabolites by selected ion-monitoring. The differentiation ratio of the d_0-methyl and the d_3-methyl group in this reaction is obtained, i.e. an apparent intramolecular isotope effect on the metabolic reaction. Data are given in Table 7.

TABLE 7 - Observed differentiation ratio of CH_3 and CD_3 groups on oxidative demethylation by rat liver microsomes of [d_3]Methoxychlor

	[d_3]Metabolite[a]/[d_0]Metabolite[a]		
	Untreated	NF-treated[b]	PB-treated[c]
[S]-[d_3] ($^D v_S$)	13.6	6.85	6.30
[R]-[d_3] ($^D v_R$)	1.06	0.71	1.34
[RS]-[d_3]	4.71	2.58	4.27

a) Metabolite = Monodemethylated product from methoxychlor.

b) NF = β-Naphthoflavone.

c) PB = Phenobarbitone.

These data lead to the conclusions that (i) microsomal oxidative demethylation of this substrate had high enantiotopic selectivity (Figure 9) and (ii) the product ratios clearly show some degree of isotope differentiation, but the value of the isotope effect is much smaller than the intrinsic value obtained from tritium data.

If a rapid exchange of competitively reacting groups occurs at the catalytic site in an enzyme, a reaction must give a degree of isotope effect as high as the intrinsic isotope effect value of the reaction. In addition, if there is an enantiotopic selectivity in the reaction, the values for R and S isomers would be geometrically averaged to derive the intrinsic value. In our study the values $\sqrt{D_{V/S} \times D_{V/R}}$ equal 3.8 (untreated) and 2.9 (NF-treated) and much smaller than 15.2 and 19.2 previously reported as the intrinsic values for the reaction. Our assumption is that the exchange of the two methoxy groups is not so rapid as anticipated.

An intramolecular exchange of competitively reacting groups

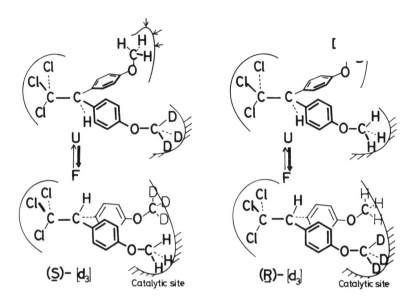

Figure 9. Possible conformations of the enzyme-substrate complex which produces mono-O-demethylated methoxychlor. (S)-[d3] = (S)-[d3]methoxychlor, in which the CH3 group will fit the catalytic site of an enzyme in the favourable conformation of ES complex. (R)-[d3] = (R)-[d3]methoxychlor, in which the slower reacting CD3 group will fit the catalytic site of an enzyme in the favourable conformation of ES complex. F = favourable and U = unfavourable conformation of ES complex.

at the catalytic site of an enzyme is usually rapid as described above. A paper recently published also reports an interesting example of a rapid exchange of two methyl groups in oxidative and stereoselective oxygenation of a chirally deuterated cumene (Sugiyama and Trager, 1986). All these substrates have two competitively reacting groups in very close proximity and this proximity seems to allow a rapid exchange of the two groups in the enzyme-substrate complex. The two methoxy groups in methoxychlor are separated by two phenyl rings and one methine carbon, and are located much further from each other than are the published examples mentioned above.

Recently, we examined insecticidal activity of the above described deuterated methoxychlor. Table 8 shows their LD_{50} values (Kurihara and Hirano, unpublished). Small but distinct isotope effects on LD_{50} values are observed especially in Blattella germanica.

TABLE 8 - Insecticidal activity of [d_0]-, [d_3]- and [d_6]Methoxychlor

	LD_{50} values (nmol/insect) relative values in parentheses	
	Male house fly (NAIDM)	Female German cockroach
[d_0]	3.73 (100)	124.2 (100)
(S)-[d_3]	3.70 (101)	97.0 (128)
(RS)-[d_3]	3.53 (106)	74.9 (166)
(R)-[d_3]	3.39 (110)	69.4 (179)
[d_6]	2.22 (168)	51.2 (243)

Kinetic isotope effects on *in vivo* degradation rates in Blattella seem to have caused such a variation in LD_{50} values, but we have not yet obtained proof of this.

An important application of isotopically sensitive branching in enzyme-catalyzed oxygenations to determining the limit of their intrinsic isotope effect has appeared (Jones et al., 1986). In this study, [1-d_3]- and [1,2,3-d_7]octane were used as substrates and the intramolecular exchange of competitively reacting methyl and [d_3]-methyl groups etc. seems again very rapid at the catalytic site of cytochrome P-450.

CONCLUSIONS

The study of isotope effects in xenobiotic metabolism, using deuterium-labelled substrates in particular, provides information on: (i) bioactivation and/or detoxification reactions of a substrate; these are detected by changes in metabolic reaction rates, and (ii) minor metabolic pathways hidden in major pathways; these may be exposed by the perturbation of a network of metabolic pathways.

As examples, we have observed remarkable increases in the insecticidal activities of lindane and methoxychlor when they were heavily deuterated. Carbon-hydrogen bond oxygenation was found to be the most critical step in the detoxification reactions of these insecticides when we compared the reaction rates of deuterated and undeuterated substrates.

If we are able to measure intrinsic isotope effects in a particular metabolic reaction by the use of its intramolecular isotope effect or its tritium isotope effect, the isotope effect value will help elucidate the precise reaction mechanism.

REFERENCES

Foster, A.B., Jarman, M., Stevens, J.D., Thomas, P. and Westwood, J.H., 1974, Isotope effects in O- and N-demethylations mediated by rat liver microsomes: an application of direct insertion electron impact mass spectrometry. Chemico-Biological Interactions, 9, 327-340.

Groves, J.T., McClusky, G.A., White, R.E. and Coon, M.J., 1978. Aliphatic hydroxylation by highly purified liver microsomal cytochrome P-450. Evidence for a carbon radical intermediate. Biochemical and Biophysical Research Communications, 81, 154-160.

Hjelmeland, L.M., Aronow, L. and Trudell, J.R., 1977, Intramolecular determination of primary kinetic isotope effects in hydroxylations catalyzed by cytochrome P-450. Biochemical and Biophysical Research Communications, 76, 541-549.

Horning, M.G., Haegele, K.D., Sommer, K.R., Nowlin, J., Stafford, M. and Thenot, J-P., 1976, Metabolic switching of drug pathways as a consequence of deuterium substitution In Proceedings of the Second International Conference on Stable Isotopes, edited by E.R. Klein and P.D. Klein (Springfield, Va.: CONF-751027, NTIS), pp.41-54.

Ichinose, R. and Kurihara, N., 1986, Deuterium and tritium isotope effects on the oxidative demethylation rate of methoxychlor in rat liver microsomes. Journal of Pesticide Science, 11, 231-236.

Ichinose, R. and Kurihara, N., 1987, Intramolecular deuterium isotope effect and enantiotopic differentiation in oxidative demethylation of chiral [monomethyl-d_3]methoxychlor in rat liver microsomes. Biochemical Pharmacology, 36, 3751-3756.

Jones, J.P., Korzekwa, K.R., Rettie, A.E. and Trager, W.F., 1986, Isotopically sensitive branching and its effects on the observed intramolecular isotope effects in cytochrome P-450 catalyzed reactions: a new method for the estimation of intrinsic isotope effects. Journal of the American Chemical Society, 108, 7074-7078.

Kurihara, N., Tanaka, K. and Nakajima, M., 1979, Mercapturic acid formation from lindane in rats. Pesticide Biochemistry and Physiology, 10, 137-150.

Kurihara, N., Suzuki, T. and Nakajima, M., 1980, Deuterium isotope effects on the formation of mercapturic acids from lindane in rats. Pesticide Biochemistry and Physiology, 14, 41-49.

Melander, L. and Saunders, W.H. Jr., 1980, Evaluation of rate-constant ratios from experimental data. In Reaction Rates of Isotopic Molecules, by L. Melander and W.H. Saunders, Jr. (New York: John Wiley & Sons), pp.91-94.

Miwa, G.T., Walsh, J.S. and Lu, A.Y.H., 1984, Kinetic isotope effects on cytochrome P-450-catalyzed oxidation reactions. Journal of Biological Chemistry, 259, 3000-3004.

Northrop, D.B., 1977, Determining the absolute magnitude of hydrogen isotope effects. In Isotope Effects on Enzyme-Catalyzed Reactions, edited by W.W. Cleland, M.H. O'Leary and D.B. Northrop (Baltimore: University Park Press), pp. 122-152.

Pohl, L.R., 1979, Biochemical toxicology of chloroform. In Reviews in Biochemical Toxicology 1, edited by E. Hodgson, J.R. Bend and R.M. Philpot (New York: Elsevier/North-Holland), pp. 79-107.

Spielmann, H., Jacob-Müller, U. and Habenicht, U., 1980, Deuterium isotope effect in teratology: changes in cyclophosphamide teratogenicity in the mouse due to 5,5-dideuteration. Archives of Pharmacology, 311, R25.

Sugiyama, K. and Trager, W.F., 1986, Prochiral selectivity and intramolecular isotope effects in the cytochrome P-450 catalyzed ω-hydroxylation of cumene. Biochemistry, 25, 7336-7343.

Tanaka, K., Nakajima, M. and Kurihara, N., 1981, The mechanism of resistance to lindane and hexadeuterated lindane in the third Yumenoshima strain of house fly. Pesticide Biochemistry and Physiology, 16, 149-157.

IN VIVO DETECTION OF FREE RADICAL METABOLITES AS
APPLIED TO CARBON TETRACHLORIDE AND RELATED HALOCARBONS

Kathryn T. Knecht[1,2], Henry D. Connor[1], Lynn B. LaCagnin[1,2],
Ronald G. Thurman[2] and Ronald P. Mason[1]

[1]Laboratory of Molecular Biophysics, National Institute of
Environmental Health Sciences, P.O. Box 12233, Research
Triangle Park, NC 27709, USA; [2]Curriculum in Toxicology,
University of North Carolina, Chapel Hill, NC 27514, USA

INTRODUCTION

As early as the 1930s, Michaelis (of the Michaelis-Menton equation) was interested in the possible importance of free radical metabolites in biochemistry. However, unlike many aromatic drugs and industrial chemicals of current toxicological interest, most biochemicals are not easily metabolized through free radical intermediates. Thus metabolism involving free radicals has historically received little attention. Because the toxicological effects of free radicals should, ultimately, have purely chemical explanations, defining radical chemistry under controlled physiological conditions in vitro is a necessary factor in understanding free radical mechanisms of toxicity. Furthermore, the characteristics and the localization of toxicities detected in vivo must be consistent with established modes of free radical production, both enzymatic and non-enzymatic. In vitro studies provide useful data by establishing the possible origins of a given free radical species, as has been shown for many different classes of compounds (Mason and Chignell, 1982; Mason, 1982, 1984). On the other hand, in vivo studies not directly involving free radicals can also be useful. Pharmacological manipulations of an animal model which alter the toxic effects of an administered compound can indirectly implicate metabolic activation, including activation via free radical formation, in the mechanism of toxic action.

THE DETECTION OF FREE RADICALS IN VIVO

Unless the free radical chemistry developed in vitro can actually be demonstrated in the whole animal, its existence in vivo and its fundamental significance in toxicity will still be subject to question. The inherent transience of free radical species and their slow rate of production in vivo makes such detection in the whole animal quite challenging. Spin trapping, a technique which integrates free radicals formed over time, is a sensitive and thus most attractive approach to this problem.

Spin trapping is a technique where a diamagnetic molecule (or spin trap) reacts with a free radical to produce a more stable radical (or spin adduct) which is readily detectable by electron spin resonance (ESR) (Janzen, 1980; Perkins, 1980). Spin adducts are substituted nitroxide free radicals which tend to be relatively long-lived and, in some cases, are stable enough for structural analysis by mass spectroscopy and NMR (Mason, 1984).

Initial applications of spin trapping to in vivo metabolism studies have involved the prototypical hepatotoxicant carbon tetrachloride. The trichloromethyl free radical, a reductive dehalogenated product of carbon tetrachloride, had been proposed as a reactive intermediate (Butler, 1961), but spin trapping with phenyl-N-t-butylnitrone (PBN) provided the first direct proof of this radical's existence in vivo (Lai et al., 1979). These workers administered carbon tetrachloride and PBN to rats and found a radical adduct in organic extracts of the liver. Its identity was assigned by comparing experimental hyperfine coupling constants with those of PBN/·CCl$_3$ generated in a microsomal system or by photolysis of carbon tetrachloride. More definitive identification of this species as the PBN/·CCl$_3$ radical adduct was later made using ^{13}C-carbon tetrachloride, which produces an additional splitting in the radical adduct spectrum (Poyer et al., 1980; Albano et al., 1982).

$$\text{Ph-CH=}\overset{+}{\underset{\underset{O^-}{|}}{N}}\text{-C(CH}_3\text{)}_3 \xrightarrow{\cdot \text{CCl}_3} \text{Ph-}\underset{\underset{O\cdot}{|}}{\overset{\overset{\text{CCl}_3}{|}}{\text{CH}}}\text{-N-C(CH}_3\text{)}_3$$

PBN

An additional radical adduct, PBN/·CO$_2^-$, was detected by Connor et al. (1986) in the urine of living rats treated with carbon tetrachloride and PBN. Use of ^{13}C-carbon tetrachloride showed that this radical adduct, like PBN/·CCl$_3$, was carbon tetrachloride-derived. PBN/·CO$_2^-$ was also detected in the effluent perfusate of livers which were perfused with carbon tetrachloride and PBN, and in the urine

of rats after administration of bromotrichloromethane and
PBN (LaCagnin et al., 1988). Bromotrichloromethane is
metabolized by the same pathway and to the same metabolites
as carbon tetrachloride, but is more readily dehalogenated
due to the relative weakness of the C-Br bond.

The U.S. Food and Drug Administration and Environmental
Protection Agency require companies to identify the urinary
metabolites derived from drugs and pesticides. As the above
studies suggest, such requirements could be usefully extended
to the spin-trapped products of free radical metabolism.
Unlike the stable products of detoxification, which can be
detected by conventional analytical techniques, the
detection of radical adducts proves the formation of highly
reactive free radical intermediates. Free radicals can
clearly react as easily with cellular constituents as with
spin trap agents and thus could cause damage in vivo.

Hepatocellular necrosis in perfused liver, as measured by
liver alcohol dehydrogenase (LDH) release from lysed cells,
occurs after the infusion of halocarbon and follows the
appearance of PBN/·CO_2^- (LaCagnin et al., 1988). Perfusion
of the liver with nitrogen-saturated buffer instead of
oxygen-saturated buffer accelerates LDH release. The
concentration of PBN/·CO_2^- in the perfusate at the beginning
of lysis is statistically correlated with the amount of time
required for LDH to be detected (LaCagnin et al., 1988).
Correlation does not imply causation; therefore, the stable
trapped product PBN/·CO_2^- can only be considered to be a
marker for some more reactive species actually responsible
for membrane damage. Nevertheless, the detection of a
radical adduct does imply production of reactive free
radical metabolites in vivo, and the correlation of radical
adduct production with an index of toxicity is consistent
with a free radical-mediated pathology. In these
experiments, PBN (5mM) did not appear to prevent membrane-
damaging free radical reactions from occurring, but this can
be explained by the very low rates of radical trapping
characteristic of PBN (Gasanov and Freidlina, 1987).

Both PBN/·CCl_3 and PBN/·CO_2^- were detected in bile samples
collected at multiple timepoints after treatment of living
rats with PBN intraperitoneally and carbon tetrachloride
intragastrically (Knecht and Mason, 1988). In vivo
manipulations such as low oxygen tension of inspired air or
phenobarbital pretreatment, known to increase the toxicity
of carbon tetrachloride, also produce qualitative and
quantitative changes in the ESR signals detected. Either
hypoxia or phenobarbital induction was required for the
detection of PBN/·CO_2^-. Both treatments also increased the
biliary concentration of PBN/·CCl_3 (Knecht and Mason, 1988).

Carbon-centered radical adducts from other halogenated hydrocarbons have also been detected in the organic extracts of livers from treated animals. Of clinical interest have been studies with the volatile anaesthetic halothane, which produces hepatitis in humans. Under hypoxic conditions, halothane produces both liver damage in phenobarbital-pretreated rats (McLain et al., 1979; Ross et al., 1979) and free radicals, which can be trapped by PBN and extracted from the liver (Poyer et al., 1981; Plummer et al., 1982; Fujii et al., 1984). The trapped radical species has not yet been identified unambiguously. Although metabolite profiles show both debromination and defluorination to be involved in metabolic activation (Ahr et al., 1982; Fujii et al., 1984), reductive defluorination forming a carbon-centered free radical is unlikely because of the relative strength of the carbon-fluorine bond. Reductive debromination is probably responsible for the reported carbon-centered free radical formation (Poyer et al., 1981; Plummer et al., 1982; Fujii et al., 1984). Chloroform, iodoform, bromoform, and bromodichloromethane are other compounds that are converted to free radicals in vivo (Tomasi et al., 1985).

Halocarbon free radical formation is thought to result in lipid peroxidation and it is logical to expect that spin adducts, which do not exhibit ^{13}C hyperfine coupling, but which are detected after administration of ^{13}C-carbon tetrachloride, are derived from carbon- or oxygen-centered lipid radicals. Such lipid-derived radical adducts have been detected in vitro by several investigators (Kalyanaraman et al., 1979; Poyer et al., 1980; Tomasi et al., 1983; McCay et al., 1984). The extracted livers of rats treated with the PBN analogue α-2,4,6-trimethoxyphenyl-N-t-butylnitrone and carbon tetrachloride yield a carbon tetrachloride-dependent, non-carbon tetrachloride-derived species believed to be a lipid-derived radical (McCay et al., 1984). However, O-demethylation of the spin trap precludes a simple explanation of these latter studies. Administration of carbon tetrachloride and PBN to gerbils results in detectable free radical adducts in organic extracts of liver, kidney, heart, lung, testis, brain, and blood, with signal intensities of spectra decreasing in the order given. In extracts of liver, PBN/·CCl$_3$ was identified, but no assignment of radical identity in other tissues could be made (Ahmed et al., 1987). Definitive identification of lipid radicals is inherently difficult, since ^{13}C-labelled fatty acids are unavailable and chromatography or mass spectroscopy of these heterogeneous biological products is a formidable task.

CONCLUSIONS

Although to date most *in vivo* investigations of free radical metabolite formation have been limited to the halogenated hydrocarbons, enough additional examples are known (Kubow et al., 1984; Reinke et al., 1987; Maples et al., 1988a,b) to demonstrate that this technique may be widely applicable. If this is the case, then the detection and identification of free radical metabolites *in vivo* should be more useful to the understanding and even to the prediction of toxicities than are traditional analytical techniques such as HPLC which are inherently limited to the detection of stable metabolites. In addition, both experience and chemical intuition would suggest that any free radicals detected *in vivo* would be of great toxicological significance. In contrast, the products of metabolism detectable by HPLC may be related to detoxification rather than to toxicity, but are not necessarily related to either.

REFERENCES

Ahmed, F.F., Owen, D.L. and Sun, A.Y., 1987, Detection of free radical formation in various tissues after acute carbon tetrachloride administration in the gerbil. Life Sciences, 41, 2469-2475.

Ahr, H.J., King, L.J., Nastainczyk, W. and Ullrich, V., 1982, The mechanism of reductive dehalogenation of halothane by liver cytochrome P-450. Biochemical Pharmacology, 31, 383-390.

Albano, E., Lott, K.A.K., Slater, T.F., Stier, A., Symons, M.C.R. and Tomasi, A., 1982, Spin-trapping studies on the free-radical products formed by metabolic activation of carbon tetrachloride in rat liver microsomal fractions, isolated hepatocytes and *in vivo* in the rat. Biochemical Journal, 204, 593-603.

Butler, T.C., 1961, Reduction of carbon tetrachloride in vivo and reduction of carbon tetrachloride and chloroform in vitro by tissue and tissue constituents. Journal of Pharmacology and Experimental Therapeutics, 134, 311-319.

Connor, H.D., Thurman, R.G., Galizi, M.D. and Mason, R.P., 1986, The formation of a novel free radical metabolite from CCl_4 in the perfused rat liver and in vivo. Journal of Biological Chemistry, 261, 4542-4548.

Fujii, K., Morio, M., Kikuchi, H., Ishihara, S., Okida, M. and Ficor, F., 1984, In vivo spin trap study on aerobic dehalogenation of halothane. Life Sciences, 35, 463-468.

Gasanov, R.G. and Freidlina, R.Kh., 1987, Application of the spin trapping method in kinetic measurements. Russian Chemical Reviews, 56, 264-274.

Janzen, E.G., 1980, A critical review of spin trapping in biological systems. In *Free Radicals in Biology Volume IV*, edited by W.A. Pryor (New York: Academic Press), pp. 115-154.

Kalyanaraman, B., Mason, R.P., Perez-Reyes, E., Chignell, C.F., Wolf, C.R. and Philpot, R.M., 1979, Characterization of the free radical formed in aerobic microsomal incubations containing carbon tetrachloride and NADPH. Biochemical and Biophysical Research Communications, 89, 1065-1072.

Knecht, K.T. and Mason, R.P., 1988, In vivo radical trapping and biliary secretion of carbon-tetrachloride-derived free radical metabolites. Drug Metabolism and Disposition, in press.

Kubow, S., Janzen, E.G. and Bray, T.M., 1984, Spin trapping of free radicals formed during in vitro and in vivo metabolism of 3-methylindole. Journal of Biological Chemistry, 259, 4447-4451.

LaCagnin, L.B., Connor, H.D., Mason, R.P. and Thurman, R.G., 1988, The carbon dioxide anion radical in the perfused rat liver: relationship to halocarbon-induced toxicity. Molecular Pharmacology, 33, 351-357.

Lai, E.K., McCay, P.B., Noguchi, T. and Fong, K.L., 1979, In vivo spin-trapping of trichloromethyl radicals formed from CCl_4. Biochemical Pharmacology, 28, 2231-2235.

Maples, K.R., Jordan, S.J. and Mason, R.P., 1988a, In vivo rat hemoglobin thiyl free radical formation following phenylhydrazine administration. Molecular Pharmacology, 33, 344-350.

Maples, K.R., Jordan, S.J. and Mason, R.P., 1988b, In vivo rat hemoglobin thiyl free radical formation following administration of phenylhydrazine and hydrazine-based drugs. Drug Metabolism and Disposition, in press.

Mason, R.P., 1982, Free radical intermediates in the metabolism of toxic chemicals. In *Free Radicals in Biology, Volume V*, edited by W.A. Pryor (New York: Academic Press), pp. 161-222.

Mason, R.P., 1984, Spin trapping free radical metabolites of toxic chemicals. In *Spin Labelling in Pharmacology*, edited by J.L. Holtzman (New York: Academic Press), pp. 87-129.

Mason, R.P. and Chignell, C.F., 1982, Free radicals in pharmacology and toxicology - selected topics. Pharmacological Reviews, 33, 189-211.

McCay, P.B., Lai, E.K., Poyer, J.L., BuBose, C.M. and Janzen, E.G., 1984, Oxygen- and carbon-centered free radical formation during carbon tetrachloride metabolism. Journal of Biological Chemistry, 259, 2135-2143.

McLain, G.E., Sipes, I.G. and Brown, B.B., 1979, An animal model of halothane hepatotoxicity: Role of enzyme induction and hypoxia. Anesthesiology, 51, 321-326.

Perkins, M.J., 1980, Spin trapping. Advances in Physical Organic Chemistry, 17, 1-64.

Plummer, J.L., Beckwith, A.L.J., Bastin, F.N., Adams, J.F., Cousins, M.J. and Hall, P., 1982, Free radical formation in vivo and hepatoxicity due to anesthesia with halothane. Anesthesiology, 57, 160-166.

Poyer, J.L., McCay, P.B., Lai, E.K., Janzen, E.G. and Davis, E.R., 1980, Confirmation of assignment of the trichloromethyl radical spin adduct detected by spin trapping during ^{13}C-carbon tetrachloride metabolism in vitro and in vivo. Biochemical and Biophysical Research Communications, 94, 1154-1160.

Poyer, J.L., McCay, P.B., Weddle, C.C. and Downs, P.E. 1981, In vivo spin trapping of radicals formed during halothane metabolism. Biochemical Pharmacology, 30, 1517-1519.

Reinke, L.A., Lai, E.K., DuBose, C.M. and McCay, P.B., 1987, Reactive free radical generation in the heart and liver of ethanol-fed rats: Correlation with in vitro radical formation. Proceedings of the National Academy of Sciences (USA), 84, 9223-9227.

Ross, W.T., Daggy, B.P. and Cardell, R.R., 1979, Hepatic necrosis caused by halothane and hypoxia in phenobarbital-treated rats. Anesthesiology, 51, 327-333.

Tomasi, A. Billing, S., Garner, A., Slater, T.F. and Albano, E., 1983, The metabolism of halothane by hepatocytes: a comparison between free radical spin trapping and lipid peroxidation in relation to cell damage. Chemico-Biological Interactions, 46, 353-368.

Tomasi, A., Albano, E., Biasi, F., Slater, T.F., Vannini, V. and Dianzani, M.U., 1985, Activation of chloroform and related trihalomethanes to free radical intermediates in isolated hepatocytes and in the rat in vivo as detected by the ESR-spin trapping technique. Chemico-Biological Interactions, 55, 303-316.

INDEX

Absorption 3
Acetaminophen
  N-acetyl-p-benzoquinonamine 236
  bioactivation 145
  bioactivation by prostaglandin
    synthetase 87
  hydrolysis 145
  nephrotoxicity 87
  oxidation in kidney 120
  protein adducts 236
Acetaminophen metabolism 250
  effect of ethanol 252
  effect of 3-methylcholanthrene 250
  effect of phenobarbitone 252
Acetanilides, bioactivation by
  sulphation 191
Acetoacetyl-CoA, in N-hydroxy-arylamine
  toxicity 159
N-Acetoxy-arylamines
  intermediates in carcinogenesis 158
  reaction with DNA 158
Acetyl mexacarbate bioactivation 141
2-Acetylaminofluorene
  bioactivation via sulphation 188
  mode of carcinogenic action 155
  protein adducts 235
Acetylation 151
  in arylamine toxicity 151
  genetic polymorphism 151
  in N-hydroxy-arylamine toxicity 159
O- and N-Acetylation, genetic regulation
  of 165
Acetylcholinesterase 99
  inhibition by organophosphates 100
Acetyltransferase
  acetyl-CoA-dependent 156
  in arylamine carcinogenesis 156
Acyl glucuronides
  acylating reactivity 197
  acylation of proteins 197
  electrophilic reactivity 197
  intramolecular acyl migration 194
  as mercapturic acid precursors 197
Acyl migration 194
  clofibryl glucuronide 196
  consequences of 196
  β-glucuronidase resistance
    following 196

Wy-18251 glucuronide 194, 196
Wy-41770 glucuronide 196
O-Acylation in arylamine
  carcinogenesis 154
Acyltransferases
  arylhydroxamic acid–dependent 155
  in carcinogenesis (O-, N- and N,O-) 151
N, O-Acetyltransferases in human
  liver 170
Adducts, with macromolecules 225
A-esterases
  in brain 111
  hydrolysis of organophosphates 111
Alcohol dehydrogenase 121, 123
Aldehyde oxidase 123
  mechanism 122
  tissue distribution 123
Alkenylbenzenes 185
  DNA adducts from 187
  hydroxylation 185
  sulphation in toxic action 186
Alkyl sulphotransferase 181
Alkylation by benzoquinoneamines 87
Alkylnitrosamines, reactive intermediates
  of 231
Allyl esters
  hydrolysis 146
  toxicity 146
Allylisopropylacetamide, suicide inhibition
  by 252
Alveolar macrophages, metabolism by 56
4-Aminoazobenzene, sulphation in toxic
  action 189
4-Aminobenzoic acid, drain on CoA
  pool 256
4-Aminobiphenyl
  binding to erythrocytes 232
  binding to serum albumin 233
6-Aminochrysene, nucleic acid
  adducts 235
2-Aminofluorene, sulphation in toxic
  action 188
Amygdalin hydrolysis 271
Antibiotics, use in study of metabolism
  28, 31
Aromatization of aryl premercapturic
  acids 218
Arylamidases in bioactivation 144

Arylamine acetylation, genetic regulation
  of 165
Arylamines
  DNA adducts 155
  N-glucuronides 198
  haemoglobin adducts 151
  metabolic activation 151
Arylhydroxylamine O-glucuronides 196
Arylsulphotransferase 181
Autoradiography 295
  carbon pool labelling 310
  of covalent binding 297
  of foetal tissue 301
  inducers used with 301
  inhibitors used with 301
  interpretation of 310
  light microscope level 296
  metabolite accumulation 297
  quantitation of 306
  sites of metabolism 308
  tape section 295
  tissue–specific metabolism 299
Azo bond reduction in sulphasalazine 269

Bacterial degradation of xenobiotics on
  skin 69, 72
Balloon catheters, use of 20
Benzodioxazoles
  oxidation inhibitors 133
  structure–activity relationship 134
Benzo[a]pyrene
  bioactivation by bronchial tissue 54
  metabolism in respiratory tract 46, 51
Benzoquinoneamines
  in acetaminophen metabolism 87
  in nephrotoxicity 87
Benzyl acetate
  mercapturic acid formation 183
B-esterases
  in brain 100
  inhibition by organophosphates 100
  in liver 100
Benzyl sulphates, electrophilic
  reactivity 182
Biliary metabolites
  autoradiography of 305
  translocation of 305
Bisulfan
  sulphonium intermediate 228
  tetrahydrothiophene formation 228
Brachymorphic mice 187
  sulphation deficiency 187
  use in mechanistic studies 187
Bromodichloromethane, free radical
  formation 378
Bromoform, free radical formation 378
o-Bromophenol 228

Bromotrichloromethane, free radical
  formation 377
Bronchial tissue
  bioactivation of polycyclic
    hydrocarbons 54
  metabolism in 54
Brush border vesicles 17
Butonate bioactivation 142
2-Butoxyethanol
  oxidation 124
  toxicity 124

Captopril
  disulphide adducts 238
  disulphide conjugates 238
  protein adducts 238
Carbon-13, NMR of tissue extracts 286
Carbon tetrachloride, free radical
  formation 375
Carboxylesterases
  bioactivation of acetanilide 145
  bioactivation of phenacetin 145
  inhibition by organophosphates 103
Carboxylic acids
  glucuronide conjugates 194
  incorporation into lipids 6
Catechol 4-dehydroxylation 270
Chemical modulators of metabolism
  245, 259
  classification 248
Chloroform
  free radical formation 378
  metabolism in kidney 82
  nephrotoxicity 82
Chromatographic translobular
  migration 345
  effect on metabolism 346
Cis-platinum, induction of renal cytochrome
  P-450 82
Clara cells
  cytochrome P-450 in 54
  metabolism by 54
Cobalt chloride
  elevation of hepatic glutathione 253
  lowering of cytochrome P-450 253
Cobalt protoporphyrin IX, lowering of
  cytochrome P-450 253
Conjugation 6
  in intestine 14
  in kidney 89
Covalent binding 225, 232
C-S lyase (β-lyase)
  bioactivation by 90
  bioactivation of
    dichlorovinylcysteine 213
  bioactivation of
    hexachlorobutadiene 213

Index

in kidney 90
Cyanatryn
  haemoglobin adduct 239
  S-oxidation 239
Cyanide ion as electrophile trap 229
4-Cyanoacetanilide
  N-glycolyl intermediate 191
  N-glycolyl sulphate 191
4-Cyano-N, N-dimethylaniline, mercapturic acid formation 191
Cycasin, hydrolytic bioactivation 271
Cysteine conjugates in glutathione conjugate catabolism 90
Cytochrome P-450 119
  in basal ganglia 105
  in brain 103
  brain in comparison with liver 105
  comparison with FMO 125
  distribution in brain 105
  in hypothalmus 105
  induction in kidney 82
  inhibition by cobalt chloride 253
  isotope effects 369
  in kidney 81
  localization in kidney 82
  in respiratory tract 46, 48
  in septum 105
  in skin 69
  suicide inhibition of 252
Cytosolic oxidation 119

Deamination
  of arylamines 230
  involvement of nitrite ion 230
Dermal absorption 65
  in dermal toxicity 65
  factors affecting 68
  in vitro methods 67
  in vivo methods 66
  species differences 68
  in systemic toxicity 65
  toxicity 65
Deuterium isotope effects 357
Diazonium ions, intermediates in deamination 230
1,2-Dibromoethane
  autoradiography 305
  binding to foetal tissue 305
  bioactivation 227
  gluthathione–dependent bioactivation 219
  metabolism 227
2,6-Dichlorobenzamide 209
  glutathione in metabolism 212
Dichloronitrophenol, inhibition of sulphation 255
Diet, effects on metabolism 250

Diesel exhaust, DNA adducts in respiratory tract 49
Diethyl maleate
  lowering of glutathione 258
Diffusion cells (for dermal studies) 67
  flow–through 67
  static 67
Dihydrodiols
  from epoxides 229
  o-quinones from 230
2,4-Dimethoxyacetophenone, selective inhibition of conjugation 256

Electrophilic metabolites, reaction with macromolecules 225
Enterohepatic circulation 8, 181, 306
  study by autoradiography 306
Enterohepatorenal disposition 193
Enzyme–substrate interactions, study by isotope effects 368
Epidermis–dermis separation 67
Episulphonium intermediates from glutathione conjugates 219
EPN 100, 106
  inhibition of acetylcholinesterase 101
Ester glucuronide conjugates
  see acyl glucuronides
Esterases in kidney 89
Estragole hydroxylation and sulphation 240
7-Ethoxyresorufin
  de-ethylation in brain tissue 105
  metabolism in respiratory tissue 47
Ethylene dibromide
  see 1,2-dibromoethane
Ethylene glycol oxidation and toxicity 124
Ethylene oxide
  DNA adducts 237
  protein adducts 237
Ethylene thiourea 227
Everted intestinal sacs for metabolic studies 13, 14

Fenazaflor bioactivation 142
First pass metabolism, study by organ perfusion 315
Flavoprotein monooxygenase (FMO) 119, 125
  in kidney 85
  mechanism 126
  properties 125
  purification 125
  role with cytochrome P-450 126
Fluorine NMR 280, 285
5-Fluorouracil, *in vivo* NMR study 285
Foetus
  autoradiography of 301

covalent binding in 301
Free radical detection 375
   in bile 377
   in liver 378
   in vivo 375
Free radicals 375
   in bioactivation 375
   membrane damage 377
   in metabolism 375
   toxicological significance 377

Galactosamine, inhibition of
   glucuronidation 258
Gastro-intestinal perfusions 325
   propachlor metabolism 329
Gastro-intestinal tract 13
   distribution of enzymes 15
   in vitro preparations 14
   metabolism 13, 14
Genetic control
   of *O*-acetylation 170
   of *N*-acetylation 151
Genetic polymorphism
   in animal models 166
   in bladder cancer 165
   in colorectal cancer 165
   in man 169
Germfree animals 28
   in metabolism studies 32
Germfree rats 28, 263
   effects on intestinal morphology 272
   effects on xenobiotic metabolizing
      enzymes 272
   relevance to man 273
   in study of intestinal metabolism 265
   in study of tissue metabolism 266
S-Glucuronides 214, 218
Glucuronide conjugates
   biliary elimination 193
   in enterohepatic circulation 193
   as intermediary metabolites 179
   reactivity 193
   as transport metabolites 198
Glucuronyltransferase 45, 180
   properties 180
   in respiratory tract 54
   substrates 180
Glutathione, protective role 225
Glutathione conjugates 205
   as intermediary metabolites 205
   catabolism 90
   C-S lyase in catabolism 90
   disposition of 210
   formation of 206
   in kidney 89
Glutathione transferase
   distribution in kidney 89

   in respiratory tract 54
Gnotobiotic rats 263
   see also Germfree rats

Halothane
   free radical formation 378
   hepatitis 378
Hepatocytes, pitfalls in use 339
Hexachloro-1,3-butadiene 228
   glutathione conjugation 90
   nephrocarcinogenesis 90
Hydrolases in kidney 89
Hydrolysis
   bioactivation by 139
   effect on toxicity 139
N-Hydroxy-2-acetylaminofluorene
   acylation in DNA binding 169
   acylation in man 169
   carcinogenic action 155
N-Hydroxy-6-aminochrysene nucleic acid
   adducts 235
N-Hydroxy-arylamine acetylation 160
   sex differences 162
   species differences 162
N-Hydroxy-arylamines, metabolic
   activation 160
5-Hydroxymethylchrysene 183
   bioactivation by sulphation 184
   DNA adducts 184

Image analysis 301
   quantitation of autoradiographs 306
Immunohistochemical detection of
   xenobiotic metabolizing enzymes in the
   respiratory tract 52
In vitro techniques 335
   formulation of substrates 337, 341
   limitations 336
   substrate buffering 337, 341
   substrate delivery 335, 341
Inhalation 41
   effect of particle size 42
   effect of polarity 44
   effect of volatility 44
   route of exposure to xenobiotics 41
Intestinal bacteria 13, 263
   concentration 23
   distribution 23
   glycosidase action 271
   role in toxicity 23, 268
   role in enterohepatic circulation 271
   role in metabolism 23
   study *in vitro* 24, 25
   study *in vivo* 28
Intestinal loops for metabolism studies 20
Intestinal metabolism in man 32
Intestinal wall, metabolism in 13

Intestine, position in general circulation 21
Intramolecular isotope effect 365
  study of enzyme–substrate
    interactions 369–71
Intrinsic isotope effect values 363
  study of enzyme mechanism 369
Iodoform, free radical formation 378
4-Ipomeanol
  lung damage 308
  metabolism in Clara cells 308
Isolated organ perfusions 315, 335
Isoprenalin metabolism in intestine 22
Isosafrole, inhibition of xenobiotic
  metabolism by 255
Isothiocyanates 227
Isotope effects 356
  in mechanistic studies 357
  in metabolism studies 356

Kidney 81
  co-oxidation by prostaglandin
    synthetase 85
  cysteine conjugate β-lyase 90
  cytochrome P-450 in 81
  metabolism in 81
  reductive reactions in 87
Kidney perfusions 315, 328
  dichlorovinylcysteine metabolism 329
  propachlor metabolism 327

Levodopa
  covalent binding 19
  intestinal metabolism 15
  metabolism to dopamine 16
Lindane
  isotope effect 360
  metabolism 359
Lindane-d_6
  insecticidal activity 359
  metabolism 360
Lipid peroxidation
  by $\cdot CCl_3$ 378
  by free radicals 378
Lipophilic substrates
  absorption 340
  behaviour in in vitro systems 338
  binding sites in blood 340
  concentration buffering 337, 341
  distribution in vivo 340
  kinetic problems 338, 342
Liver perfusion 323
  propachlor metabolism 329

Macromolecules, covalent reaction
  with 225
Membrane damage by free radicals 377
Menaphthyl sulphate, reaction with
  glutathione 182

Mercapturic acid formation 205, 211
  alternative to C-S lyase action 211
Metabolism
  biphasic nature 5
  in brain 99
  capacity of organs 7
  factors affecting 9
  in gastrointestinal tract 13
  in intestinal wall 13
  in isolated organs 315
  in kidney 81
  non-enzymic 5
  options for 4
  in respiratory tract 41
  in skin 65
  in vitro techniques 335
Methanol bioactivation 123
Methoxychlor
  demethylation 368
  enantiotopic selectivity 369
  isotope effects 366
  mechanism of oxidation 366
Methyl paraoxon 101
  inhibition of acetylcholinesterase 100
Methyl parathion 100, 106
Methylation of thiols 214
Methylazoxymethanol, from hydrolysis of
  cycasin 271
Methylenedioxyphenyl compounds 132
  biphasic action 133
  inhibition of oxidation 132
  mechanism of action 132
N-Methylformamide
  bioactivation to methyl isocyanate 226
  mercapturic acid formation 226
Methylthio metabolites 216
  demethylation 217
  disposition 216
  formation via C-S lyase action 214
Metronidazole metabolism in intestine 31
Microautoradiography 296, 300
Miserotoxin
  hydrolysis 147
  toxicity 147
Monooxygenase
  cytochrome P-450 119
  flavoprotein 119
MNFA toxicity via hydrolysis 140
  via fluoroacetate 140
Molybdenum hydroxylases 121
  distribution 123
  mechanism of action 122

Naphthalene, cysteine and glutathione
  conjugation 210
Nasal carcinogenesis of
  acetaminophen 299
Nasal tissue, metabolism of xenobiotics

Nasal–pharyngyal region, exposure to particles 42
Nephrotoxicity via metabolism 81
4-Nitrobenzoic acid, model for study of metabolism in intestine 265
Nitropyrene metabolism in respiratory tract 47, 51
Nitroreduction
  mechanism 268
  studies using germfree rats 265
Nitrosamines, DNA adduct formation in human respiratory tissue 55
Nitroso group in nitroreduction 268
NMR 277
  carbon-13 natural abundance studies 286
  5-fluororuracil studies in vivo 285
  localization methods 281
  signal acquisition 281
  suitable nuclei 278
  in xenobiotic metabolism 277
Non-enzymic transformations 5
Nuclear magnetic resonance 277
  see also NMR

Olefactory mucosa
  autoradiography of 300
  phenacetin metabolism in 300
Organophosphate oxons 103
  formation in brain 106
Organophosphates
  bioactivation in brain 99
  detoxification in brain 99
  mode of action 99
Organophosphorothionates, bioactivation in brain 103, 106
Oxidation 119
  induction 132
  inhibition 132
  multiple pathways 130
Oxidative metabolism 119, 121
Oxidative transamination in metabolism of cysteine conjugates 212
Oxprenolol, metabolism in intestine 23

Paracetamol, oxidation in kidney 120
  see also acetaminophen
Paraoxon, inhibition of acetylcholinesterase 100
Parathion 100, 106
  activation in brain 108
  distribution in vivo 340
  metabolism in hepatocytes 339
Particles
  fate of pollutants on 51
  inhalation of 42, 43
Pentachlorothioanisole 217
Perfusin techniques 315

Phenacyl glutathione, further reaction with glutathione 219
Phosgene
  in $CHCl_3$ metabolism 83
  reaction with glutathione 83
Piperonyl butoxide 254
Phenacetin
  effect of glutathione on binding 303
  olefactory mucosal binding 300
Phenyl-N-t-butylnitrone (PBN) 376
  for detection of free radicals
  PBN/·CCl_3 radical 376
  PBN/·CO_2^- radical 376
Phenytoin teratogeneticity 122
Phorate, stereochemistry of oxidation 128
Phalate esters, hydrolysis and toxicity 146
Polychlorinated biphenyls
  autoradiography 298, 307
  in lung 307
  metabolism in Clara cells 309
  methylsulphone metabolites 307
  tissue distribution 298
Polymorphism in acetylation 151
Polynuclear benzyl sulphates 183
  carcinogenic action 183
  mutagenic action 183
  properties 183
^{32}P-Post-labelling assay, DNA adducts in respiratory tract 49
Propachlor conjugates 217
  gastro-intestinal metabolism 329
  hepatic metabolism 329
  renal metabolism 329
Propanil bioactivation by hydrolysis 143
Propionyl-CoA in N-hydroxy-arylamine toxicity 159
Prostaglandin synthetase 119
  bioactivation by 86
  co-oxidation of xenobiotics 120
  in kidney 85
  distribution 120
  mechanism of action 86
  oxidation of arylamines 122
  oxidation of paracetamol 120
  oxidation of polycyclic hydrocarbons 122
Pulmonary macrophages 57
  cytochrome P-450 in 58
  enzymes in 58
  pollutants on particles 58
Pulmonary tissue 57
  metabolism in 57

Quinoline adducts with proteins 234
Quinoxalines 234

Reactive intermediates 225
Reduction 119

Index

of azo groups 269
in intestine 27
of nitro compounds 265
of nitroheterocyclics 267
of sulphoxides 27
Reductive dehalogenation 219
glutathione–dependent 219
reductive metabolism 119, 121
Renal nitroreductase 88
Renal reductases 87
Respiratory tract 41
deposition in 42
distribution of enzymes in 45
metabolism in 41

Safrole hydroxylation and sulphation 240
Simetryn haemoglobin adduct 239
SKF-525A
inhibition of cytochrome P-450 254
non-selectivity on inhibition 254
Skin 65
absorption and disposition of xenobiotics 66
contribution to xenobiotic metabolism 70
toxicity 65
Skin enzymes
activities 70
induction 71
inhibition 71
location 70
Spin trappers 376
of free radicals 376
metabolism *in vivo* 378
Stable isotopes in metabolism studies 355
advantages 356
limitations 356
Substrate buffering 341
in *in vitro* studies 341
in perfused liver 345
Sulindac 27
metabolism in intestine 34–36
reduction in kidney 88
Sulphamethazine N-acetylation, genetic polymorphism 169
Sulphate conjugates 179
formation 181

as intermediary metabolites 179
reactivity 181
S-Sulphation 229
Sulphinpyrazone 24
metabolism in intestine 24–28
Sulphotransferase, properties and substrates 180
Sulphoxides 27
reduction in intestine 33

2,3,7,8-Tetrachlorodibenzo-*p*-dioxin
induction of renal cytochrome P-450 92
Thioacylation in nephrotoxicity 92
Thiols 213
from glutathione conjugates 213
methylation of 214
as reactive metabolites 213
Thioridazine S-oxygenation 130
Tri-*o*-cresyl phosphate, inhibition of hydrolases 144
Tracheal tissue metabolism 54
Translocation of metabolites
between tissues 305
within tissues 307
1,2,4-Trichlorobenzene
formation of acetylated glutathione conjugate 209
Trichloromethanol in chloroform metabolism 83

Vinyl chloride-DNA adducts 237

Xanthine oxidase 119
distribution 123
mechanism of action 122
Xenobiotic lipids 6
Xenobiotic metabolizing enzymes 3
in brain 99
in gastro-intestinal tract 13
in kidney 81
in respiratory tract 41
in skin 65
Xenobiotics
life cycle in the mammal 9